LIFE ON EARTH

A Picture History

LIFE
ON EARTH

A Picture History

Text by
G. Pretto
A. Minelli
M. U. Tanara

NEWSWEEK BOOKS

New York

Art Director
Enzo Orlandi

Editorial Director
Giorgio Marcolungo

Editor
Italo Bosetto

Picture Editors
Ettore Mochetti
Adolfo Segattini

Picture Researchers
Paola Brunetta
Luciana Conforti
Gioietta Gioia
Franco Minardi
Nicoletta P. Tanucci
Katherine Silberblatt

1977 First American Edition

© 1975 Arnoldo Mondadori Editore, Milano

English translation © copyright 1977 Arnoldo Mondadori Editore, Milano

Library of Congress Cataloging in Publication Data
    G. Pretto, A. Minelli and M. U. Tanara
    Life On Earth
    Index
1.Life—Worldwide.   2.Authors, Italian—Picture History.
3.Life Forms—Ecology.
Library of Congress Card Number 77–78800
ISBN 0–88225–251–8

Printed and Bound in Italy
by Arnoldo Mondadori Editore, Verona

CONTENTS

LIVING MATTER

THE CELL

When we approach the study of living matter, we find such a multiplicity of forms and phenomena that it seems unlikely that we shall ever be able to get down to a study of all the individuals which make up the animal and vegetable kingdoms. When the Swedish naturalist Carolus Linnaeus completed his systematic classification of living things in 1758, he had catalogued approximately 8,500 plant species and 4,200 animal species that were known at that time. Today we know over 300,000 plant species and approximately one million animal species. The multiplicity of phenomena in this vast field were listed with exemplary detail and patience by the earliest students of botany and zoology. While such classifications grew progressively more precise, they neither told us anything about living matter nor revealed the distinguishing characteristics that set living matter apart from non-living things. Many of the early authors were unable to disguise their wonder upon finding that adult individuals, of considerable size, could develop from a small egg or a tiny seed. In the absence of other possible explanations, almost everyone believed that tiny but completely formed animals or plants existed inside the egg or seed. This left unanswered the question of what had formed that tiny living thing inside the seed or the egg.

A genuine advance was the discovery of the basic units that make up all living matter. This discovery evolved from the observation of extremely tiny things made possible when the microscope was invented in the mid-seventeenth century. In 1665 the English astronomer and mathematician Robert Hooke published a book, *Micrographia*, that described the microscope and some of the things he had observed through it. One of these was the pattern of tiny rectangular holes he saw in cork and other plant tissues. Hooke named these tiny poles *cells*, from the Latin word for little rooms. But almost two centuries passed before anyone recognized the significance of cells in the structure of living things. In 1838 the German botanist Mathias Jacob Schleiden declared that all plants have cellular structure. He described the exact nature of cells and their precise functions – to act as tiny laboratories where living matter is produced. Within a year the cell theory for plants was supported by Schleiden's friend and compatriot, physiologist Theodor Schwann, who extended the cell theory to the animal kingdom.

All living organisms are formed of cells that have basically the same structure. This does not mean there is a 'typical' cell repeated in identical form, but that there are different cells which share common characteristics. Cellular form varies considerably: it may be cylindrical, cubic, prismatic, conical, or star-shaped. The isolated cell, without any membrane and at rest, has a spherical shape. To form some idea of the dimensions of a cell, we have to find a comparison. Let us assume that a man were as big as a whale. Then one of the more than a million million cells that form the man's body would be the size of the tip of a ball-point pen.

The cell consists of two basic parts, the *cytoplasm* and the *nucleus*; together, they are called the cell's *protoplasm*. The nucleus is embedded in the surrounding cytoplasm, has a rounded shape, and is encased in a *nuclear membrane*. Inside the nucleus is a *nucleolus* (or several nuceleoli), and a number of *chromosomes*. The latter are long, curled filaments of *deoxyribonucleic acid* (DNA), which play a vital role in the reproduction of cells. The cytoplasm is a semi-fluid substance rather like gelatin. It contains other highly important *organelles*, or 'little organs', that have very definite functions.

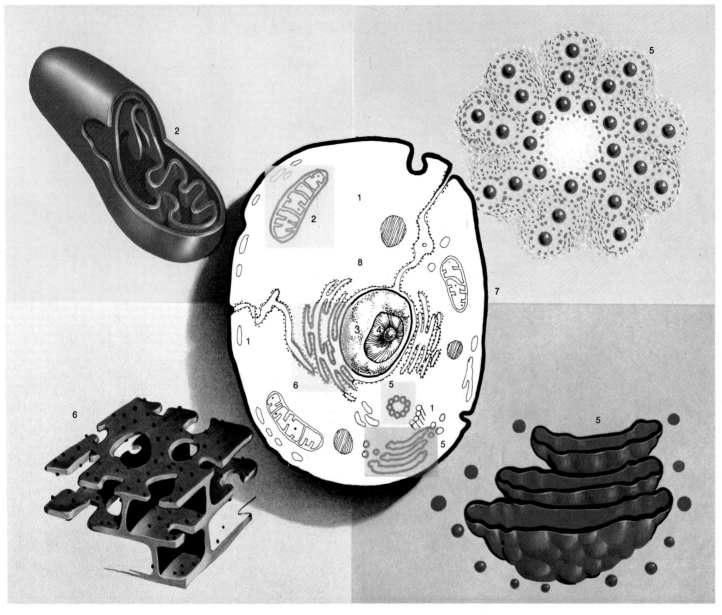

Diagram of a generalized cell as it would appear through an electron microscope: (1) cytoplasm; (2) mitochondria; (3) nucleus; (4) nucleolus; (5) Golgi complex; (6) endoplasmic reticulum; (7) cell membrane; (8) nuclear membrane. Detail diagram numbers relate to those inside cell.

(1) The *mitochondria* are rod-shaped organelles in which respiration takes place, combining food and oxygen to produce most of the energy needed by the cell.

(2) The *chloroplasts*, found only in plant cells, are indispensable in the process of *photosynthesis*, which uses light energy to change carbon dioxide and water into starch.

(3) The *centrioles*, always paired, are in charge of cellular division, the means by which cells reproduce.

(4) The *endoplasmic reticulum* consists of membranes that form a network of tiny, tiny canals where there is intensive metabolic activity (chemical reactions).

(5) The *ribosomes* are very tiny spheres that contain mainly *ribonucleic acid* (RNA) and play an important rôle in protein synthesis (the assembly of *amino-acids*, the 'building blocks' of living matter, to form protein).

(6) The *Golgi complex* seems to 'package' proteins in membranes for delivery to the other parts of the cell.

The cells of micro-organisms and unicellular, or single-celled, organisms, such as protozoa and many types of algae, are self-sufficient. For this reason

they may have, in addition to the organelles described above, other highly complicated structures: devices for moving the cell through a liquid or for adhering to a substratum, small visual organs, organs capable of eliminating excess water, and so on. Cells in many-celled organisms are rarely self-sufficient, but in compensation such cells may be highly specialized to perform different specialized functions within the plant or animal.

In plants we find cells that have covering and protective functions, cellular structures that convey the sap (water and substances dissolved in it) that circulates in the organism, cells that perform photosynthesis, and others that store nutritional matter. Most plant cells have cellulose walls which adhere to the walls of adjoining cells and support the plant in its upward growth.

In animals there are four principal cell types.

(1) *Epithelial* cells cover the external surfaces or line the digestive duct and structures connected with it.

(2) *Neurons*, or nerve cells, receive and transmit sense stimuli, analyze such stimuli, and produce responses to them.

(3) *Muscle* cells contract and relax to move different parts of the animal's body and make locomotion possible.

Left: equipment for a biological research laboratory. Top right: Dutch naturalist Anton van Leeuwenhoeck, the first to observe micro-organisms, using a small hand microscope he designed and made. Below: French chemist Louis Pasteur used these two glass vessels to show that micro-organisms in the air may contaminate exposed food.

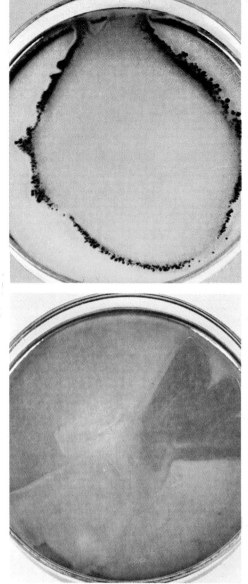

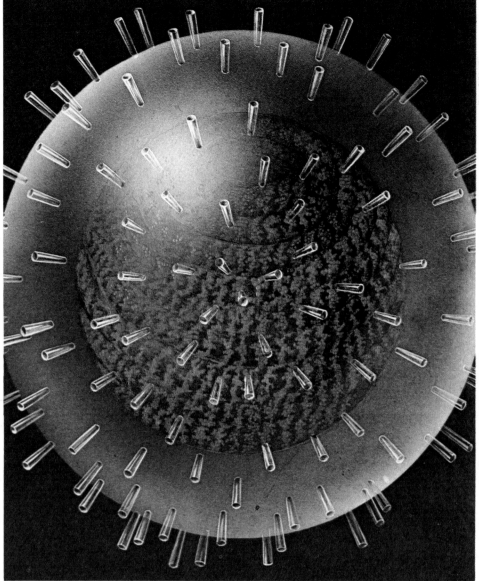

Top and below left: germ samples, cultured on agar in petri dishes, were stained for identification and observation. Right: diagram of an influenza virus as it appears through an electron microscope.

(4) *Connective* cells provide support and consistency for the organism.

Epithelial cells are shaped something like the many-sided tiles used for floor paving or wall covering; they fit together closely, forming a layer of tissue. Some layers are very thin; others are about as thick as their cells are wide or several times that thickness. The cells near the surface of the layer are connected to keep materials (from the stomach, for example) out of the space between the cells. Tiny indentations in the exposed faces of the cells help them absorb fluids.

Nerve cells (neurons) vary considerably in shape. Those that sense olfactory or visual stimuli are elongated and have at one end several *flagella*, or thread-like filaments, that, respectively, react upon contact with odoriferous molecules or when light falls on them. At the other end of the cells is a long filament which establishes contact with other nerve cells along the line of transmission. These cells are more or less star-shaped, and each has two extensions – fibres with tassels at the tips – where the cell connects with other neurons. Along these extensions, faint electric impulses move, analogous to those which are conveyed over the wires of a telephone network. Some extensions of nerve fibres link up with cells that form muscle fibres. Under the

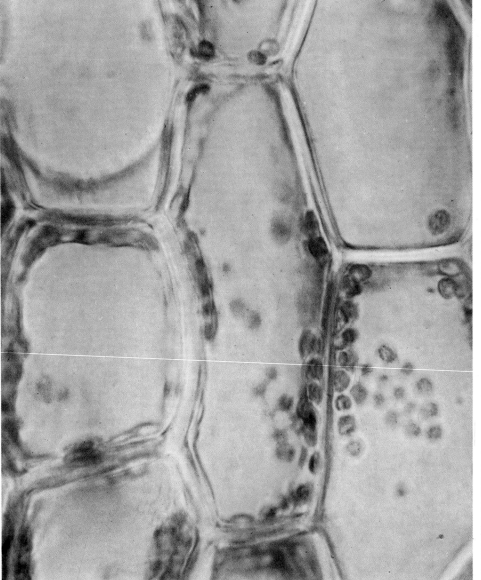

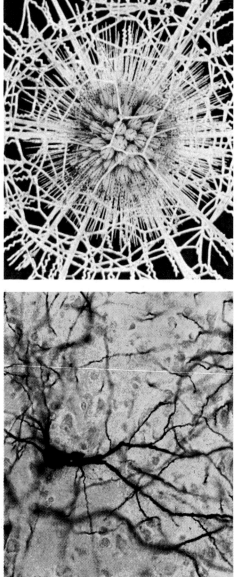

*Left: rectangular plant cells are visible in this highly magnified photograph of leaf tissue in waterweed (*Elodea canadensis*), often used as an aquarium plant. Top right: radiolarian, one-celled marine animal whose siliceous skeletons form much of the sedimentary deposit on the ocean floors. Below: a nerve cell (neuron) as seen through a microscope. The dendritic branches connect the cell to other neurons.*

influence of a nerve impulse (and, on rare occasions, without such a stimulus), chemical energy in these cells is transformed into mechanical action which enables the organism to move part or all of its body. However, if the muscular fibres were not supported in rigid structures formed by the connective and epithelial cells, they would not function effectively. Although the connective cells are more or less insignificant in appearance, they produce the different forms of matter in which they are finally confined. This matter, known as *matrix*, is elastic and compact in the cartilages (of the nose or ear lobes, for example); stiff and rather brittle in the bones; elastic or inelastic, as required, in the fibres that hold soft tissues in place, connect muscles with bones or bones with joints, and so on.

The cell has a very special characteristic – the ability to reproduce and divide itself so as to form two daughter cells. The complex process of cell division begins in the nucleus. The German anatomist Walther Flemming in 1882 called this process *mitosis* (*mitos* is Greek for 'thread'), because the thread-like chromosomes play a very important role. Mitosis occurs in substantially the same way in every cell, and it is usually described in four phases: *prophase, metaphase, anaphase,* and *telophase.* Some time before the prophase begins,

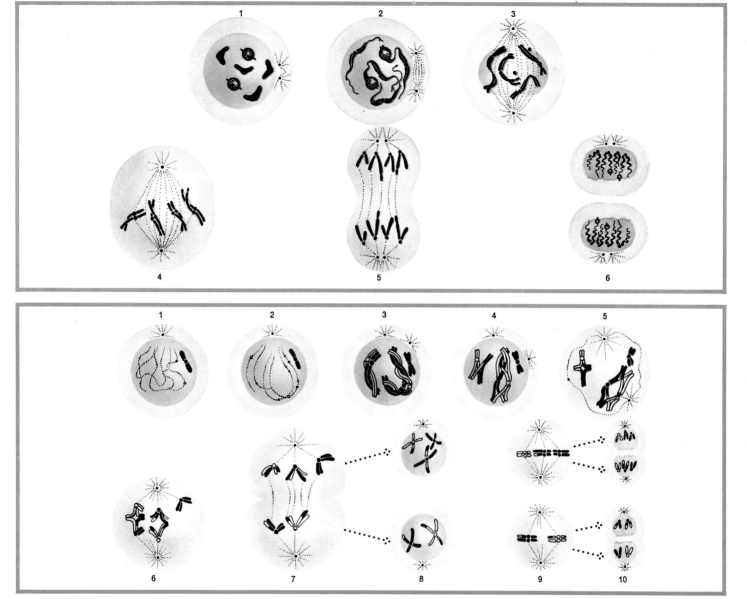

Top: stages in the cell division known as mitosis: (1,2) prophase; (3) metaphase; (4,5) anaphase; (6) telophase (for explanation, see text). Below: stages in meiosis, or production of the reproductive germ cells: (1,2) pairing of chromosomes; (3) duplication of chromosomes; (4) division and exchange of chromosomes; (5,6) orientation towards two poles; (7,8) origin of cells with half of chromosomes; (9,10) second division, resembling normal mitosis.

each chromosome duplicates itself and the two resulting strands, called *chromatids*, are connected near their centres by a tiny *centrosome*. At prophase, the thread-like chromatids coil up to form short, thick rods. The paired centrioles divide and each projects filaments to form a star-shaped *aster* (Latin for 'star'). The metaphase starts with the dissolution of the nuclear membrane. The asters have moved to opposite poles of the cell, stretching their connecting filaments into a *spindle*. The connected chromatids line up across the cell's 'equator', each pairs with its centromere on a spindle filament. As the anaphase begins, each centromere splits in two and the two parts carry their daughter chromosomes along the spindle to the opposite poles of the cell. Meantime the centriole in each aster duplicates itself and the cell itself begins to divide. In the telophase, a nuclear membrane forms around the complete set of chromosomes at each pole; the chromosomes resume their tangled-thread shape, and the cell divides into two daughter cells, each with half of the parent cell's cytoplasm and a set of chromosomes identical to those of the parent cell.

A different process of cell division, called *meiosis*, produces the *gametes*, or germ cells, that are required for sexual reproduction. An ordinary cell contains two chromosomes of each type, one from the organism's male parent and one

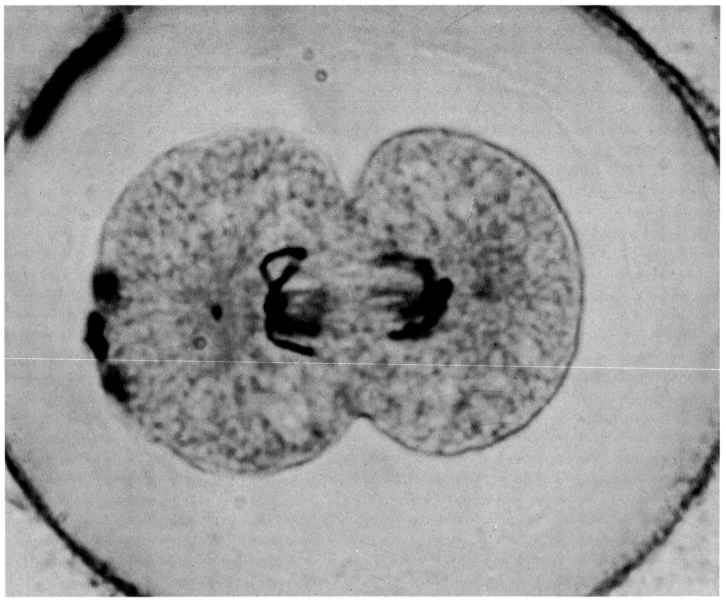

from the female parent. In meiosis, two chromosomes of the same type come together side by side, duplicate themselves (as in mitosis), then divide and form two nuclei within the cells. Each nucleus now has the normal number of chromosomes, but none of the same type from both of the organism's parents. Next the two nuclei divide, as in mitosis, but the cell then divides into four daughter germ cells. Each germ cell has just one chromosome of each type (instead of two of each type, as in a body cell). Each germ cell contains a different combination of the chromosomes inherited from the organism's male and female parents, so each cell carries a different genetic message. When a male gamete (*sperm*) unites with a female gamete (*egg*), they form a cell called a *zygote*, with the normal number of chromosomes, half from each parent cell. The *zygote* is a new individual; if it is formed by the union of gametes from two multicellular organisms, repeated mitosis develops the zygote, its daughter cells, their offspring, and so on, into a multi-celled organism like its parents.

The smallest one-celled organism is the *bacterium*, which has no cell nucleus and reproduces by simply dividing into two. Like the plants called fungi, most bacteria are *parasites*, getting their food from other living organisms, or *saprophytes*, getting their food from non-living organisms. They live in and on

Photomicrograph of a cell in the telophase of mitosis, showing the formation of two daughter nuclei with complete sets of chromosomes. Soon the single cell will divide, producing two daughter cells with the same genetic heritage.

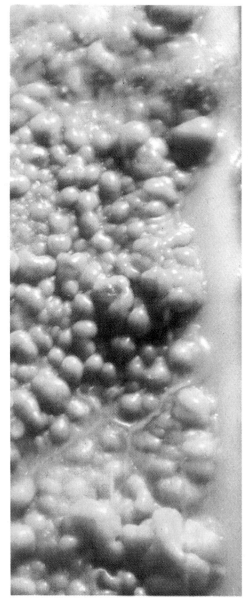

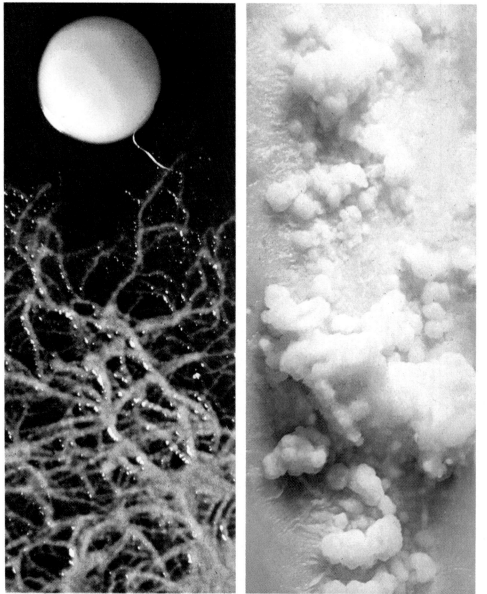

When unicellular organisms such as bacteria are placed on a culture medium like agar, they proliferate rapidly, forming colonies. Left: tuberculosis bacteria after thirty days of incubation. Centre: a colony of Bacillus mycoides *shown below a spherical colony of staphylococci. Right: colony of* Microbacterium phley, *commonly called dung-heap bacterium.*

other living things or their remains, in soil and water. Some bacteria are harmful to plants and animals, including humans, but most kinds serve vitally useful functions in the web of life.

On the border between living and non-living matter are the *viruses*. These are smaller than bacteria and have no cellular organization. A virus is simply a strand of nucleic acid with a coat of protein. Only when it has injected its nucleic acid into a living cell can the virus 'come to life' and replicate, by using the cell's mechanism to produce hundreds or thousands of new viruses. In so doing, viruses cause human ailments such as the common cold, influenza, measles, poliomyelitis, and certain types of cancers.

Only recently have these disease-producing agents become visible to the investigator. Identifying them had to await the development of the electron microscope. Viruses come in various shapes—some are spherical, others are rod-like. And they affect not only man and animals (foot-and-mouth disease and rabies are of viral origin and can be transmitted to man); plants, too, have viral diseases, including tobacco mosaic virus, the study of which has opened investigation into a number of diseases that hitherto have resisted cures or prevention.

HEREDITY

When we speak of heredity, we refer to all those characteristics that one living creature hands down through its direct descendants to successive generations.

When primitive humans first started to breed livestock and grow their own food plants, they began, in an empirical way, to discover and use some of the principles of heredity. However, the man who really discovered the laws regulating heredity was Gregor Mendel, born in 1822 in a small village in Silesia (then part of the Austro-Hungarian Empire). At a youthful age, Mendel became a novice in a monastery of Augustinian monks at Brünn, Austria (now Brno, Czechoslovakia), and was ordained a priest in 1847. He dedicated himself to teaching mathematics and natural science, spending his free time in the monastery orchard.

In 1857, he began to experiment by cross-breeding plants, with special emphasis on peas. Continuing these experiments with most painstaking care over a number of years, he obtained some interesting results, which he recorded very intelligently, preparing the first known statistical data based on successive cross-breeding. Unfortunately the article in which Mendel reported his findings in 1866 did not attract any attention among his contemporary biologists. But Mendel persevered in his research, studying bees and certain plants, although these studies did not yield any very interesting results. In the meantime he had become the abbot of his monastery, and as a result had less free time to pursue his research work. He died in 1884 without having gained any recognition for his important contribution to science.

Mendel's findings lay neglected for some thirty-five years. Then, about 1900, three botanists who did not even know each other – Hugo de Vries in the Netherlands, Karl Correns in Germany, and Erich von Tschermak in Austria – independently worked out the principles of heredity at about the same time. In checking their results against the literature on the subject, each of the three found Mendel's papers and brought them to the public's attention. Moreover, all three credited Mendel with the original discovery and said their work confirmed his findings. In a scientific climate already favourably disposed towards biological theories that were supported by ample experimental data, Mendel's laws gained quick acceptance, and Mendel was acknowledged as founder of the scientific study of heredity, now called *genetics*. Mendel's findings can be summarized as follows:

Law of unit characters. A plant contains certain 'factors' that control its inheritance of such characteristics as height, colour, and so on. A pair of factors – one inherited from each parent – controls each characteristic of the offspring.

Law of dominance. If a red-flowered pea plant is cross-bred with a white-flowered pea plant, the first generation of offspring plants (called the F_1, or first filial, generation) all have red flowers: the red factor is the *dominant factor*, because it prevails; the white factor is the *recessive factor*, because it is unable to control the colour of the first-generation plants. However, for certain types of characteristics, neither of the two controlling factors dominates. For example, if two plants of the genus *Mirabilis* (commonly called Four o'Clocks), one having red flowers and the other having white flowers, are cross-bred, the resulting hybrids are pink.

Law of segregation. The second generation (F_2), obtained by cross-breeding F_1 plants, has about three red-flowered plants to each white-flowered plant – a three-to-one ratio. In this second generation we observe the phenomenon of

Transmission of two hereditary characteristics in the pea plant – colour of flower and colour of seed. Cross-breeding a plant having red flowers (R) and yellow seeds (G) with a plant having white flowers (r) and green seeds (g) will produce first-generation plants with red flowers and yellow seeds. In the second generation only nine plants out of sixteen still have red flowers and yellow seeds (both dominant characteristics) ; three out of sixteen will have white flowers and yellow seeds (one recessive and one dominant characteristic) ; another three will have red flowers and green seeds (one dominant and one recessive characteristic), and one out of sixteen will have white flowers and green seeds (both recessive characteristics).

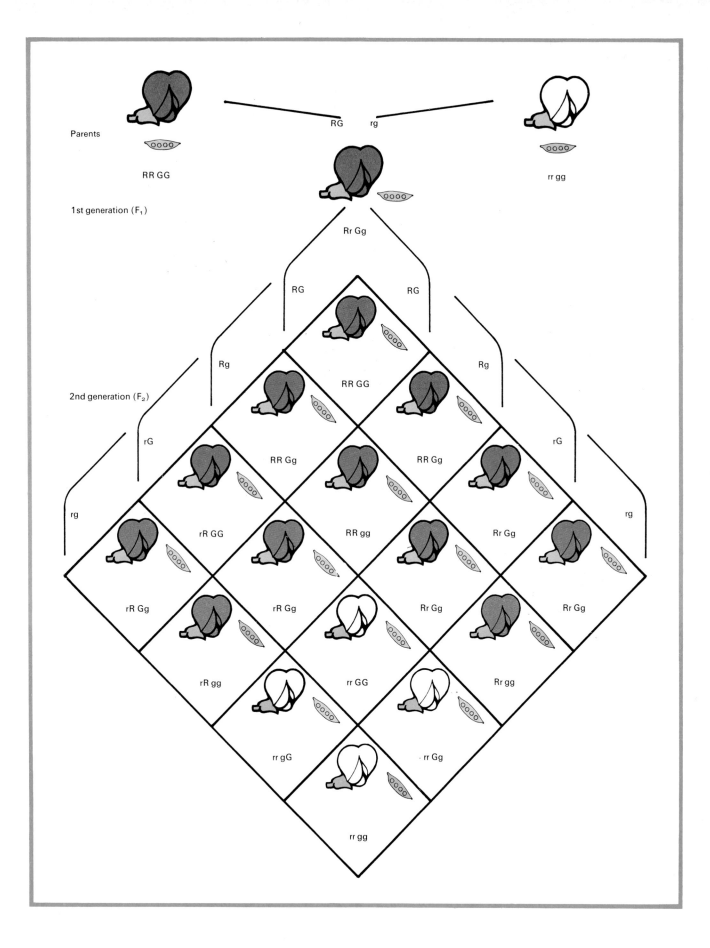

Parents

RR GG

rr gg

RG   rg

1st generation (F₁)

Rr Gg

2nd generation (F₂)

RG   RG

Rg   Rg

rG   rG

rg   rg

RR GG

RR Gg   RR Gg

rR GG   RR gg   Rr Gg

rR Gg   rR Gg   Rr Gg   Rr Gg

rR gg   rr GG   Rr gg

rr gG   rr Gg

rr gg

disconnection of the two characteristics; that is, once again we have flowers of both colours.

The third generation (F_3), obtained from F_2 plants, yields both red-flowered plants and white-flowered plants. All the F_2 plants having white flowers produce peas that grow into F_3 plants with white flowers. This derivation is constant and produces pure individuals. However F_2 plants with red flowers produce peas that grow into F_3 plants with the following characteristics: about one-third have red flowers, and the remaining three red-flowered plants. There are no exceptions to this law of segregation, which is regarded as a fundamental principle of genetics.

A plant, then, may have recessive factors that do not manifest themselves in its physical appearance but can be transmitted to its offspring. Because of this, scientists call the complex of a plant's *apparent characteristics* its *phenotype*; the complex of the plant's *inherited characteristics* is called its *genotype*. For example, a hybrid plant may have the phenotype 'red flowers' but the genotype 'red flowers and white flowers', because it has inherited a recessive white-flower factor as well as a dominant red-flower factor.

Law of independent segregation. This concerns the simultaneous inheritance of

Hereditary characteristics of the Hapsburg family – full lower lip, long chin, and prominent nose – are evident in these portraits of Maximilian I, his daughter Margaret of Austria, his grandson Charles V, and his great-grandson Philip IV. Such characteristics are transmitted by the chromosomes in the gametes produced by meiotic cell division. Top right: condensing into rods, chromosomes begin to pair off for first meiotic division. Below: paired chromosomes move towards poles of the cell.

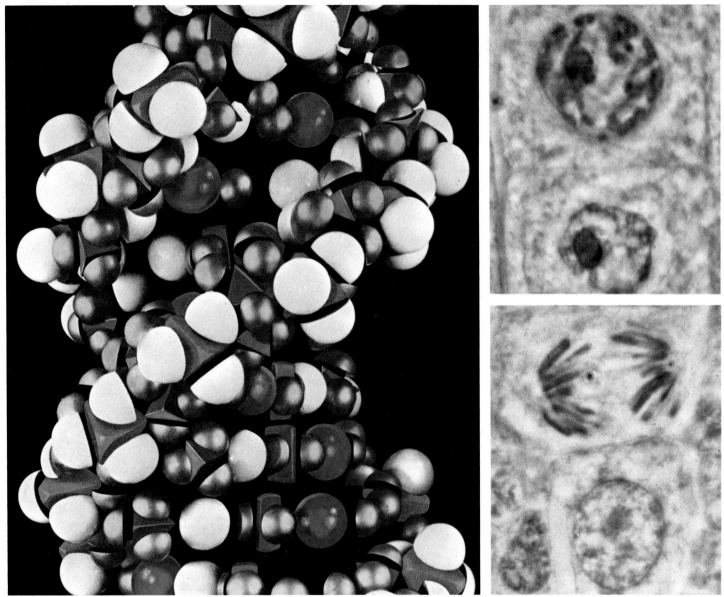

Reproduction of individual organisms and renewal of cells in a tissue (top and below right) are achieved by the reproduction of long protein molecules assembled according to the pattern encoded in the DNA molecules that make up chromosomes. Left: model of a portion of a DNA molecule, showing how the two spiral coils are joined by rungs of substances whose arrangement determines the code. Displacement of a single sphere (atom) in the complex can cause a mutation in the daughter cells.

two or more pairs of characteristics. If we cross-breed pea plants having red flowers and yellow seeds with plants that have white flowers and green seeds, the first generation (F_1) plants will show identical characteristics – red flowers and yellow seeds, produced by the two dominant factors. But in the second generation (F_2), we find the so-called *independent segregation* of the two sets of characteristics. The phenotype of the F_2 plants will show characteristics according to this arrangement (see page 19): of sixteen plants derived from the hybridization process, nine retain the dominant characteristics (red flowers and yellow seeds); three show one dominant characteristic (red flowers) and one recessive characteristic (green seeds); another three show one recessive characteristic (white flowers) and one dominant characteristic (yellow seeds); and one plant shows two recessive characteristics (white flowers and green seeds). This illustration of segregation permits us to ascertain something very important. The two pairs of factors producing these characteristics are independent of each other, so a plant of the F_2 generation may have any combination of the characteristics. This law of independent segregation is valid not only with two pairs of characteristics, but also in cases where there may be three or four such pairs. The variety of possible combinations increases in

proportion to the number of characteristics involved.

After 1882, when the German anatomist Walther Flemming described the behaviour of chromosomes in a cell undergoing mitosis, biologists realized that the chromosomes probably had something to do with heredity. In 1902, after Mendel's laws had been rediscovered, the young American geneticist Walter S. Sutton reasoned that Mendel's hereditary 'factors', now called *genes*, were strung together in linear sequence on the chromosomes. And by 1911 a scientific team led by the American biologist Thomas Hunt Morgan had established the role of chromosomes in carrying heredity.

In the early 1950s American and British biochemists James Watson and Francis Crick figured out the complicated structure of deoxyribonucleic acid (DNA). Every chromosome contains a long DNA molecule, formed of two long strands intertwined in a shape something like that of a dual-flight spiral staircase. Each of the 'rungs' connecting the two strands is a linked pair of two of the three basic substances: *cytosine, adenine, guanine,* and *thymine.* The function of this structure is to convey genetic information from one cell to the next generation of cells, except in certain cases where this duty is performed by ribonucleic acid (RNA).

Top: example of the hereditary transmission of skin colour in mice. Below left: eye colour is a very obvious hereditary characteristic in humans. Centre: beautifully striated flower obtained by repeated cross-breeding of plants with different flowers and colours. Right: brightly coloured male and duller coloured female exemplify dimorphism in fish.

Albinism – the absence of pigment in skin or fur caused by a genetic mutation – produced this colourless peacock, kangaroo with its young, and gorilla. Albinos occur in almost every species of animal, including humans; in plants it is usually a lethal mutation, because a plant without the pigment chlorophyll cannot make food to nourish its cells.

If an organism's genetic information is changed before or during transmission to the organism's offspring, the latter will have one or more 'new' inheritable characteristics, called *mutations*. This involves a change in a chromosome (which may occur accidentally in the process of cell division) or – more frequently – damage to a particular gene. Such damage can result from the organism's exposure to high temperature, to high-energy radiation (X-rays, gamma-rays, beta-rays, even ultraviolet rays), or to a number of mutagenic chemicals. Most mutations are harmful, tending to lessen the organism's ability to survive and reproduce. But some mutations improve an organism's ability to cope with its environment, and such favourable mutations are vital to the process of *evolution*.

EVOLUTION

From the time of Aristotle, scholars wondered what life was like at the dawn of the world and how living things originated. On one point they all seem to have agreed, from the earliest naturalists to the great Linnaeus; no one doubted that every plant and animal species on earth had begun in the identical form the observer saw in his own day and age.

When the English naturalist Charles Darwin revealed his revolutionary opinions in 1859, in his *Origin of Species*, they had the impact of an exploding bomb – not only in the domain of science, but also in those of religion and philosophy. It was not the first time there had been talk among scientists of evolution, the idea that living creatures might have undergone anatomical and physiological changes over a period of many generations. The evolutionary theory of the French naturalist Jean-Baptiste Lamarck preceded Darwin's theory by about half a century. Lamarck maintained that in time a species undergoes modifications resulting from environmental factors and the use or non-use of the organs with which members of the species are endowed, and that these modifications are genetically transmitted. For example, he maintained that the long neck of the giraffe resulted from a continuous stretching of the cervical region by these animals in their attempts to reach the leaves on the highest branches of trees on which they fed, and that this extension of the original neck was then passed along to their offspring. Unfortunately, the theory was never supported by irrefutable evidence. But the Darwinian hypothesis, shared by Darwin's compatriot Alfred R. Wallace, had a definite advantage, because Darwin had accumulated a good deal of evidence in support of his theory, and that evidence has withstood the test of time.

One observation fundamental to Darwin's theory was that plants and animals are produced in a far greater number than those which actually survive; in science this phenomenon is known as *biotic potential*. To explain the concept of biotic potential better, let us cite a few significant examples. When oysters lay their eggs, they produce hundreds of thousands at a time, but those that both hatch and survive are limited in number. The common housefly has an even higher biotic potential. A female housefly produces an average of a hundred and twenty eggs at one time. In a warm climate, one female can produce seven generations a year, and half of the eggs develop into females. If every one of the flies survived, and the females in each generation reproduced to the maximum of their capability, by the end of the year there would be more than six billion flies.

Darwin also called attention to the variations in characteristics among living things, even among organisms of the same species. He suggested that any individual plant or animal endowed with a variation of physical characteristics that gave it a greater probability of survival than its fellow organisms also had a greater probability of transmitting to offspring the physical advantages with which it was itself endowed. Darwin called this process *natural selection*. Unfortunately, Darwin was unable to explain how some individuals of the same species came to have 'new' characteristics that would be transmitted to their offspring. We know now that twins born from the same egg have identical genetic heritages, yet one may be taller and more robust if the twins have grown up under different living conditions. But such non-genetic variation is not passed on to the children of either twin. Only the variations called mutations, caused by changes in the chromosomes or genes of a mutant's parent, can be passed on to the mutant's offspring. These genetic variations provide the new

Four examples of plant fossils formed by the process of carbonization. Organic matter such as plant leaves or stems slowly decomposes under water or when covered by sediment, losing most of its oxygen, hydrogen, and nitrogen, and leaving only a thin residue of carbonaceous material. This may produce a hardened imprint in the solidified sediment (rock), as shown. When large amounts of organic matter are fossilized in this way, they form deposits of peat or soft or hard coal.

inheritable characteristics on which Darwin's theory of evolution by natural selection depends.

Among the sciences that have contributed to the consolidation of the theory of evolution, one of the most important is *palaeontology*, the study of primeval life forms. The material of palaeontology consists of *fossils* – the preserved remains of prehistoric plants and animals. Fossils (the word means 'something obtained by digging') are usually found embedded in the earth or in rock. They provide the evidence that, in the remote ages, the earth contained plant and animal life very different from our contemporary flora and fauna. In a way, we could compare fossils to illustrations in a book, in which the pages represent geological stratifications. Fossil relics have been found that reproduce original organisms in such minute detail that the organisms seem to be merely dead and embalmed. Significant traces of other organisms have been found that can 'tell' the story of those organisms with ample clarity. Of still others only a few traces pertaining to their biological activity have been found, and we do not know what the organisms looked like.

A plant or animal reaches the stage of fossilization only when two specific, coincidental factors exist. First, some mineral residue (sediment) capable of

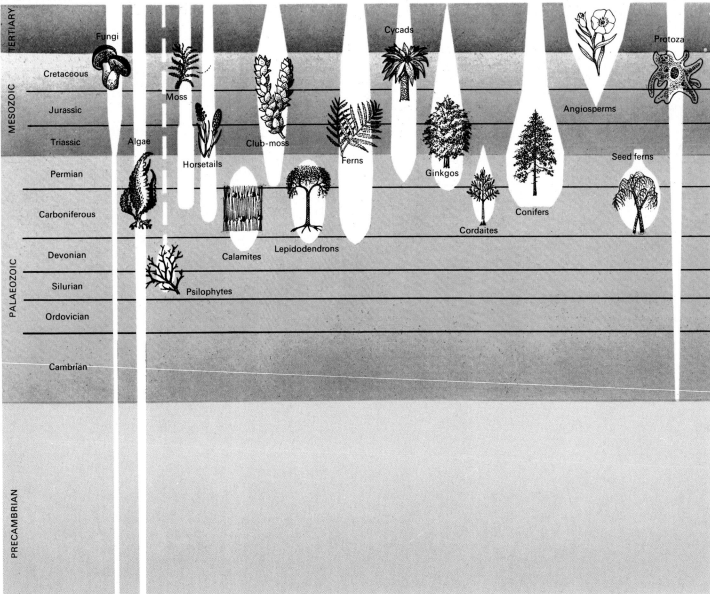

This chart shows the intervals of geological time (see next caption) over which the main groups of plants and animals evolved. The importance of a particular plant or animal group in successive geological periods is indicated by the width of the group's white band in that period.

preserving the remains from bacteria, erosion, scavenging animals, and other external factors that would otherwise destroy or damage them must settle on the organism in question before it decomposes. Second, it is essential that the organism's mineralized parts – those parts which do not decompose and can therefore remain intact – are not crushed, heated, or damaged in other ways that would impair their integrity. If all goes well, the slow process of fossil formation will begin. The sediment which has covered the organism will make a mould, or *external contour*. The inner part will be filled progressively as its softer parts decompose, and in this way a negative imprint, or *internal contour*, will form. In rocks formed more than three thousand million years ago, much earlier than life was once thought to have existed, microscopic traces have been found of fossil cells that resemble present-day species of blue-green algae, a unicellular plant.

Arranging fossils according to their age, we are able to reconstruct the history of living things on earth – how they were slowly transformed, underwent abrupt changes, experienced tranquil progression, and even inexplicable disasters. However, we do not have all the pages in this great book of life, so many problems are still unsolved, particularly those concerning

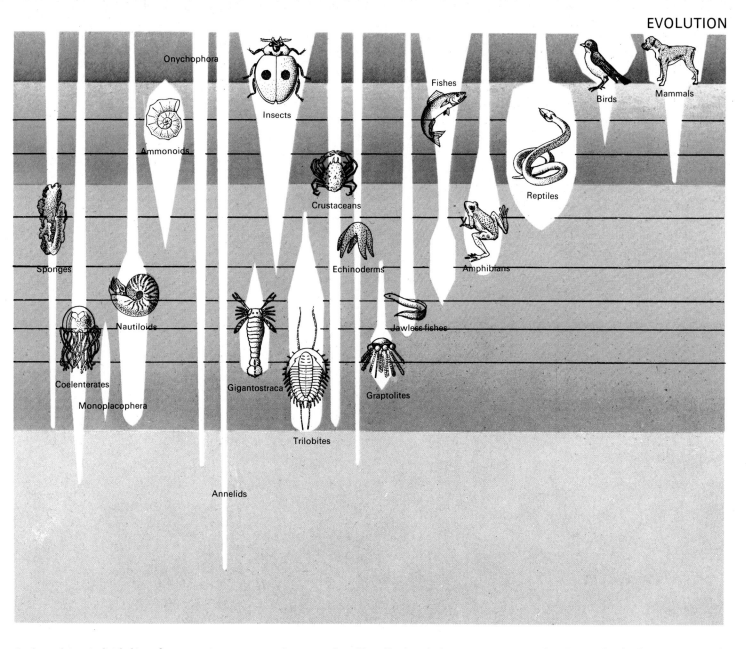

Geological time is divided into five eras and a number of periods, according to the times when certain layers of sedimentary rock – and the fossils within them – were deposited. The Precambrian era began about 4,500 million years ago; the Palaeozoic era, about 600 million years ago; the Mesozoic era, about 230 million years ago; the Tertiary era, about 65 million years ago; and the Quaternary era, about one million years ago.

extinct species. Fossils that help us reconstruct the chronological sequence and duration of the different ages of life on earth are called *index* fossils. They must be fossils of organisms that were great in number at one time, widely scattered over the earth, and particularly suited to the process of fossilization. They must also be fossils of organisms that lasted only a short span of time. Examples of index fossils include: for the early Palaeozoic era, the trilobites, so-called because their bodies were divided into three parts, or lobes; and for marine deposits of the Mesozoic era, the ammonites, distinguished by a shell that was subdivided into different chambers by seven spirals.

A second category is that of the *facies* fossils, which help us to establish environmental characteristics in the different geological periods. Unlike the index fossils, facies fossils are the remains of organisms that resemble present-day organisms living in limited but well-defined geographical environments. Assuming that the fossil organisms required the same environmental conditions as their modern counterparts require, we are able to reconstruct those ancient environments. Typical among facies fossils are the corals; from our knowledge of contemporary corals we deduce, by analogy, that the fossil corals must have lived in an environment where sea-water was pure but rough,

in places that were far away from river mouths, and at temperatures that never dropped below 68°F. (20°C.).

The third category is that of *climatic* fossils, which help us reconstruct the sequence of cold and hot periods in a single area. The discovery of cold inhabitants and hot inhabitants in different geological stratifications of the Mediterranean area enable us to deduce that, during the Lower Quaternary era, periods when the Mediterranean climate was cold (corresponding to glaciations) alternated with periods when the climate was hot. The cold inhabitants were Lamellibranchia, which still thrive today in cold northern waters; the warm inhabitants were the Gastropoda, which can still be found along the coasts of Senegal and Gabon.

But how is it possible to put dates to these ancient forms, when we are dealing in tens of millions of years? That question was answered some years after radium was discovered, although the palaeontological use of the element was not suspected then. By using the presence of radium and radioactive elements, modern scientists have found that what has been called a 'clock of utmost precision' exists in the earth itself. Radioactive elements break down, or decay, on a rather precise time basis. The age of rock formations can thus be

Left: fossil shell of an extinct sea animal called an ammonite, which flourished widely over the Mesozoic era and now serves as a useful index fossil for that era. Right: a well preserved trilobite fossil. Named for its three-lobed body, this animal sometimes reached 18 inches (45 centimetres) in length and was one of the two dominant animals, both in number and size, of the Palaeozoic seas.

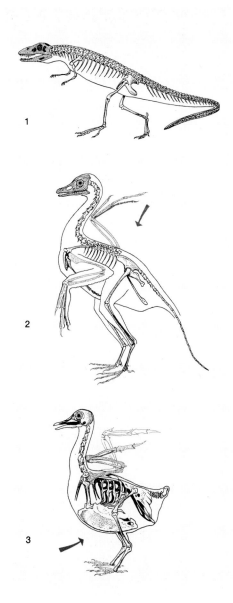

The drawing of Archaeopteryx *(2) shows how its long tail, teeth, and skull resembled those of Triassic reptiles (1,* Saltopsychus*), while the feathers, wings, and claws resemble those of a modern bird (3). Scales on the claws of modern birds remind us that they evolved from reptiles. Right: fossil remains of* Archaeopteryx lithographica, *the earliest known bird, which lived in the Jurassic period; it was about the size of today's crow.*

timed by examining the stages of various products of the decay process that can be found in the various kinds of rocks.

By using this method it was discovered, for instance, that the shark, among the oldest of the vertebrates, first appeared in the Silurian period (320 to 440 million years ago) and has remained relatively unchanged during the enormous time span that has followed. Man, of course, is millions of years younger than that primitive life form.

Since the fossil species classified to date run into many thousands, it is easy to realize how essential palaeontology is for the study of evolution. Another science vital to this study is the discipline called *comparative anatomy*, which analyzes parallels between the structure and function of the organs of living creatures and those of extinct species. This kind of analysis is particularly helpful when the researcher has access to fossil remains of an animal's forbears that were deposited over a long period of time and can be arranged in the sequence of their deposition. From such a series of fossils, one can reconstruct the developments in the animal's evolution. A well-known example is the evolution of the horse. Fossils deposited over the past sixty million years show that the horse has evolved in that time from a dog-sized animal with a short

muzzle and four toes on each foot to the large, long-muzzled, single-toed animal of today. More astonishing comparisons are those that show the evolutionary relationships between the limbs of organisms very different from each other. The hand of a human being, the front paw of a dog, the flippers of a whale, and the wing of a bat all have more or less the same bones. But there is still more: the essential bone structure is the same in the limbs of a salamander (amphibian), a lizard (reptile), a pheasant (bird), and a cat (mammal). It is easy to deduce that these individual animal species once had a common ancestor long, long ago.

Other evidence revealed through the comparative analysis of fossils is furnished by *embryology*, the study of organisms in the earliest phases of their lives. For example, embryologists have ascertained that many organs that are plainly perceptible in the embryo no longer exist in the adult organism, or else have been transformed through evolution until they are no longer recognizable. For example, the embryos of reptiles, birds, and mammals have branchial fissures very similar to those of fishes and amphibians; in the latter these develop into respiratory organs, whereas in the former they develop into quite different organs. Such *affinities*, or likenesses, between living things that

Skeleton of a mammoth, reconstructed and exhibited in the Paris Museum. Members of the elephant family, the mammoths roamed North America and Eurasia during the Pleistocene period, about a million years ago. Reaching 9 feet (3 metres) or more in height, they had evolved a thick layer of fat and long, woolly hair that protected them against the bitter Ice Age cold. They became extinct about ten thousand years ago, but parts of their bodies have been found in northern lands, fully preserved in ice.

Skeleton and reconstruction of the Brontosaurus, *a peaceful, four-legged dinosaur of the Jurassic period. Some reached a length of 85 feet (26 metres) and weighed 35 or 40 tons (30 or 35 tonnes), making them the largest land animals ever to roam the earth. The* Brontosaurus *and its related dinosaurs of the Sauropod sub-order are usually described as having lived in swamps and fed only on water plants, but some scientists believe their grinding teeth were adapted to eating tougher, dry land plants.*

have undergone evolutionary transformation can often be pinpointed only through embryological research.

The geographical distribution of living animals and plants can also be interpreted in the light of their evolutionary development. On some islands – New Zealand, for example – there were no mammals until humans brought them there; in the Arctic region there are polar bears but no penguins; in the Antarctic region there are penguins but no polar bears. In other words, various geographical zones had become isolated and became home for species that could adapt to each specialized environment.

In the process of reconstructing the degree of affinity between living and fossil organisms, whether plants or animals, we find new evidence supporting the theory of evolution. Discovering such affinities is the role of *taxonomy*, the science of classifying living things systematically. Today the taxonomists put two groups of plants into the same higher classification if it can be shown that the plants of both groups evolved – or probably evolved – from a common ancestor.

ADAPTATION AND BEHAVIOUR

No living thing can remain isolated from its surrounding environment. It cannot help being affected by factors such as temperature, humidity, air pressure, light, availability of food, and so on. Nor can it be entirely free of all relationship with other organisms of the same or different species. Individual organisms must be *adapted* to their environment – equipped to cope with it – if they are to survive, and a species survives only if enough of its individual members survive long enough to reproduce. The evolutionary process of natural selection promotes *adaptation* by favouring reproduction by those members of a species best equipped to survive in their environment. Thus adaptation is a continuing process in which genetic characteristics that prove advantageous to individuals in one generation of a species tend to be transmitted to their offspring. Every environment is constantly changing, if only very gradually; but so, too, is the genetic stock of a species, as mutations introduce new characteristics, some of which favour the mutant's – and thus its species' – survival in a changed environment.

Each animal or plant, then, is a product of the continuous adaptation of its species to changing environmental conditions. Adaptations can account for the organism's shape, size, colour, organs, senses, and so on. In so far as such characteristics determine the organism's capabilities, they also determine the ways in which the organism can react, or behave, in its contacts with its environment. While each organism behaves generally in a way typical of its species, the so-called higher animals have ranges of innate capabilities that permit individuals of the same species to react to the same conditions in different ways. The study of animal behaviour is called *ethology*.

We can analyze any particular behaviour in terms of a *stimulus* and a *response*. The stimulus is the external factor that confronts the subject; the response is what the subject does about it. Since external factors are constantly changing, and responses to them are shaped by changes within the subject organism, it is easy to understand why one rarely observes uniform stimuli and responses. Under the spur of hunger, for example, a cat will devour food that it would refuse on a full stomach; the males of some species woo the females only in the so-called mating seasons; some birds decline food and pine away until they die, solely because they are enclosed in a cage.

When individuals of the same species react in identical ways to identical stimuli, ethologists describe this as *stereotype* behaviour. In such behaviour there is no trace of experience responsible for the reaction; the individual's DNA is programmed to produce that response, therefore the individual acts that way. The simplest organisms – tiny one-celled life forms, for example – react with their whole bodies, in stereotype responses to external stimuli. Their behaviour is limited to two elementary movements, which we call *positive* when the organism turns or moves towards the stimulus and *negative* when it turns or moves away from the stimulus. If the organism merely turns, leans, or grows towards or away from the stimulus, its movement is called a *tropism*. Since plants cannot move around, they exhibit only tropisms, such as turning toward a source of light (*phototropism*) or growing in the direction opposite to the downward pull of gravity (*geotropism*). A movement that changes the location of the organism is called a *taxis*; movement toward a light is *phototaxis*; movement toward or away from a heat source is *thermotaxis*.

Animals with more complex structures have specialized nerve and muscle systems that can receive a stimulus and automatically transmit an impulse that

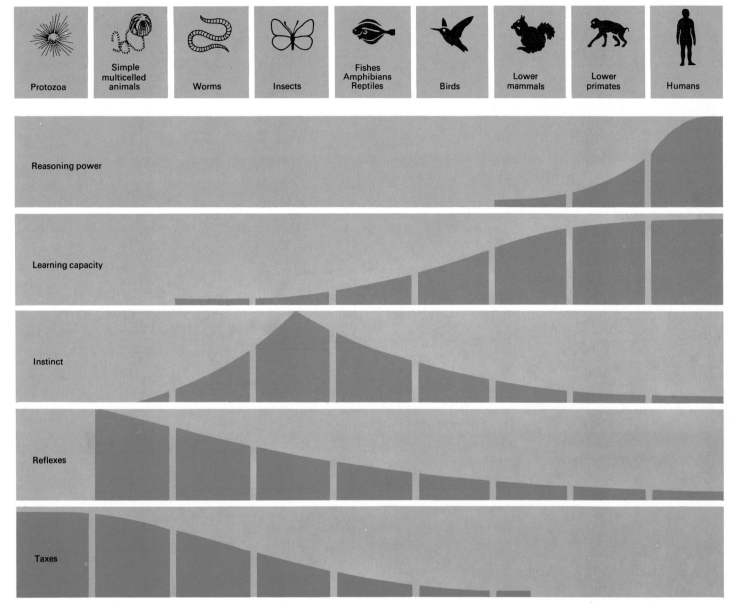

| Protozoa | Simple multicelled animals | Worms | Insects | Fishes Amphibians Reptiles | Birds | Lower mammals | Lower primates | Humans |

Reasoning power

Learning capacity

Instinct

Reflexes

Taxes

This chart gives some idea of the degree to which certain groups of animals display various types of behaviour. For example, the dark green curves show that the protozoa and simple multi-celled animals behave mainly by taxes and reflexes, whereas humans and other higher animals have higher capacities for learning and reasoning, although neither instinctive nor reflex behaviour is lacking in them.

moves one or more particular parts of their bodies. This happens when you touch a hot object and pull your finger away automatically, without any thought or decision on your part. Such a response is called a *reflex*. Many kinds of reflexes, including the example above, are inherited and qualify as evolutionary adaptations, or *instincts*, as some prefer to call them.

Much behaviour that was once considered 'instinctive' has proved under scientific investigations to be what ethologists call *acquired* behaviour. For example, we now know that cats do not have an innate 'instinct' for killing mice; a kitten must observe adult cats chasing mice and be induced to imitate them in order to acquire this particular behaviour. When this does not occur, kittens and mice have been observed to grow up together in perfect harmony.

To acquire behaviour, animals must be able to make use of experience and derive from it a behavioural pattern which they subsequently follow. This is already a far cry from the innate behaviour involved in taxes, tropisms, and reflexes. However, there are significant differences in the ability of animals of different species to acquire behaviour. The first step consists of *habit*; this is acquired when an individual ceases to make any response to a certain stimulus that is repeated over a period of time and is found to be *neutral* – without

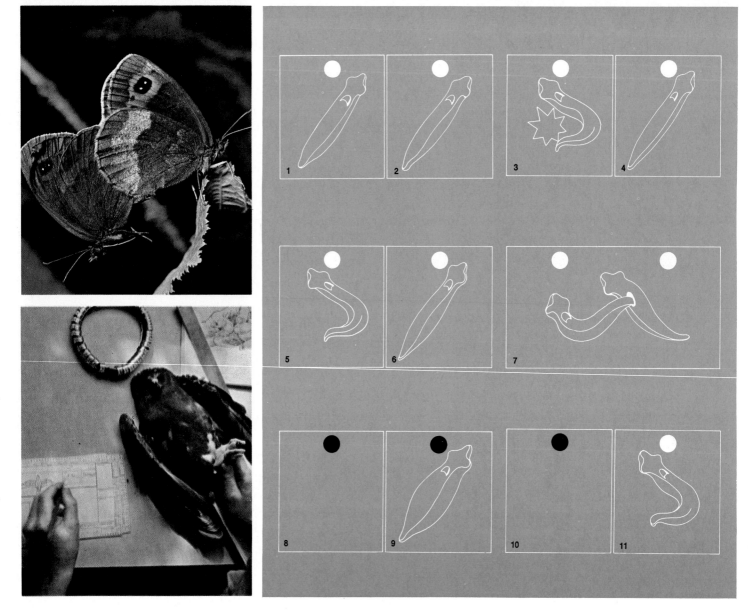

advantage or disadvantage for the animal. A common example is the behaviour of game animals that do not react to the thousand innocent sounds in nature but flee at the sound of an enemy. A second level of learning is acquired by *imprinting*. Usually this dates back to the first days in the life of an individual or during the time of its development. It is frequent among some birds, which tend to follow consistently whatever object they first see in motion. For example, baby chicks born in an incubator may consistently follow the person who took care of them during their first days of existence, just as they would have followed the hen if she had hatched them.

Another way of acquiring behaviour involves a so-called *associative* memory and has many applications in training domestic animals. Many living things – from insects to the larger mammals – are able to associate two or more phenomena when these occur repeatedly and in rapid succession. It is associative memory that guides domestic pets and helps them to find their way home. In the particular form of associative memory that we call *conditioning*, the subject is exposed to two stimuli simultaneously, one neutral, one agreeable. After a certain lapse of time, the neutral stimulus alone will recall to the animal's mind the agreeable stimulus. For example, if a dog is fed

Top left: mating behaviour of two butterflies. Below: indexing a banded bird for migration studies. Mating and migration are both instinctive behaviour. Right: two planarians, or flatworms (1 and 2). No. 1 is trained to receive a shock when a light is turned on (3), so contracts at the light signal even if not given a shock (5). No. 2 is left undisturbed (4 and 6). In phases 7,8,9, No. 2 devours No. 1. Now No. 2 contracts at the light signal, apparently because it has somehow acquired No. 1's reflexes.

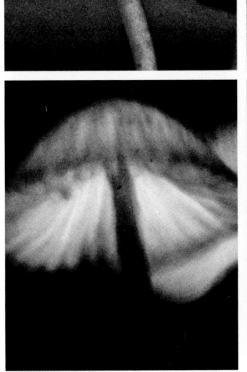

Left: photographed in light (top) and in dark (below), a fungus of the genus Mycena *displays the phenomenon of bioluminescence, produced by part of the energy released in the plant's respiration. Top right: bioluminescence in siphonophores, sea animals similar to jellyfish. Below: luciferin and luciferase, substances extracted from fireflies, slowly oxidize when combined in water, producing relatively cold light. In nature, energy from the firefly activates the reaction.*

repeatedly subsequent to the ringing of a bell, eventually the sound of the bell alone will activate the dog's salivary glands, because the dog associates the sound of the bell with the pleasure of eating. This was, of course, one of the famous conditional reflex experiments of the Russian physiologist Ivan Pavlov.

Still another type of acquired behaviour is known as learning by *trial and error*. Through making repeated attempts, unsuccessful and successful, to fulfil a particular need, a subject eventually learns to use the course of action that has proved most successful in the past. The speed with which the subject acquires this ability is one measure of its capacity to learn by this particular method. When an animal learns by *observation*, a typical process in younger animals, it is because it has observed over a period of time the behaviour of other animals of its own kind.

The third and final way of learning behaviour is called *intelligence*. We use the word *intelligence* almost exclusively in connection with humans, so it might seem improper when applied to other animals. However, if we interpret intelligence as the ability to respond adequately to a new and unexpected situation, we have ample experimental data to indicate that other animals, too,

are endowed with a degree of intelligence. The most common test for assessing an animal's intelligence is to put food in front of it so that to reach the food the animal must overcome certain obstacles or use objects with which it has had no previous experience. Except for humans, the only animals able to pass such a test the first time they are subjected to it are certain types of more evolved apes. Perhaps cats, dogs, and rats will not succeed at the first attempt, but they learn very quickly. Last on this scale come the birds and fishes and, in a few rare instances, some invertebrates.

Acquired behaviour may or may not contribute to an individual animal's survival; in any event, it is not transmitted genetically to the animal's offspring. On the other hand, the kinds of behaviour, and the physical characteristics affecting behaviour, that are most vital to a species' survival are the evolutionary adaptations, which are transmitted genetically. A fascinating physical adaptation that helps some species survive is called *mimicry*. When a species has evolved in colour, and sometimes shape, to resemble something else in its natural environment, this is called *deceptive mimicry*. For example, the snowshoe rabbit's coat changes from brown in summer to white in winter, making the animal less visible to predators and thus giving it a better chance of

Top left: a chameleon ambushing insects in a green plant takes on the plant's coloration as pigments in the animal's skin react to its surroundings. Below: rest position of the geometrid moth's caterpillar camouflaged it as a twig. Right: a scarpena on the sea-bed is camouflaged by algae and other marine organisms living on the fish's speckled skin – an example of mutualism, or mutually beneficial symbiosis.

Top left: treehoppers, parasites that suck the sap from the plants, have evolved a thorn-like shape that protects them from insectivorous birds. Below: this leaf insect has evolved a shape and colour to match the leaves it feeds on. Top right: the bumble-bee hawkmoth has evolved a protective shape and colour that mimic those of its stinging namesake. Below: a threatened caterpillar draws in its claws and activates two eye-like markings that give it the appearance of a small poisonous snake.

survival. Many insects that live on plants have evolved to resemble leaves, twigs, or thorns. Deceptive mimicry gives some animals an aggressive, as well as a defensive advantage. The chameleon's body colour changes somewhat to camouflage the reptile, improving its ability to sneak up on prey and decreasing its visibility to predators. Bad-tasting butterflies and stinging bees have evolved bright colours that warn experienced predators to avoid them; this adaptation is called *advertising* mimicry. Other species of butterflies that predators savour gain some protection from their adaptive resemblance to the bad-tasting species.

Another interesting physical adaptation is *bioluminescence*, the light produced by certain plants (such as some fungi), by certain sea fish and one-celled sea animals, and by fireflies. This light is produced by chemical processes that give off almost no heat at all. In some fish the luminosity serves as a bait to lure prey; in others it may serve, like advertising mimicry, to warn off predators. Some fireflies are known to flash their lights in coded patterns that attract fireflies of the same species for mating.

Evolutionary adaptation shapes such behaviour as animals' mating habits, certain animals' ability to find their way over long distances, and the

adjustments most animals make to rhythmic changes, such as the day and night and seasonal cycles. Some species make the latter adjustments by *migrating* – moving individually or in large groups from one place to another. Tiny sea plants and animals called plankton migrate upwards in the ocean at night, and downwards in daytime. Cooling weather affects the glands of certain birds and butterflies, triggering their migration – often in vast flocks – to warmer climes; some species fly thousands of miles between their summer and winter environments. Salmon travel similar distances from the oceans to the headwaters of rivers to breed in the same places that they were hatched.

Eels are particularly well-known migrants. Both American and European eels spawn in the Sargasso Sea, an area in the Atlantic Ocean between the West Indies and Bermuda. The European larvae reach Europe when they are about three years old. The American eel young move north when they are about a year old. Bats are migrants, too, leaving their summer quarters for caves or tree hollows where they can hibernate.

Then there is another movement that might be termed migration, albeit loosely. That pattern is age-dependent. For instance, larvae of some insects live in the water – even on stream or lake bottoms. When they are grown they come

Top left: salmon migrate from the open sea through rivers to mountain streams to deposit their eggs where they themselves were hatched. Below: monarch butterflies massed at a rest stop on their migratory flight. Top right: vast flocks of birds migrate thousands of miles each spring and autumn. Below centre: flightless penguins migrate on foot, without fish to feed on, to breed hundreds of miles inland in Antarctica. Below right: banding birds helps scientists to study their migration and navigation behaviour.

Zebras, like many other species of animals, are adapted to live together in groups that provide defence against predators. For predators such as wolves and hyaenas, group living provides the advantages of group hunting. Animal communities are often dominated by a single strong male or female; some operate as a social hierarchy, with a definite 'pecking order', in which each adult dominates those 'below' him and is dominated by those that have proved their superiority, by combat or threats.

to the surface and fly off, never to return to their watery birthplace except to lay their own eggs. Some scientists believe this phenomenon is related to water's being the original home of all life.

Social behaviour can be defined as the relations between individuals of the same species. Zoologists often describe animal 'societies' as either *individualistic* or *collective*. In individualistic societies, such as wolf packs, many herds of ruminants, and certain insect groups, each adult individual is completely autonomous in providing for its own vital necessities. A collective society, such as those of the termites, ants, and 'social' bees and wasps, is distinguished by the division of work among the members of the group. A *polymorphic* collective society is one composed of individuals of the same species, but of different forms, such as fertile and sterile. Since social behaviour varies from species to species, it, too, seems to be chiefly the product of evolutionary adaptation.

ECOLOGY

The word *ecology* was first defined by the German biologist Ernst Haeckel in 1869. He described ecology as the study of the relationships between living things and the environment in which they live. Ecologists view the plant and animal populations within a given area as a *community* in which different organisms perform different rôles. A community of living things, together with the non-living parts of their environment, forms what is called an *ecosystem*. The whole earth is an ecosystem; so is a forest, an ocean, a pond, or a rotting log. The living things in any ecosystem can be classified according to their functions. The *producers* are the green plants, which make food; the *consumers* are the animals, which eat plants or other animals; the *decomposers* are the bacteria and other tiny organisms that break down animal wastes and the remains of dead plants and animals into raw materials that are used by the plants to make more food. Each species of organism in the ecosystem has a specific task, each plant species producing a different kind of food, each animal species consuming different food, and each decomposer performing a different operation on wastes. Ecologists call a species' task its *niche*, and if two species compete for the same niche in an ecosystem, one species or the other eventually dies out.

All these living things are constantly affecting and being affected by each other, as well as by the non-living things in their environment – the soil, the water, and the air. A very simple ecosystem, such as a small, still-unpolluted mountain lake, tends to maintain a *dynamic equilibrium*, or balance of energy, among the producers, consumers, and decomposers, even though their populations vary somewhat with the seasons. But everything in nature is constantly changing – usually very gradually, but sometimes rapidly – so no ecosystem can remain completely stable over a long period of time. A change of climate, or the entry of humans into our ideal mountain lake ecosystem, may shake its equilibrium drastically.

The *physical*, or non-living, environmental factors in an ecosystem include the light, temperature, radiations, pressure, water, air, climate, gravity, and soil. For each of these factors the organisms have minimum and maximum acceptable values in the environment in which they live. For example, the maximum temperature tolerated by an animal is about 126°F. (52°C.), but for certain bacteria the maximum temperature may be as high as approximately 185°F. (85°C.). Spores and the seeds of some plants tolerate even higher temperatures. For some living things the lowest tolerable temperature levels are close to 32°F. (0°C.); some die when the liquid within their cells congeals (at about 28°F. [−2°C.]), while others may die even at higher temperatures. Each organism has an optimum temperature that is closer to the maximum than to the average between the two limits. Pressures also vary to an incredible degree in some environments; a fish at deep sea levels is exposed to pressure a thousand times greater than that at which a high-mountain bird may live.

The biological factors in an ecosystem are even more complex than the physical factors. Two of these factors that tend to limit animal populations are *predation* and *competition*. Predation is the killing of animals by other animals for food. Predatory animals are usually much fewer in number than the animals they prey on, and their hunting helps to keep the prey animals from increasing beyond the capacity of their food supply. Competition is the striving of two or more organisms for a necessity of life (such as food, water, minerals, living space, or sunlight for plants) that is inadequate to fill the needs of all.

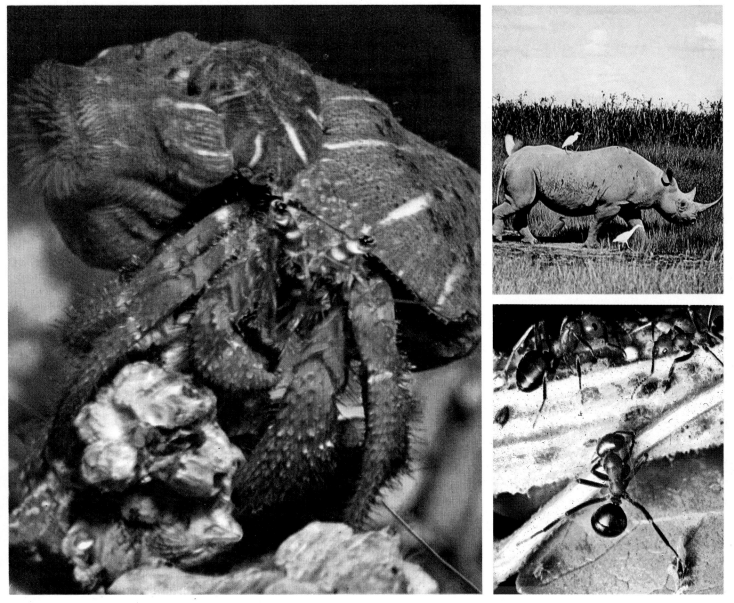

Examples of symbiosis. Left: a sea-anemone (animal) rides on a hermit crab, whose mobility brings food to the anemone while its stingers help protect the crab from predators – a case of mutualism. Top right: cattle egrets hunt parasitic insects on a rhinoceros's tough hide, mainly benefiting the egrets – a case of commensalism. Below: ants 'milk' aphids (plant lice) for a sugary secretion that nourishes the ants – commensalism. But some ants also 'feed' the aphids by placing them on plants – mutualism.

Competition often involves organisms of the same species, because they have the same requirements. It seldom results in bloodshed, but the 'losing' competitor may have to migrate or die.

A relationship in which two individuals of different species live together is called *symbiosis*. A symbiotic relationship that benefits both *symbionts*, or participants, is called *mutualism*. For example, the sea-anemone (an animal despite its floral name) often becomes attached to the back of a hermit crab; scraps from the crab's food feed the sea-anemone while its stinging cells help protect the crab. When one symbiont benefits from cohabitation without helping or hunting the other, we call it *commensalism*. Ants, for example, often associate with aphids, or plant lice, to feed on the saccharine substances produced by the latter.

When one symbiont lives entirely at the expense of its host, we call it *parasitism*; this is often harmful to the host-organism, but seldom lethal. *Ectoparasites*, such as lice, ticks, and fleas, lodge on the surface of their host and usually feed on its blood; they may harm their host by transmitting internal parasites through their blood-sucking organs. The *endoparasites* install themselves in the host-organism's blood, intestines, liver, lungs, or

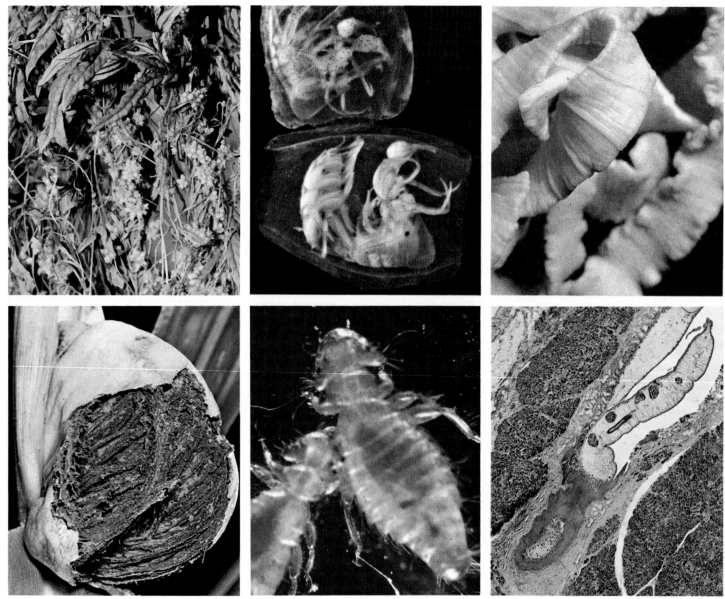

muscles. Particularly lethal are those parasites that deposit their eggs inside other insects, which are devoured by the parasite's larvae when they hatch.

Some plants, as well as animals, are parasites to plants. Many of the parasite plants lack *chlorophyll*, the substance that enables green plants to make food by photosynthesis. One such example is the dodder, a leafless yellow stem that climbs the stalks of green plants and puts out special roots that pierce the host-plant and suck nourishment from it.

The part of an ecosystem where an organism lives is called its *habitat*. A fish's habitat could be a whole pond, and thus the pond's entire ecosystem; a micro-organism's habitat might be a tiny area on the pond floor, and thus only a small part of its ecosystem. Some typical habitats in nature are forests, ponds, marshes, grasslands, oceans, rivers, and so on. A barren island occupied only by birds of the same species would be an *exclusive* habitat; a forest teeming with populations of different species would be a *heterogeneous*, or mixed, habitat.

One very unusual habitat is the cave, which accommodates different animal and plant species in an environment of almost total darkness. There are two basic kinds of cave-dwelling organisms. The *troglophiles* (*troglos* is Greek for

Top left: a leafless, parasitic plant, the dodder, winds around a plant and sucks food from its bark. Below: the 'carbon' parasite in corn – a microscopic fungus that resembles black powder. Top centre: the crustacean Phronima sedentaria, *a parasite of several marine invertebrates. Below: parasitic fleas of the chicken. Top right: the* taenia, *or tapeworm, is a parasite that lives in the human intestine. Below: human bile duct damaged by an oriental parasitic worm,* Chlonorchis sinensis.

Top left: predatory snake about to swallow a bird's egg. Below: vultures devour a dead animal's carcass. Vultures are mostly scavengers, rather than predators. Centre: waves breaking on the shore produce an ecosystem abundant in microscopic plant and animal life. Top right: fumes spewed from a steel mill pollute the atmosphere and adversely affect the ecosystem for miles around the mill. Below: indiscriminate hunting of coyotes results in an increase of the rodents on which coyotes feed.

'cave') prefer a dark environment inside caves, though they may also live elsewhere; they include certain gastropod molluscs, some reptiles, such as a species of salamander, and also a few mammals, such as bats. The *troglobioms* never leave the cave; their structure and physiological make-up are totally adjusted to darkness, scanty ventilation, almost constant temperature, and maximum humidity. They include insects, worms, spiders, crustaceans, myriapoda, molluscs, and a few vertebrates, such as fish and amphibians. Cave plants vary in species according to the amount of light reaching them. An area that receives abundant light is tenanted by mosses, ferns, and lichens; in dimmer light there are sterile ferns and fungi; only a tiny bit of light suffices for some mosses and a few small algae; in total darkness, the only plants are microscopic fungi.

The forest, with its abundance of trees of different heights, bushes, climbing vines, and so on, is one of the environments richest in plant and animal variety. Ecologically it is one of the finest, because the soil retains all its natural richness and the water system and climate are regulated by the trees and smaller plants. The forest is a genuine 'explosion' of vitality, and provides an ideal habitat for the most widely diversified species of living things, with all kinds of living

habits. Unfortunately, humans have destroyed many forests and converted the land to farms.

Marshes, or wetlands, are extremely rich in living organisms. They form in the shallow parts of a lake where water plants can take root and multiply, along rivers in places where the land is too flat for the water to drain off, and especially in estuaries – the places where rivers slow down and deposit their sediment before emptying into the sea. The latter places are called *tidal marshes*, because the tide pushes salty ocean water into the marsh twice a day. Inland marshes provide favourable habitats for fresh-water fishes and plants, amphibians, reptiles, insects, and especially birds, which feed on the other marsh animals. Tidal marshes are filled with plants that are adapted to both salt water and the brackish fresh water left when the tide ebbs. Many ocean fishes, crabs, lobsters, and so on lay their eggs in tidal marshes, where the plants and their varied insect populations provide food for the young sea animals when they hatch. These 'nurseries of the ocean' contain many times as much living material as the open sea. Only recently has the value of wetlands been recognized and an effort made to keep them from being filled in with rubbish and soil as sites for homes or factories.

Left: reptiles, mammals, birds, fishes, and insects all share this forest-pond habitat. Top right: a polar bear hunts for fish in its typical Arctic habitat. Below: the water-lily, with its large, floating leaves, is adapted for survival in a marshy habitat.

Cave-dwelling organisms. Top left: millipede. Below: moth. Top right: moss growing in an area almost devoid of light. Below: blind and unpigmented salamander. The moth may spend part of its time outside the cave; for the others, the cave is their permanent home.

Ecologists refer to a large area in which the same climate, plants, and animals predominate as a *biome*. Similar biomes are found around the earth at the same latitudes, north and south, and at comparable altitudes above sea level. The *tundra* (Russian for 'marshy plain') is a vast, treeless land that surrounds the Arctic ocean. The *taiga*, or northern coniferous (evergreen) forest, forms a wide band across North America, Europe, and Asia just south of the tundra. Neither of these two biomes has a counterpart in the southern hemisphere, because the land areas there do not extend into comparable southern latitudes. In the middle latitudes are the *temperate deciduous forests*, composed mainly of trees that shed all their leaves periodically, usually in autumn; the *temperate grasslands*, or prairies; and *deserts* in the dry climates. In tropical areas where rain falls daily, are *tropical rain forests*; and where dry seasons are especially long and severe, the *tropical grasslands*. Mountain areas do not qualify as biomes, because they have complex zones of plant life ranging from deciduous trees and meadows in the valleys, upward through coniferous forests, then tundra-like vegetation, to no plants at all on the high peaks.

Biomes of the same type have the same kinds of ecological niche, though a specific niche may be filled by plants or animals of different species in different

biomes. For example, the primary consumers in grasslands are grazing animals; bison and pronghorns (antelope) have yielded this niche in North American grasslands to cattle and sheep. In Africa it is occupied by zebras and gazelles, in Asia by wild horses, and in Australia by kangaroos. Ecologists believe that by studying a particular biome intensively, they can use their findings to draw conclusions about how other biomes of the same type work.

Analyzing the *energy budget* of the biome is an important part of such a study. All of the energy in an ecosystem comes originally from the sun. By drying and weighing all the plants from a measured area of land, scientists can estimate how much of the radiant energy reaching the area has been changed by green plants into chemical energy, or food. Some of that energy is transferred to the prime consumers – grazing mammals, birds, insects, even parasitic plants – that feed on the green plants; in turn, some of their stored chemical energy is transferred to the predator animals that prey on them. Finally, some of the energy reaches the bacteria and other micro-organisms as they decompose the wastes and remains of the plants and animals. Energy cannot be destroyed; but whenever it is used, some is changed into heat and transferred to the surrounding air, where it is no longer available to power the

Types of forest. Left: coniferous forest, where only a few species of deciduous trees are found. Top centre: fluvial, or gallery, forest, following the course of a river and forming a vegetative 'roof' over it. Below: temperate rain forest with numerous small trees and a thick undergrowth of mosses and ferns. Right: tropical rain forest, dense with hundreds of species of broadleaved and evergreen trees, abundant lianas, and thick undergrowth of arborescent ferns and tall grasses.

View of a marsh environment: new trees grow from the hulk of a 'drowned' tree; aquatic plants such as grasses, rushes, and water-lilies flourish. Aquatic birds find the marsh relatively safe and well stocked with food – an ideal habitat for resting on migration or nesting and raising young. Many marine fishes are hatched and spend the early part of their lives in the salt-water marshes, where the tides mix sea water with the fresh water from rivers and streams.

growth or activities of living organisms. A portion of the energy stored in the green plant is used and lost in this way by each organism in the *food chain* – the plant itself, the prime consumer, the one or more predators in the chain and finally the decomposers. Thus the population of any species of organism in the food chain is limited by the energy available to it from the organisms lower on the food chain, on which it feeds. Humans are no exception to this rule. Since our forbears learned to raise plants and animals for food about ten thousand years ago, the human population has been growing – slowly at first, but at an ever-increasing rate. In the two hundred years between 1650 and 1850 the earth's human population doubled from five hundred million to a thousand million. This increase was made possible by the development of new ways to raise food crops, of machines to use the energy stored in the earth's deposits of coal, gas, and oil, and of ways to extend human lives by curing certain diseases and preventing others. Such developments have continued to increase, making possible a world population that is nearly four thousand million today and growing at a rate that could double that figure by the year 2000.

Man's need for living space and food has increased in proportion to his numerical increase. In addition, his desire for objects and services that can only

be provided by exploiting the earth's limited supply of forest, stored minerals, and stored energy has skyrocketed over the past hundred years. To satisfy these needs and desires, we have been encroaching more and more on natural areas, destroying their plant and animal populations as we grow. In the past century the millions of bison that once roamed the North American prairies have been reduced by ruthless hunting to a few small herds that would disappear without human efforts to keep them alive. Many species of whales – the earth's biggest mammals – have been hunted to near-extinction and will not survive unless all nations forbid their destruction. Many other species of mammals, birds, and fish have been destroyed or require an intense human will and effort to ensure their survival. In many parts of the world the destruction of forests for farming, timber, and fuel have left immense areas without plant cover, exposed to continuous erosion by wind and water. Fortunately, nearly a hundred years ago, some far-sighted persons realized the importance of preserving at least some natural areas. Thus was born the world's first national park, Yellowstone, in Montana, Idaho and Wyoming; many more such parks have since been established in the United States and other countries. At first these parks were set aside to protect the wild animals and plants therein from

Top left: ibex in Grand Paradise National Park, Italy. Below: Tokasaki Park in Japan, where a colony of seven hundred macaques lives. Centre: skeletal remains of large bald cypress trees in Everglades National Park, largest remaining sub-tropical wilderness in the United States. Top right: Amboseli Park in Kenya, where numerous animal species are protected. Below: the few remaining European bison survive in Oka Park, USSR.

Unique and beautiful geological formations are also protected from human exploitation in parks such as Zion National Park in Utah (top left); Geological Park in Ojcow, Poland (below); Yellowstone National Park in Idaho, Montana, and Wyoming (top right); and Wind Cave National Park in South Dakota, where a herd of the near-extinct American bison is protected (below).

human depredation and to serve as 'showcases' for visitors, in the hope of developing respect for nature and understanding of its processes. This effort has become so successful that some parks now must limit admissions to prevent further deterioration of their natural environments by the sheer number of visitors. Scientists now realize that our need to preserve natural areas is not merely aesthetic or educational in character. The extinction of any plant or animal species withdraws its unique genes from the earth's 'gene bank', thus diminishing its capacity to fund the evolution of new species or to supply genes that scientists might use in developing new and better breeds of plants and animals for human needs. As a result some areas little frequented by humans are now being preserved as 'wilderness areas', to be kept 'ever wild'.

Man's ability to change the environment to suit his needs and desires is an important reason for the immense growth of the human population. But we have been pouring our waste gases, liquids, solids, and heat – many of them toxic to living organisms, including ourselves – into the earth's atmosphere, waters, and soil without regard for the capacity of natural forces to make the wastes harmless or useful as raw materials. In this way we have polluted the physical part of our environment and thus threatened our own survival as well

as that of other living things. Furthermore, we have only recently recognized that our supply of fossil fuels – energy from the sun that was stored underground in plant and animal remains millions of years ago and gradually converted into coal, oil, and natural gas – is not endless. It seems clear that we must adapt our behaviour to the requirements and capabilities of our natural environment, if man is to continue being a 'successful' species.

Biogeography is the study of the geographical distribution of animals (*zoogeography*) and plants (*phytogeography*). By comparing the living and fossil flora and fauna of the earth's land areas, biogeographers have sought to explain the differences and similarities between the species that evolved in different regions. (See maps on pages 52–53.) There is little doubt, for example, that certain animals – including humans – came from Asia to North America via a land bridge that connected Siberia and Alaska in the Ice Ages, when considerable ocean water was frozen in glaciers. Biogeographic evidence also suggests that animals and plants spread into the various regions when the continents were together in a single land mass, then evolved into similar but different species when the continents separated and were isolated by the

Left : many fish die each year as sewage, agricultural and industrial wastes pollute streams, rivers, lakes, and eventually the seas. Centre and right : particles of ash and unburned fuel, plus waste gases that are often toxic, pour into the air from factories, domestic fires, and car exhaust pipes. In big cities the sunlight often transforms these waste gases into photochemical smog, a mixture of smoke and fog harmful to living things.

Early detergents (man-made 'soaps') in sewage water blanketed rivers with foam that kept air from the water, suffocating the fish, and polluted the water used by towns downstream. This problem was met by changing detergents so that bacteria can decompose them. But, like the fertilizers that wash off farmlands, some detergents still contain phosphorus; in streams, ponds, and lakes, this causes great growths of water plants, which use up so much oxygen in decaying that many fish suffocate.

oceans. This theory seems to be confirmed, at least in general terms, by the recent discovery – resulting from drilling into the oceans' floors – that the continents actually did drift apart, beginning about sixty million years ago.

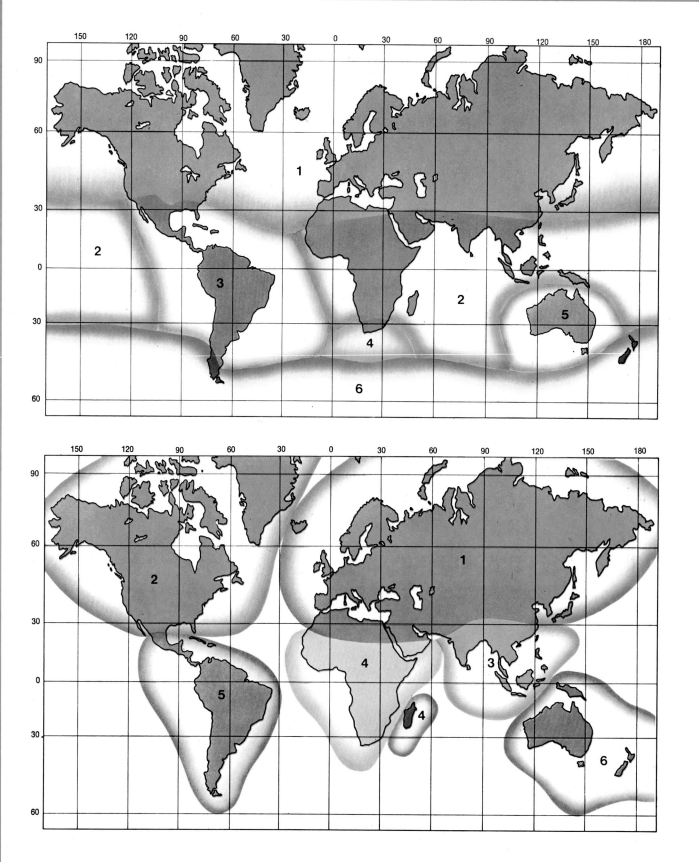

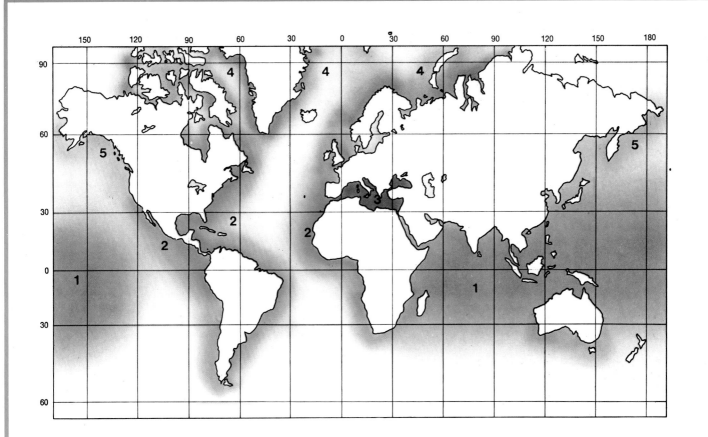

Scientists divide the earth into floral and faunal regions on the basis of the present distribution of plants and animals. A region includes areas that, in general, have the same or similar plants or animals. The map at the top of the opposite page shows the world's floral regions: (1) Holarctic region; (2) Palaeotropical; (3) Neotropical; (4) Cape; (5) Australian; (6) Antarctic. The bottom map shows the faunal regions: (1) Palaearctic region; (2) Nearctic; (3) Oriental; (4) Ethiopian (some scientists list the island of Madagascar as a separate region); (5) Neotropical; (6) Australian. The map at the top of this page shows a simplified version of the geographic distribution of coastal marine animals: (1) Indo-Pacific region; (2) Atlantic and western American; (3) Mediterranean; (4) Arctic; (5) North Pacific. The map on the right shows the present distribution of the Camelidae family (camel, dromedary, llama) in red; areas occupied by their ancestors in brown; and routes of their migration in red lines.

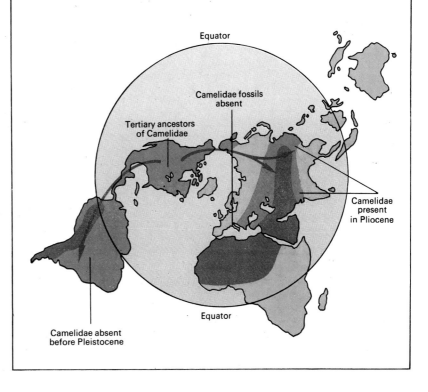

REPRODUCTION

Strive as it may to survive, each living organism must eventually die. But every living thing is capable of reproducing itself. So as long as the individual members of a species live long enough to mature and reproduce, thus replacing the individuals that die, the species itself will survive. The large number of individuals that an animal or plant is capable of producing makes species survival highly probable – though not completely certain. The processes of reproduction are many and varied; however, we can make a definite distinction between two forms: the *asexual* process and the *sexual* process. Each has its place in the two kingdoms of nature, plant and animal.

Sexual reproduction always involves the union of two gametes, or sex cells, to form a new cell, the zygote. Since the gametes usually (but not always) come from two different parent organisms, the zygote develops into an individual with some of the genetic characteristics of each of its parents. Asexual reproduction, on the other hand, always involves only one parent organism; from it a single cell or a many-celled fragment separates and forms an independent organism with the same genetic characteristics as its parent. Many plants and some animals reproduce both sexually and asexually; some can reproduce both ways simultaneously.

Asexual reproduction usually takes one of two forms. In one form, called *vegetative* reproduction, both one-celled and many-celled organisms reproduce by *fission*, or splitting, of the organism's body, each part of which develops into an individual organism. When a single-celled organism reproduces this way, the maternal cell simply divides into two identical daughter cells. Many plants can be made to reproduce from a *cutting* – a piece of leaf planted in soil or a twig grafted on to the stem of another plant. Some species of planaria (flatworms) 'drop' the tail, which then *regenerates*, or develops into a complete planarian. Some plants (yeast for one) and animals (hydra for one) reproduce by *budding* – growing a new organism that separates from the parent and becomes autonomous. Many plants grow special organs of vegetative reproduction, such as the tubers of potatoes, the bulbs of onions, and the runners of strawberries, which grow into individual plants. The foregoing methods are called vegetative reproduction because they involve cells from vegetative parts of the organism – leaf, stem, or root, for example.

The second form of asexual reproduction occurs in plants that produce *spores*, or cells that are specialized for asexual reproduction. Since this phenomenon is limited to plants, we will discuss it later along with their sexual reproduction processes.

Sexual reproduction is the most widespread form, occurring in all animals and many plants. This is the union of two gametes to form a zygote, the first cell of a new organism. The evolutionary advantage of sexual reproduction over asexual reproduction lies in the fact that the former produces individuals with different combinations of their two parents' genetic characteristics, while the latter produces individuals with genetic characteristics identical to those of the single parent organism. Even the protists and other micro-organisms that usually reproduce by fission occasionally come together, at least temporarily, and exchange some of their genetic material. But biologists usually describe reproduction as sexual only when *fertilization* – the union of two cell nuclei within a single cell – takes place. Before this can happen, cells with only half the normal number of chromosomes (gametes) must be produced by the process of meiosis. (See Chapter 1.)

For the rhinoceros and other mammals, reproduction alone is not sufficient to ensure the survival of the species. Unlike animals such as fish, reptiles, and amphibians, a newly hatched bird or a newborn mammal needs protection and care, usually provided instinctively by the mother or both parents. Once young birds have learned to fly, they leave the nest; young mammals take from a few weeks to many years (in the case of humans) to learn the techniques of survival.

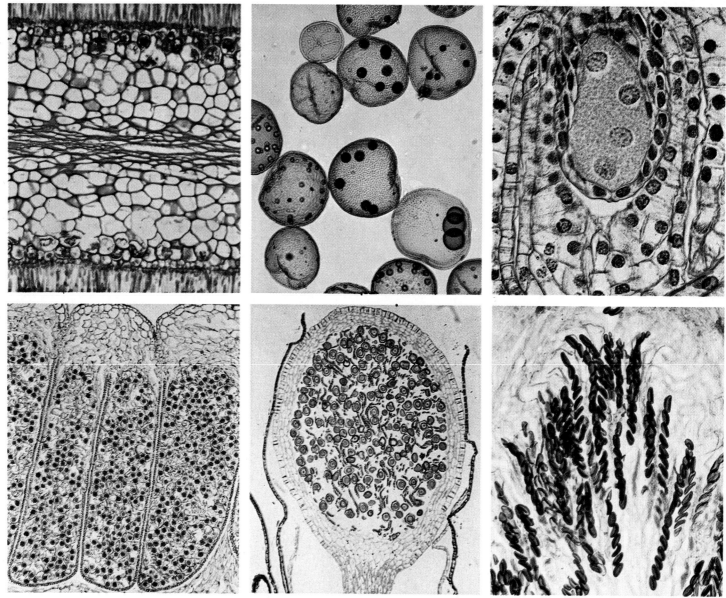

In the less evolved kinds of many-celled animals, the location of the gametes in the organism's body is not sharply defined. In some cases – sponges, for example – all parts of the body may at different times accommodate gametes, which are always on the move.

In the more highly evolved animals, gametes are produced in organs called the *gonads*. Gonads that produce *ova*, or egg-cells, are called *ovaries*; those that produce *sperm*-cells are called *testes*. An animal that has only ovaries is a female; one that has only testes is a male. An animal that has both ovaries and testes is called a *hermaphrodite*. Some common hermaphroditic animals are earthworms, most flatworms and some crustaceans. In most such species the sperm from one individual will unite only with the egg from another individual. But others, such as the tapeworm, have sperm that can unite with an egg from the same individual – a process called self-fertilization.

A very special type of reproduction, called *parthenogenesis*, occurs in certain animals and under certain conditions when non-fertilized eggs develop into individual animals. In this way aphids (plant lice) produce generation after generation of females during the summer. When autumn comes they produce males, permitting fertilization of the zygotes, which lie dormant until spring.

*Examples of spores, asexual plant reproductive bodies. Top left: kelp (*Laminaria*) blade in section, showing spores growing on blade surfaces. Below: section of horsetail with spores in sporangia. Top centre: volvox colonies with spore clusters (black spots) and daughter colonies (green spots). Below: section of sporangium in liverwort plant. Top right: lily ovule (centre) containing four megaspores, one of which will become a female gamete, or egg. Below: spores in sac of Ascomycetes fungus.*

A seed is an embryo plant with a seed coat and usually a store of food. Top left: milkweed seeds with tufts of silky hairs that facilitate dispersion by the wind. Below: winged seeds of the maple tree. Centre: sweet sorghum seeds (above) and rye seeds – both food grains. Top right: winged seeds from a pine tree cone. Below: seeds of Sterculia tragacantha, still encased in their pods.

Honey-bees combine parthenogenesis and regular sexual production; eggs laid by the queen bee hatch males if they have not been fertilized and females if they have been fertilized. The males, having only half the usual number of chromosomes in their cells, tend to be less robust than the females. This may explain why the eggs laid at the end of summer are all fertilized; they lie dormant through the cold winter and in spring hatch females that reproduce by parthenogenesis. Embryologists can produce this phenomenon artifically by subjecting the eggs of certain animals to weak acids or changes in pressure; usually the process is incomplete, but sometimes it produces an individual that can be raised to maturity.

The male and female gametes produced by most animals are specialized and quite different from each other. The egg-cell is relatively large, approximately spherical, and contains more stored food than the sperm-cell. The sperm is relatively tiny, consisting mainly of a rounded head containing the nucleus and either a tail or a whip-like *flagellum* that propels the cell rapidly through liquids. Many fish and other aquatic animals simply shed their gametes in the water, where some of the swimming sperms, which vastly outnumber the eggs, meet and fertilize the eggs. The large numbers of eggs and sperm ensure that at

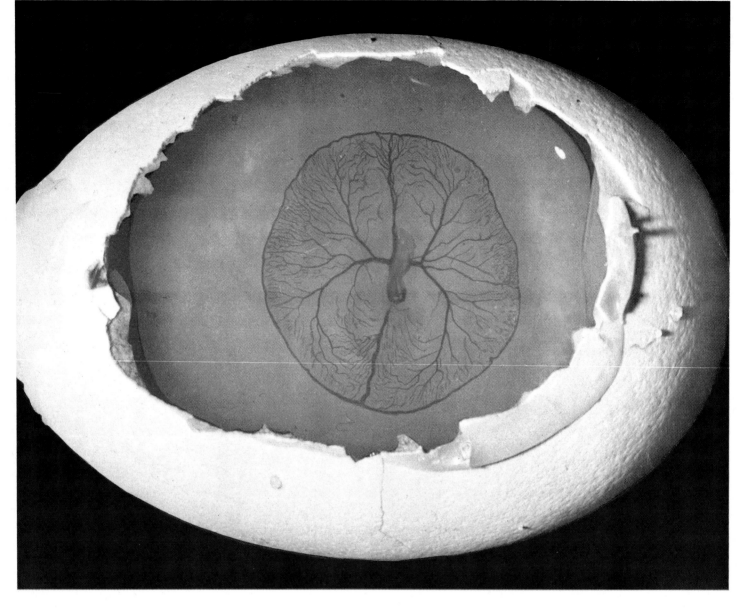

least enough get together to perpetuate the species. In a number of cases, a species has seasonal periods of sexual activity, or behavioural patterns, such as courting, that help make the eggs and sperm available to each other at the same time and place. In many crustaceans and certain fish, such as sharks, fertilization takes place inside the female. Since sperm require a watery environment to move around in and eggs die quickly when exposed to air, internal fertilization is also the rule for land animals – mammals, birds, reptiles, and insects. In the greatest number of species, internal fertilization is facilitated by the act of copulation, in which the male inserts his penis into the genital duct of the female and deposits his sperm there. Attracted by a chemical substance emitted from the egg, the many sperm race towards it: as soon as one male gamete penetrates the double membrane around the egg with its head, the space between the two membranes fills with a clear, sticky gel that forms a *fertilization membrane*, which no additional sperm cells can penetrate. As soon as the sperm nucleus contacts the egg nucleus, they merge; this union of gamete nuclei, called *karyogamy*, completes the fertilization process. The zygote, or fertilized egg-cell, now has a complete set of chromosomes carrying the genetic heritage of its species, and it immediately enters the first phase of its

Eggs produced by some animals have a solid calcareous shell; others are enveloped in a protective membrane; birds' eggs have both. Eggs vary considerably in size, depending mainly on the amount of material stored inside for nourishment of the embryo. This material is called the 'yolk'. Bird eggs, like the one shown here, have additional nutritive material called 'albumen'.

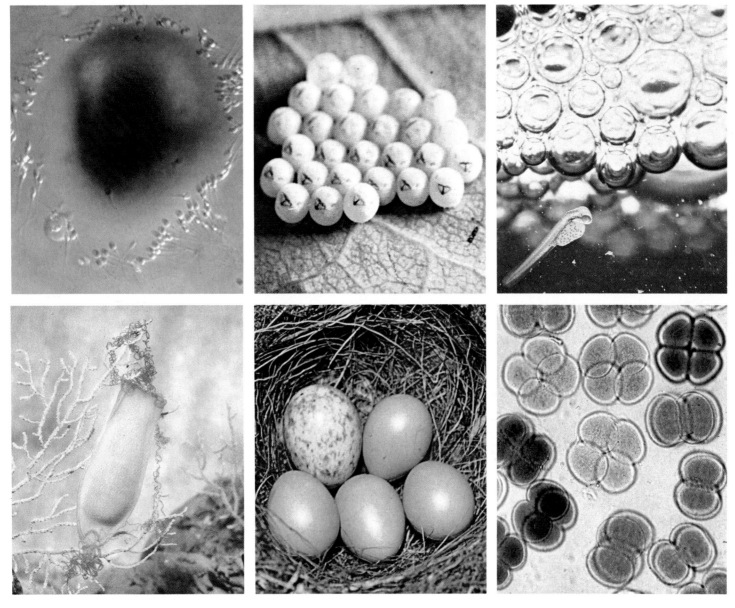

Top left: enlarged photomicrograph of a human egg surrounded by sperm. Below: dogfish shark's egg in horn-like egg case. Females of some shark species retain eggs and bear live young. Top centre: enlarged photograph of insect eggs. Below: the cuckoo often lays its egg in another bird's nest. Top right: developing embryo of a fighting fish (Beta). Below: eggs of sea-urchin, shown in process of division.

development as an *embryo*. Some animals lay their eggs as soon as they are fertilized; others retain them for various periods of time. In birds and reptiles the yolk, albumen, and shell are added to the egg as it travels down a long tube before being expelled. The large yolk provides food for the developing embryo. In humans, usually only one egg is fertilized; the rare cases of two or more produce twins, triplets, and so on. The fertilized human egg attaches itself to the wall of the female *uterus* and receives food from the mother's body during its nine month *gestation*, or development period.

There are many phases in embryonic development, and they vary widely from one species to another. We will mention only the principal ones. After fertilization of the egg, *segmentation* begins, with the egg dividing by mitosis into two *blastomeres*, each of which divides in two, and soon, when segmentation is completed, the blastomeres have formed a *blastula* – a kind of hollow ball. Next, in a process called *gastrulation*, the multiplying cells form a second layer within the blastula; the resulting bowl-shaped structure is called the *gastrula*. The outer layers of cells is the *ectoderm*, or outside skin, while the inner layer is the *endoderm*, or inside skin. This marks the beginning of *differentiation*, the development of unspecialized cells to serve specialized

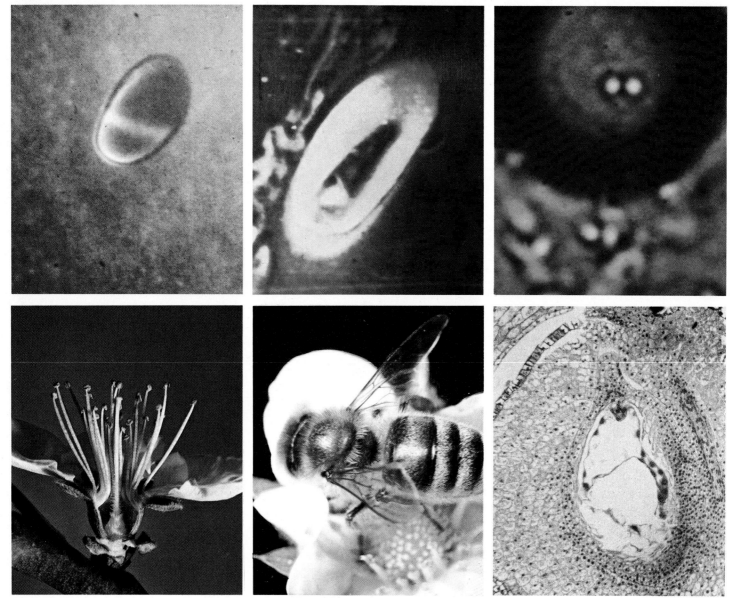

functions. In many-celled animals, a third layer, called the *mesoderm*, forms between the ectoderm and the endoderm. When this is complete, the cells begin to form different organs and parts of the body, such as muscle, bone, nerve, and so on, depending mainly on the location of the cells in the embryo.

When it is mature the new organism emerges from the egg or is expelled by the mother. If it emerges in the shape of its parents and can fend for itself or rely on its parents for early feeding, we say that the individual will undergo *direct development*, experiencing only relative changes during its growth. There is, however, an *indirect development*, in which the individual that emerges from the embryonic phase, known as a *larva*, has characteristics far different from those it will have as an adult. The process of transformation – at times long and complex – which a larva must pass through to acquire its definitive adult outward appearance is called *metamorphosis*. The complexity of this phenomenon is due to the fact that a larva is different from the adult organism in both its external appearance and its internal structure, habits, type of nourishment and even *habitat*. The larva that emerges from the egg will eventually change, frequently in a most radical way. The larvae of frogs and toads are the fish-like creatures we call tadpoles, which have a tail, no limbs,

Top left: a human egg-cell, about six thousandths of an inch in diameter. Diameter of the sperm-cell head is about a hundred times smaller. Centre: the sperm has penetrated into the egg-cell. Right: nuclei of the sperm and egg are about to unite, producing a zygote. Below left: section of a hermaphroditic flower. Centre: a bee in search of nectar transfers pollen from flower to flower, cross-fertilizing plants. Right: photomicrograph of a fertilized plant ovule.

Top left: using a device known as an automatic bee, a scientist gathers pollen from one variety of tomato plant to fertilize a plant of another variety. Below: fruit of the hybrid plant obtained by cross-breeding with the automatic bee. Right: hybrid iris of unusual coloration, grown by cross-breeding two different species.

and breathe through gills; in a drastic metamorphosis they become long and springy-legged amphibians that breathe with lungs. The dragonfly is aquatic in the larval state, but leads an aerial life when adult. Many animal species have this characteristic metamorphic development, from sponges, flatworms, and molluscs to vertebrates such as amphibia and fishes, but it is most prevalent among insects.

For many animals, the reproductive urge remains dormant for periods of time, often determined by unfavourable environmental conditions such as frigid temperature, insufficient food, or shortened daylight hours. But as soon as conditions become favourable again, the mating season sets in for many animals. Then the males of the species tend to grow noticeably more aggressive and instinctively compete for the conquest of the female. An important factor in the success of such contests may be the external appearance of the male contenders, because the males of many vertebrate species display all their beauty during such periods, accentuating what is most spectacular about them and flaunting their most vivid colours. This evolutionary difference in secondary sexual characteristics between males and females of a species is called *sexual dimorphism*. It may be either permanent or temporary. In the

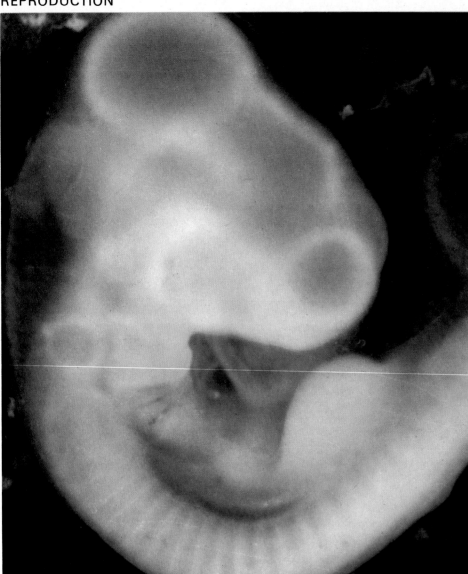

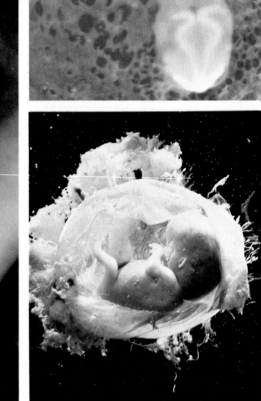

pheasant, for example, it is permanent, as it is in the lion. In certain fishes, like the stickleback, in many insects, and in the crested newt among reptiles, the secondary sexual characteristics are known as *nuptial livery*, and they disappear at the end of the reproductive season.

Sexual reproduction is also widespread among plants, but in almost all plants it alternates with asexual reproduction. The latter phase is reproduction by spores, which we mentioned earlier. A spore is a reproductive cell that – unlike gametes – can develop into a new plant without uniting with another cell. In brief summary, the sexual–asexual reproduction cycle works like this: Two gametes unite and develop into a *sporophyte*, the spore-producing plant you see growing in the garden. This plant produces *microspores* (tiny spores) and *megaspores* (larger spores) that develop into *gametophytes*, or gamete-producing plants. The gametophytes formed from microspores produce sperm-cells, while the ones formed from megaspores produce egg-cells. When a sperm and egg unite, the zygote becomes an embryo sporophyte, better known as a *seed*.

In *angiosperms*, or flowering plants, the sexual organs are contained in the flower. The long thin *stamen* is the male reproductive organ; in the *anther* at its

Left: chicken embryo on third day of incubation. The developing heart can be seen in the centre. Top right: trout embryo on third day of incubation. Below: human foetus approximately eighty days after fertilization.

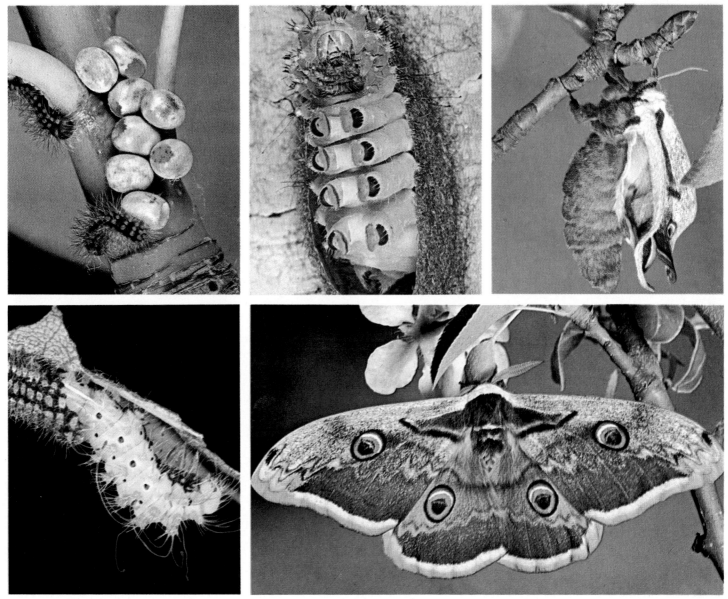

Phases in the metamorphosis of the pear moth. Top left: the moth's larvae emerge from eggs. Below: as the caterpillar grows, it undergoes minor transformations. Top centre: the silky cocoon the caterpillar wove around itself has been cut to show the animal on its back, in the pupal phase, where it undergoes radical metamorphosis. Top right: the newly transformed moth begins to emerge from its cocoon. Below: wings and antennae unfolded, the insect now displays its definitive form.

tip, the male gametophytes divide by meiosis, and several clump together to form *pollen grains*. The *pistil*, or female reproductive organ, contains the *ovule*, where the egg-cells are produced. *Pollination* occurs when pollen grains are transported – usually by the wind or by a nectar-seeking insect – to the *stigma* at the top of the pistil. There one of the male gametophytes grows a tube down the *style*, or neck of the pistil, to the *ovary* at the bottom. Two male gametes travel down this tube, and one of them unites with the egg cell, forming a zygote – the beginning of a new sporophyte. The second male gamete unites with the *embryo sac* surrounding the zygote and develops into the *endosperm*, or food supply of the embryo plant. The ovule tissues surrounding the embryo harden to form a tough *seed coat*. Some plants shed their seeds in this condition; others retain their seeds in the ovary, which then develops into a fruit.

Because plants are immobile, the survival of a species depends on *dissemination*, or transportation of the seeds from the mother plant to suitable places where the seeds can germinate and develop into mature plants. There are many different ways in which dissemination is assured. Many familiar plants, such as dandelions and milkweed, release seeds with tiny 'parachutes' that may drift with the wind for miles; maples and some other trees drop winged seeds

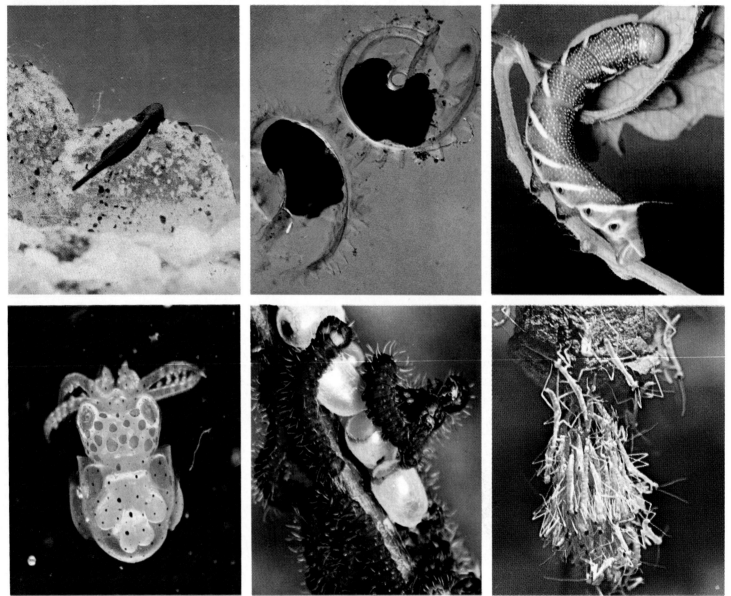

that float down and away like tiny helicopters. Seeds in fruit or nuts are eaten by animals and dropped elsewhere in their body wastes. (In this connection, it is interesting to note how many of the seeds eaten by birds are red or bluish red. Birds' eyes are especially sensitive to that colour. Hence, such fruits as mountain-ash berries, cherries, currants, and so on rank high on birds' favourite-food list. By eating the red fruits, which are easy to see, the birds have become prime agents of seed distribution.) Seeds like those of the cockleburr and beggar's tick have tiny hooks that catch on animal fur and are thus transported from one place to another. Other plants expel their seeds like catapults. The touch-me-not is one of the best known of these. The pods split down the sides and the seeds they contain fly out for surprising distances. This action can also be triggered by animals or people brushing the ripe pods. Water, of course, is a prime mover of seeds, which can float long distances down rivers and even across broad expanses of ocean. Man, too, contributes to seed dissemination, particularly with his technologically increased mobility. Motor-car tyres and grills and aeroplane tyres can pick up seeds and transport them for thousands of miles. Often such seeds will germinate in countries far from their place of origin and, not infrequently, will flourish,

Several forms of larvae. Top left: frog larva, or tadpole. Below: cuttlefish larva. Top centre: tiny, egg-shaped, free-swimming larvae typical of water worms and molluscs are called trochophores. Below and top right: grubs – soft, worm-like larvae of certain insects. Below: praying mantis larvae.

Examples of sexual dimorphism. Top to bottom and left to right: the male of a genus of Echiuroidea worm is almost microscopic in size, but the male and female live together symbiotically, ensuring reproductive capacity. In pairings of lions, mallard ducks, frigate-birds, scarlet tanagers, and praying mantises, the male is more spectacular in form and colour than the female, an evolutionary adaptation closely related to courtship behaviour, and thus to reproduction.

crowding out the native flora.

Many plants are hermaphroditic, capable of self-fertilization. However, fertilization between different plants occurs more frequently, thanks to the wafting of pollen grains by the wind. Some plants have anthers that mature before the stigma matures, or vice versa – an evolutionary block to self-fertilization. Other plants are *dioecious*, growing male and female flowers on different individual plants.

In order to improve certain plant species; to produce certain characteristics of colour, shape, or flower fragrance; and even to produce new species, two plants of different species or two of the same species but different varieties can be cross-bred by artificial means. The resulting plants are *hybrids*.

PLANTS

BOTANY

When we think of the immense variety of life forms, the task of making a definite distinction between the animal and plant kingdoms would not appear easy. In fact, certain micro-organisms with characteristics of both kingdoms, are often jokingly called 'plantimals', and some scientists now assign them to a *protist* kingdom of their own. Except for the simplest forms, however, we usually think of plants as organisms attached to the soil or the sea-bed, and mostly green. This idea is not too far from actual fact. Most plants are attached to some kind of substratum, and thus are immobile, while most animals are mobile throughout their lives. And most plants are indeed green in colour, because their cells contain an important green pigment called *chlorophyll*. This substance enables the plant to absorb sunlight and use its energy to convert carbon dioxide (from the air), water, and certain mineral salts into the organic matter that makes up its tissues. This important process is called *photosynthesis*. The chemical energy stored in this manner powers the plant's own living processes and also those of any animal that eats the plant. Thus most plants are *autotrophic*, or 'self-feeding'; animals are all *heterotrophic*, or 'other-feeding', because they depend for food on plants or on other animals that eat plants. Certain plants, such as fungi, are also heterotrophic, because they lack chlorophyll and thus cannot make their own food.

Another difference between plants and animals is that most animals grow to a maximum size and typical form, then stop growing, while most plants usually keep growing and changing their overall shape throughout their lives. Because of this difference, and because plant stems are relatively slender compared with their height, plants need strong structural support. Cell walls made of a strong, yet elastic, substance called *cellulose* provide this support. Most plants contain cellulose; only a small group of marine animals have it.

Plants get their raw materials from the air, water, and soil, so they do not have to move around or search for food. So they need neither muscles nor specialized sensory organs and nerve tissue to receive environmental stimuli, transmit them, and trigger muscular contractions. Consequently, as we noted in the chapter on cells, plant organs and tissues are much simpler and less specialized than animal organs and tissues. Plant and animal cells are very similar and they function in much the same ways; the main differences are the plant cell's cellulose wall and its *chloroplasts*, the organelles that contain chlorophyll.

Chlorophyll is actually a mixture of several slightly different green pigments, two of which are basic – chlorophyll *a* and chlorophyll *b*. Together with these in most plant cells are two other pigments – pale yellowish or brownish *xanthophyll* and yellowish-orange *carotene*. The green chlorophylls mask the colours of the other two pigments except in the autumn, when the chlorophylls seem to decompose faster than the xanthophyll and carotene. It is the unveiling of the latter, sometimes accompanied by the formation of red *anthocyanin* pigments in the leaves, that produces the brilliant display of yellow, brown, and crimson leaves each autumn in some parts of the world.

The *a* and *b* chlorophylls in each cell are responsible for photosynthesis, but the xanthophyll and carotene pigments may assist by transferring some energy from sunlight to the chlorophylls. Xanthophyll and carotene also provide the colour for some of the plant's organs, such as petals, pollen, roots, and so on. In addition carotene has a nutritional value, since it forms vitamin A in humans and some other animals. The chemical composition of chlorophyll is

CLASSIFICATION OF PLANTS

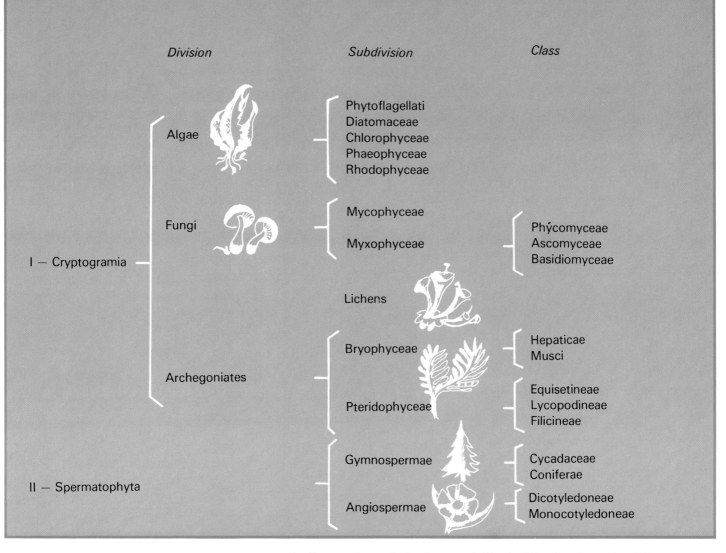

Division	Subdivision	Class
Algae	Phytoflagellati Diatomaceae Chlorophyceae Phaeophyceae Rhodophyceae	
Fungi	Mycophyceae Myxophyceae	Phycomyceae Ascomyceae Basidiomyceae
	Lichens	
Archegoniates	Bryophyceae	Hepaticae Musci
	Pteridophyceae	Equisetineae Lycopodineae Filicineae
Gymnospermae		Cycadaceae Coniferae
Angiospermae		Dicotyledoneae Monocotyledoneae

I — Cryptogramia

II — Spermatophyta

very similar to that of the haemoglobin in blood. The only difference is a magnesium atom at the core of the chlorophyll molecule and an iron atom at the core of the haemoglobin molecule.

To study a variety of objects, compare their similarities and differences, and establish some kind of order in our concepts about them, it is necessary first to give each different kind of object a distinctive label, and then to *classify*, or group, them according to their similarities. Early botanists labelled plants with descriptive names that grew longer and longer as the discovery of more kinds of plants in new-found areas of the world complicated the problem of distinguishing one kind from another. The earliest classification system grouped plants by shape and size: as herbs, shrubs, and trees. In the Middle Ages, classification by use – edible plants, medicinal plants, poisonous plants, dye plants, and so on – became popular. But this labelling method was unwieldy, and it grew increasingly apparent that such classification systems were 'artificial' – not based on any inherent similarities between the different kinds of plants. In 1753 the Swedish Linnaeus introduced a system of labelling and classifying plants that serves as the basis for the plant and animal classification systems now used by scientists all over the world.

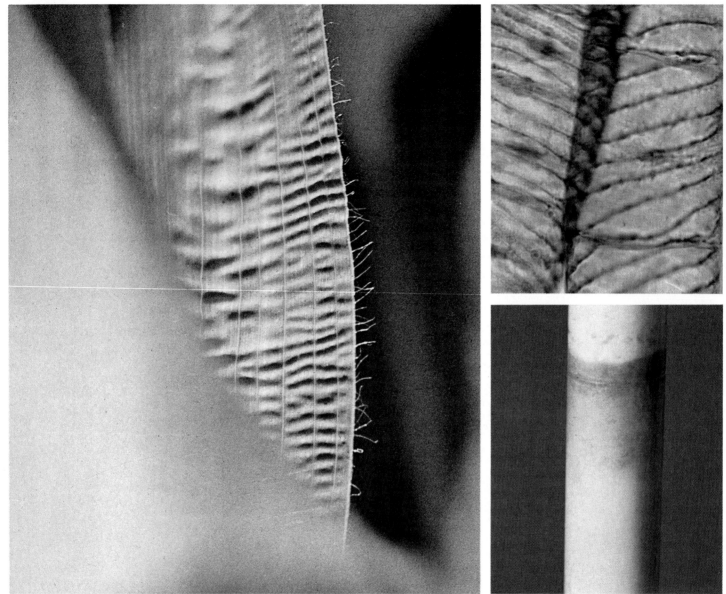

Linnaeus defined the specimens awaiting classification – plants such as the daisy, the sunflower, the almond, the apple, and so on – as *species*. Different species that closely resemble each other are classified in the same group called the *genus* (plural: *genera*). For example, the different oak species – commonly called white oak, chestnut oak, live oak, cork oak, and so on – all belong to the same genus. To simplify the problems of scientists the world over, whatever their native language might be, Linnaeus suggested that species and genus names always be either Latin names or written in the Latin form. This ensures against mistakes and misunderstandings about the particular species with which one is concerned. Each species is described by two such words, the first word indicating the genus, the second word indicating the species of that genus. All oaks, for example, belong to the genus *Quercus*; the white oak has the scientific name *Quercus alba*; the chestnut oak, *Quercus prinus*; the live oak, *Quercus virginiana*; the cork oak, *Quercus suber*. Note that the genus and species names always appear in different type from that of the surrounding text. The genus name always begins with a capital letter, and when the genus-species name has appeared once in a manuscript, the genus initial letter may be substituted for the full genus name; thus *Q. alba* means *Quercus alba*.

Left: a green leaf – a natural factory in which the pigment chlorophyll, powered by the energy it absorbs from sunlight, helps transform inorganic matter into organic matter by the process of photosynthesis. Top right: enlarged view of portion of Spirogyra, *a filamentous alga common in stagnant waters. Below: extract from a green leaf, separated into its component substances by a process called chromatography, displays the two shades of green of chlorophyll a and chlorophyll b.*

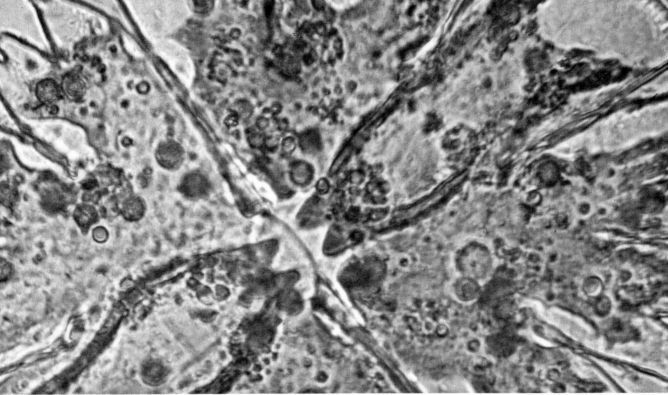

Grass-green colour of a green alga (Chlorophyta) is produced by the same proportion of chlorophyll and carotinoid pigments found in higher plants. In converting carbon dioxide and water into carbohydrates, this tiny plant and all others capable of photosynthesis produce food for themselves and – directly or indirectly – for all of the world's animal life.

Genera that resemble each other are grouped in the same *family*; the name of the family is derived from the name of the most important genus, to which is added the ending '-aceae' in the case of plants. In this way, the genera *Quercus, Fagus* (beech), and *Castanea* (chestnut) all belong to the Fagaceae family. Several related families form another natural systematic group called an *order*. The families Fagaceae and Corylaceae belong to the order Fagales.

Several orders having similarities constitute a *class*. The Fagales, Salicales, Rosales, Umbellales, and so on all belong to the class Dicotyledoneae. As you can see in the chart on page 69, the classes Dicotyledoneae and Monocotyledoneae belong to the *sub-phylum* (or subdivision) Angiospermae, and the subphyla Angiospermae and Gymnospermae belong to the *phylum* (or division) Spermatophyta. The phyla (or divisions), taken together, make up the plant *kingdom*.

Linnaeus's binomial designation of genera and species is still in use; so are his group names, which only went from 'species' up to 'class'. But his actual classifications of plants have undergone many, and often drastic, changes. In Linnaeus's time, the belief was still prevalent that a species never changes – that a daisy plant was created the way it was and would never be anything different.

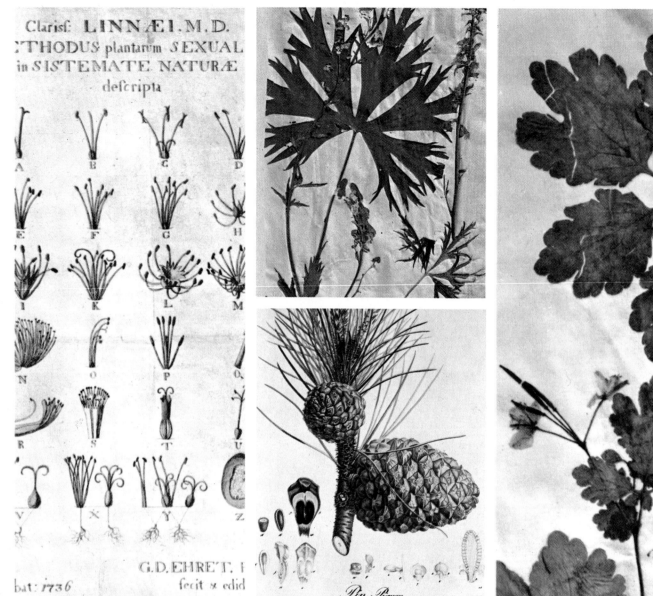

Any daisy that seemed a bit different from other daisies was considered a 'deviation' from the 'ideal' daisy to which the species name applied. Aware of the need for a more 'natural' way of ordering plants, Linnaeus based his classification system on the number of stamens in a plant's flower.

In the early nineteenth century the Swiss-French botanist Augustin de Candolle re-classified plants according to similarities in their structure and physiology; he also extended Linnaeus's hierarchy of groupings from classes to phyla. This system was more 'natural' than its predecessors and thus provided a more solid point of departure for the changes in classification that were to come.

Darwin's theory of evolution upset the concept of unchanging species. As evolution gradually won acceptance in the later nineteenth century, the principle of grouping species according to their similarities to archetypes, or 'ideals', gave way to the principle of grouping species according to their common ancestry. The definition of a species according to an archetype persisted, however, until the discovery of Mendel's laws of heredity led scientists to begin studying genetics in animal and plant populations. The modern concept that emerged treats a species as a population of individuals

Left: a page from Systema naturae *by Linnaeus, published in 1735, with drawings comparing flower stamens and pistils, on which the great Swedish naturalist based his system of classifying plants. Top centre and right: pages from the herbarium, or collection of dried plants, compiled in 1532 by the Italian Gherardo Cibo. It may be the oldest collection of this kind. Below: a page from* Les arbres fruitiers, *published in 1853.*

Left: flora typical of temperate zone deciduous forests, in autumn colour. Centre: flora of the savannah, a type of dry grassland with occasional trees and shrubs among the prevailing grasses, which feed such grazing animals as bison, antelopes, and kangaroos. Right: tropical forest, where the tangle of trees and lianas rises out of a mass of plant forms in the undergrowth, and where animal life – birds, insects, mammals, reptiles, and amphibians – is as varied as the plant life.

that evolved from common ancestors, live together, and interbreed in similar environments, and maintain similar ecological relationships with other living organisms. On this basis the similarities in structure and physiology of specimens is only one of many factors to be considered in classifying them. Knowledge of the behaviour and ecology of a plant population provide the only direct evidence for classification in species. Information that indicates the evolutionary line of a species guides the grouping of species in higher classifications.

The determination of common ancestry depends on comparative analysis of both living species and fossils of their possible forbears. For example, botanists must decide whether similar structures in two groups of plants are *homologous* – inherited from the same ancestors – or *homoplastic* – evolved the same way but from different ancestors. The vast expansion of the sciences during the present century has uncovered more and more evidence for making such decisions, along with new techniques for studying the evidence and making the decisions. Such developments have complicated the process of classification beyond anything Linnaeus could have imagined, yet they have also put the process on a more solid and natural foundation than it ever had before. The

complexity of the process, the lack of some key pieces in the puzzle, and the tendency of scientists in related but different fields to use slightly different approaches – together, these factors explain why some organisms are still classified in different ways by different scientists. More important, however, classification is no longer just a system for cataloguing and identifying plants and animals; it is now a vital tool for helping us to understand how living things got to be the way they are.

Thanks to Linnaeus, the complex of plants that are characteristic of a given geographic area is known by the name of the Roman goddess of flowers, *Flora*. The terms *flora* and *vegetation* are frequently and mistakenly used as synonyms; they have different meanings to a botanist. By 'vegetation' we mean all the plant growth in a particular place, whereas 'flora' refers to the species of living plants found in a particular area. For example, it is possible to find an area with abundant vegetation that is singularly deficient in flora, which may consist of only a limited number of species. One does not always find similar flora in places with similar climate and other environmental conditions, perhaps because the plants in the different areas had different origins or evolved in different ways.

Left: vegetation of the steppe is mainly grasses that grow shorter and farther apart than on the more humid prairie and savannah grasslands. The steppe gradually thins out into desert. Centre: desert vegetation consists of plants adapted to survive with a minimum of water, like the cacti of the Arizona desert. Right: the taiga, extensive coniferous forest, covers vast regions of northern America, Europe, and Asia.

Botanical garden of the University of Padua, Italy. Started in 1544, it is believed to be the first of many such gardens devoted to the scientific study of plants.

Botany is one of the most ancient of sciences. Records carved in stone show that six thousand years ago the Sumerians and Egyptians were cultivating many kinds of plants and had at least practical skills in working with them. However, the title of first true botanical scholar is usually accorded to a Greek, Theophrastus, who about 340 BC wrote a *History of Plants* describing some five hundred food and medicinal plants.

Like the other early sciences, botany generally regressed during the Middle Ages. In the sixteenth century it began to move ahead with the establishment of *herbaria*, or botanical gardens, usually in connection with universities and medical schools. Botany began to flower as a true science after the budding sciences of physics and chemistry had laid to rest the myths and superstitions that had squelched curiosity and investigation for so many centuries.

THE INFERIOR PLANTS

For convenience, botanists often distinguish members of the plant kingdom as either *inferior* or *superior* plants. Like other aspects of classification, there is some difference of opinion about how these two groups should be defined. Some limit the term 'inferior' to the *thallophytes*, plants without specialized tissues such as roots, stems, and leaves and whose seeds do not develop in a complex structure such as an ovary (see page 69). In this system, all the other plants are considered 'superior'. Some botanists group the thallophytes and *bryophytes* as plants 'inferior' to the *pteridophytes* and *spermatophytes*, which have evolved internal vascular systems adapting them to life on dry land. Still others call the thallophytes, bryophytes, and pteridophytes 'inferior' to the spermatophytes, or seed plants; this is the basis on which we shall distinguish between inferior and superior plants. There are so many different kinds of inferior plants, and they vary so extensively in shape and function, that we can at best highlight some of the main characteristics of the various groups.

The *algae* consist of seven unrelated groups of thallophytes that possess chlorophyll and photosynthesize their own food. Ranging in size from one-celled plants to the giant kelps, which grow to lengths of 95 feet (30 metres), the algae are found in fresh and salt waters, from cold to exceedingly hot, over the entire earth. It has been estimated that more than 90 per cent of the world's oxygen is produced by the many species of marine algae.

The Cyanophyta, or blue-green algae, are one-celled plants lacking any well-defined nucleus. They reproduce by fission, and live either isolated or in colonies, in water or soil, and in places such as the sides of damp stones and flower pots. In addition to chlorophyll, they contain a blue pigment, *phycocyanin*, which produces their characteristic colour, and sometimes a red pigment, *phycoerythrine*, which in abundance makes the algae red. (One of these species gives the Red Sea its colour.)

The Euglenophyta are one-celled organisms with a definite nucleus and one to three *flagella*, or whip-like organs, to propel them through water. Some species have a gullet and ingest solid foods, which leads zoologists to classify them as protists, rather than plants. However, chlorophyll makes them green and they carry on photosynthesis, so botanists treat them as primitive plants. They reproduce by mitosis and are found mainly in fresh water.

The Chrysophyta, or 'golden' algae, include yellow-green algae, golden-brown algae, and diatoms. The first two groups live predominantly in fresh water; the last in fresh or salt water or soil. Most are one-celled organisms living individually or in colonies; a few are multi-celled. Diatoms make up the large part of the *phytoplankton*, the plant life that floats on the surface of waters and is food for all aquatic animals. Diatoms are one-celled organisms enclosed in two shells of transparent silicon that fit together like a pill box. Their beautiful shapes have won them the title of 'jewels of the sea'. Diatomaceous earth, formed of diatoms deposited on the sea-bed over millions of years, is used as an abrasive in metal and tooth polishes, for insulation, and in filters.

The Pyrrophyta, or 'flame-coloured' algae, are actually yellow-green or yellow-brown. Most are one-celled organisms with two flagella. They form much of the fresh- and salt-water plankton.

With the dark brown Phaeophyta, we enter the realm of superior algae. All multicellular and mainly marine organisms, they vary from a few cells long to more than 100 feet (33 metres) in the giant kelps. In the latter, the *thallus*, or vegetative body, is considerably differentiated: a large, rootlike *holdfast*, often

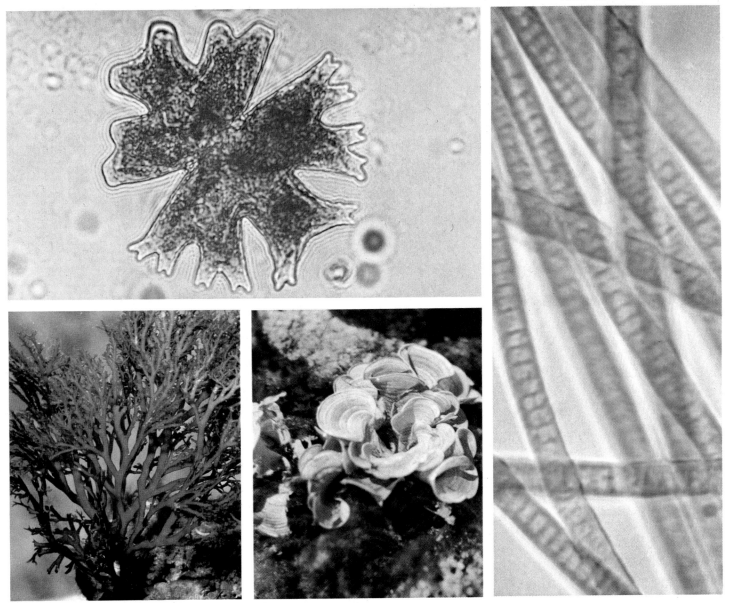

Top left: desmid, a one-celled green alga with cellulose walls, is found in fresh water, singly or in filamentous or amorphous colonies. Below: a typical red alga, Rodococcus coronopifolius, source of agar, the laboratory medium for culturing bacteria. Below centre: brown algae, Padina pavonia, source of viscous alginate, an industrial chemical. Other brown algae supply iodine or potassium and food for both humans and cattle. Right: highly enlarged multicellular filaments of green algae.

equipped with suckers, secures the organism to the substratum (rocks or reefs). A cylindrical, stemlike *stipe* leads to one or more leaf-like *blades*. Some kelps have small gas-filled bladders, which help the blades to float on the water. A substance called algin obtained from brown algae is used to thicken ice cream, toothpaste, shaving cream, and certain cosmetics, giving them a creamy appearance.

The Rhodophyta, or 'rose' plants, are also known as red algae and as sea mosses. Particularly profuse in warm seas, they are also present in fresh water. Some of the more than three thousand species live at depths of 200 feet (66 metres). Many rhodophytes have important medicinal properties, and some provide a substance known as *agar* on which micro-organisms are grown for laboratory study.

The Chlorophyta, or green algae, are usually regarded as the most-evolved algae because of their abundant chlorophyll, carotene, and xanthophyll and their reserve starches, similar to those of the superior plants. Though found in the sea, they live prevalently in fresh water, but are capable of living in conditions very different from those typical of algae life. They thrive, for example, in humid terrain, on the bark of trees with rich water content, and at

times on leaves or in the snow. A few live within the tissues of other plants, sometimes as parasites.

Perhaps the best known of the inferior plants are the *fungi*. Like the algae, fungi are thallus plants, lacking roots, stems, and leaves. Unlike algae, almost all fungi lack chlorophyll and are thus heterotrophs – parasites feeding on live organisms or saprophytes feeding on dead organisms, on wastes from live organisms, or in some cases on inorganic substances. The fungi include three groups of plants which modern botanists consider unrelated to each other.

The Schizomycophyta, or bacteria, are microscopic one-celled organisms that – like the blue-green algae – lack a definite nucleus and reproduce by simple fission. A few species of bacteria are autotrophic, capable of synthesizing organic compounds from carbon dioxide and other simple inorganic substances. These species are widespread and perform important functions, such as converting sulphur and nitrogen compounds into sulphates and nitrates needed by green plants. Some biologists believe the world's iron ore deposits were produced by bacteria that oxidized iron from water in which it was dissolved. A few species contain pigments that enable them to photosynthesize organic compounds. Parasitic bacteria cause many serious

Top left: Amanita caesarea, *known as 'good ovule', a mushroom delight for gourmets. Below: meadow mushrooms of* Psalliota *genus, like those raised commercially for market. Top centre:* Coprinus comatus, *only edible in its early stage. Below: edible 'drumstick' mushroom,* Lepiota procera. *Top right: edible yellow morels,* Morchella rotunda. *Below: edible mushrooms of the* Armillariela mellea *species. Only experts should determine the edibility of wild mushrooms.*

Many fungi are poisonous, and while some of these are not dangerous, others are lethal. The ones shown here are all poisonous to varying degrees. Top left: Russula foetens. *Below:* Amanita muscaria. *Top centre:* Amanita phalloides *(lethal). Below:* Entoloma lividum. *Top right:* Russula emetica. *Below:* Clitocybe dealbata.

diseases in humans, other animals, and plants; yet vast numbers of bacteria live on and inside our bodies and those of other organisms without harming their hosts and often contributing to their well-being. Bacteria are immensely helpful in processing many foods and producing organic compounds with various industrial uses. Most important, saprophytic bacteria are one of the two principal groups of decomposers that convert plant and animal wastes and remains into the inorganic substances needed to sustain all forms of life.

The Myxomycophyta, or slime fungi, are of interest mainly because of their life cycle. A thin, slimy mass of naked protoplasm, called the *plasmodium*, contains many nuclei and spreads out in a thin film on soil, grass, dead leaves, a fallen tree or a stump, sometimes covering an area of several square metres. Flowing, often at noticeable speeds, over the substratum, it absorbs food from non-living materials. Eventually the modium gathers together and produces a number of *sporangia*, or spore cases, resembling certain mushrooms in structure. Each sporangium breaks open and releases one-celled spores which, upon reaching a favourable substratum, germinate and produce up to four *swarm-cells*. These move about with the aid of their flagella, divide, and then fuse in pairs in sexual reproduction that results in a new plasmodium.

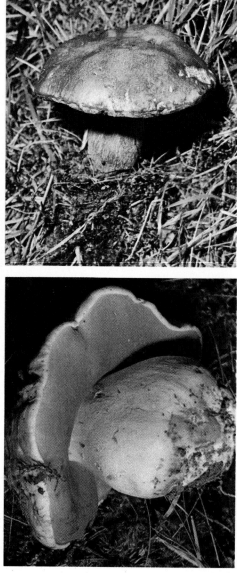

The Eumycophyta, or true fungi, include about seventy-five thousand species varying widely in form, size, physiology, and reproductive methods. All are heterotrophic, and many are saprophytes – principal decomposers, along with the saprophytic bacteria. Most true fungi are composed of filaments called *hyphae,* one of which may be a single long cell or a multi-celled structure; a mass of hyphae is called a *mycelium.* All true fungi reproduce at least part of the time by releasing microscopic spores that are usually abundant in the air, water, and soil. There are four classes of true fungi.

The Phycomycetes, or algae-like fungi, range from one-celled organisms to those composed of hyphae. Some species, found only in water, have flagella to move them around, like certain algae. Others live as parasites on animals and plants of all kinds, including other fungi; such species cause mildew diseases in many crops. Another species is the black mould that grows on bread left in a damp, warm place. Other species are used to make industrial chemicals.

The Ascomycetes, or sac fungi, are named for the sac-like structures that grow out of the hyphae and contain the reproductive *ascospores.* The Ascomycetes include some forty thousand widely diverse species, among them the economically important yeasts, which are micro-organisms of the

Left: common pore fungus, Boletus edulis, *also known as 'little tree trunk' because of its shape and colour, is one of the best known and most highly prized edible mushrooms. Top right:* Boletus luteus, *or yellow boletus, edible mushroom that grows mainly at the foot of pine and birch trees. Below:* Boletus satanas, *or 'malefic pore fungus', is the most poisonous mushroom of the* Boletus *genus.*

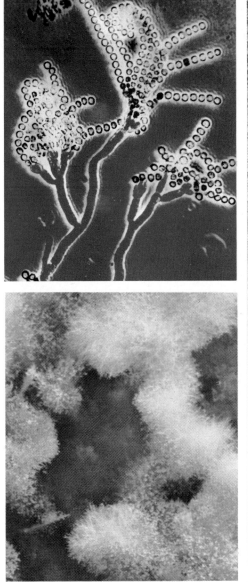

Top left : Penicillium roqueforti, *a fungus of the Aspergillacae family, provides the characteristic aroma of Roquefort cheese. Below : a wood-invading mould in the bark of an elm tree. Right :* Penicillium notatum *culture, source of the antibiotic known as penicillin.*

Saccharomycetaceae family and usually consist of isolated cells. The yeasts are useful in baking and brewing, and as vitamin and protein synthesizers. Another famous member of the family is *Morchella esculenta*, the morel, whose fleshy spore sac makes a delicious food. Certain mildews of this family are parasites of plant leaves and stems, and some are highly destructive. *Penicillium notatum*, a mould fungus, yields the famous bacteria-killing penicillin, which has been so helpful to humans. Other moulds of the same genus give Roquefort and Camembert cheeses their characteristic odours and flavours.

The Basidiomycetes are named for the *basidium*, a club-shaped and usually enlarged cell that grows from the end of a hypha and forms reproductive spores. The large family Agaricaceae, or gill fungi, includes most of the familiar fungi known as mushrooms or toadstools. Of these some species are highly edible, whereas other species of the same genus are poisonous – even lethal – to humans. Eaters of wild mushrooms need expert knowledge of the fungi to be safe. Certain smut fungi and rust fungi of this family are extremely destructive to plants, especially those producing grains.

The Deuteromycetes, or imperfect fungi, consist of specimens that are not

yet properly classified, either because they have lost their sexual reproductive stages or because they are not well enough known biologically. One group, familarly called the ringworm fungi, causes a number of skin diseases, including athlete's foot, on humans and other animals.

Certain fungi, often Ascomycetes but sometimes Basidiomycetes, maintain an unusual symbiotic relationship with some of the green and blue-green algae in forming the 'compound plants' known as *lichens*. The lichens grow on rocks, on trees and other wooden objects, and on certain types of soil in places such as high mountains, the Arctic tundra, and deserts, where conditions are unfavourable for other types of plants. Lichens vary widely in colour and form; they may look like crusts, leaves, scales, or even shrubs. Strictly speaking, lichens are not plants, but associations of algae and fungi that have evolved, apparently, to their mutual benefit. The fungi obtain food from the algae, and the fungi may help the algae absorb and retain water; however, the fungi cannot live without the algae, although the algae can survive without the fungi. Lichens are eaten by certain insects, snails, slugs, and, in the tundra, animals such as caribou and reindeer. By excreting acids that break down rocks, lichens help form soil in which other kinds of plants may grow.

Left: Parmelia coperata, *a lichen usually found on broadleaf tree trunks. Top centre: reindeer lichen (*Cladonia rangiferina) *is the main food source for many tundra animals. Below: Icelandic lichen (*Cetraria islandica), *used in herbal medicine as a respiratory decongestant. Right: geographic lichen (*Licidea geographica), *so-called because it resembles a relief map, is widespread on siliceous rocks.*

Left: Marchantia polymorpha, *a bryophyte of the Hepaticae, or liverwort family. Bryophytes are green plants, and thus autotrophic, but lack true roots, stem, and leaves. Top centre: liverworts and mosses, both bryophytes, require moist habitats and are often found growing together. Below: sporophyte of moss plant, with spore case at tip, grows out of the 'leafy' gametophyte. Released spores fall to earth and grow into new plants. Right: moss clinging to rock in an inclement environment.*

Moving up the evolutionary ladder, we come to the *bryophytes*, or moss plants. All contain chlorophyll and are thus autotrophic. Primarily land plants, they have root-like, stem-like, and leaf-like structures called *rhizoids*, *stemlets*, and *leaflets*, but no vascular system to transport water and nutrients. The bryophytes mark a transition from the aquatic algae to land plants, such as ferns, that require a humid environment. The Musci, or true mosses, grow vertically, though often only a centimetre high, in dense strands resembling carpets of vegetation. The Hepaticae, or liverworts, usually grow flat on the ground. Some have flat, lobed thalli; others have stemlets and tiny leaflets. Bryophytes grow on moist rocks, trees, and bare soil, which they help protect from erosion.

The *pteridophytes* are divided into three classes, the most common being the Filicineae, or ferns. The others are the Equisetineae, or horsetails, and the Lycopodineae, or club mosses. All have chlorophyll, making them autotrophs; true roots, stems, and leaves; and vascular systems – all of which leads many botanists to class them as superior plants. However, none is a seed plant. In each case the large sporophyte produces spores that fall to the earth and develop into small, inconspicuous gametophytes; these produce immobile eggs

and swimming sperms that can only move through water to reach and fertilize the eggs. The horsetail species are herbaceous to shrubby, and grow mostly in damp places, to heights of about 3 feet (1 metre). The sporophyte has two distinct kinds of stems, one colourless and bearing at the tip a cone in which spores are produced, the other green and quite bushy, earning the plant its name. The club mosses are small, herbaceous plants with small, thin, crowded leaves, resembling fir needles; the stems run along the ground at or just below the surface, putting up forked, vertical branches, some tipped by narrow spore cones. The club mosses are widely distributed, and fossils show that they once included tree-sized plants.

The ferns are distinguished by their leaves; arranged in well-developed fronds, they have visible veins and are often divided, *pinnate* (leaflets on each side of a common axis, or *palmate* (leaflets radiating from a single point). They usually grow in humid places without much exposure to the sun, as in undergrowth, along river banks, in swamps, and so on. The familiar fern plant is the sporophyte; dots on the underside of the leaves are *fruit dots*, or *sori*, consisting of several to many spore sacs. In many ferns, a spring-like layer of cells called the *annulus* reacts to humidity changes and flips the spores out of the

Left: Ceterach officinarum, *similar to the American spleenwort fern, grows on rocks and walls in humid places. Top right: elkhorn fern (*Platycerium alicorne*) is an epiphyte, a plant that grows on other plants. Below: close-up view of new, uncurled fern frond, often called a 'fiddlehead'.*

Undergrowth of a coniferous forest. This frequently consists of plant communities in which different species of fern predominate. The ideal habitat of these plants is an environment where light is limited and humidity is high.

sac with considerable force. A spore germinates on the ground and develops into a small, roughly heart-shaped gametophyte, called the *prothallus*, which produces both sperm and eggs; the sperms usually swim to other prothalli to fertilize their eggs. Ferns may be terrestrial or aquatic, epiphytic (living on other plants) or upright, creeping or twining. In tropical zones some grow to heights of 30 feet (10 metres) or so. Widely cultivated as ornamental plants are the maidenhair fern, *Adiantum capillus-veneris*, with leaves that form a delicate aerial arrangement, and the lady fern, *Athyrium filix-foemina*, which is cultivated in many gardens particularly near fountains, which supply humidity.

THE SUPERIOR PLANTS

The so-called superior plants are the *spermatophytes*, or seed plants. They form the largest, most varied, and most economically useful group of plants existing today. In general, the seed plants are larger than the inferior plants. But this is not the reason they are called superior. Rather, it is because they have evolved further than the inferior plants, developing structures and reproductive systems that are better adapted to survival on land and in a wide variety of environments. The seed plants all have three vital elements: true roots, stems, and leaves. Many have woody stems or trunks, composed of a combination of cellulose and a substance called *lignin* that is secreted by certain cells; the woody plants are mostly *perennials* – plants that keep on growing year after year. The *annual* and *biennial* seed plants, which live for only one and two years, respectively, are generally *herbaceous*, or non-woody. The spermatophytes are composed of two sub-groups – the *gymnosperms* (meaning 'naked-seed') and the *angiosperms* ('seed-in-vessel'). The former are mostly cone-bearing trees, such as the pines, spruces, and cedars. The latter are the flower- and fruit-bearing trees, shrubs, and herbs.

As a rule the trunk or stem is the most highly developed part of any plant. It supports the branches, leaves, flowers, and fruit and also serves as a conduit between these elements and the roots. If you examine the top of a freshly sawed tree stump, or the sawed end of the trunk, you can see how the trunk is constructed. Working in from the outside towards the centre, you see first the bark, which may be fairly smooth or quite rough, depending on the species of tree and its age. Next comes a thin layer of *phloem*, and inside that the lighter-coloured *xylem*, or wood. If the central part is light in colour, it is *pith*; if it is dark, the pith has become *heartwood* (as opposed to the xylem, which is called *sapwood*). The most important layer, called the *cambium*, lies between the phloem and the xylem, but it is invisible to the unaided eye. The cambium is a thin layer of living cells that form new cells, mostly of xylem, each growing season. The new xylem cells produced in spring are larger than those produced in summer, adding one light and one dark ring of wood each year. Not only do these rings reveal the age of the tree, but the relative width of the rings indicates whether the rainfall and other climatic conditions were favourable for growth the year a particular ring was formed. In the outer part of the bark a tissue called *cork cambium* grows, but not as rapidly as the xylem and phloem expand, and this explains why thick bark is usually split and furrowed. The xylem and phloem both consist of networks of cells: some are long, narrow, vertical, and connected to similar cells above and below them in the trunk; other cells are connected horizontally and radiate from the centre of the trunk outwards to the bark. The latter are called *vascular rays*, and they transport liquids horizontally through the xylem and phloem. The vertical cells in the xylem conduct water with dissolved salts from the roots to the leaves. Interwoven between the various conduit cells of the angiosperms are fibres that give the trunk elasticity combined with mechanical strength. The gymnosperm trees lack these fibres, which is why they are called *softwood* trees.

The inferior plants might seem to hold the record for absorbing and utilizing water; most of them require either an aqueous or very humid environment. Yet most superior plants absorb and give off into the atmosphere so much water – using only about 10 per cent of it for their growth processes – that they might be considered profligate. They take in water, of course, through the roots, and lose most of it by evaporation from the aerial parts of the plant. This

An example of superior plant life: a powerful trunk sustaining many ramifications rich in leaves, which, taken together, form the head. Only in autumn and in certain areas do the broad leaves of deciduous trees display such a variety of coloration.

evaporation, called *transpiration*, is a continuous process. Transpiration occurs through *stomata*, the pores in the stem and leaves through which the plant breathes, 'inhaling' carbon dioxide and oxygen from the atmosphere and 'exhaling' oxygen produced in photosynthesis. As a fairly typical example, a single corn plant loses more than 50 gallons (220 litres) of water through transpiration during its hundred-day life span. When transpiration is slowed by a cool and humid atmosphere, as at night, some plants exude water (liquid, not vapour) through special pores along the leaf margins; this process is called *guttation*.

Superior plants have evolved a complex system for taking in and transporting nutrients and excreting wastes. Among the inferior plants only a few forms of algae and the pteridophytes have such a system. The prevalence of sexual reproduction among the superior plants has undoubtedly given them an evolutionary advantage over the inferior plants, which reproduce asexually more often than sexually. By blending the genetic heritages of two different plants, sexual reproduction provides variations in characteristics that facilitate and accelerate the processes of evolutionary change. One result of this is the wide variety of *semination*, or seed distribution, devices found in superior

plants. Their seeds carry ample food stores to allow very slow germination, even under unfavourable conditions. And the seeds have protection against cold and dryness that enables them to survive under exceptional conditions. Cereal seeds found in tombs thousands of years old have proved capable of germinating. Inferior plants play a vital role in making and keeping fertile the soil in which superior plants grow. While dependent on the inferior plants for their soil, superior plants tend to limit the living space, water, nutrients, and light available to the smaller plants. However, when a superior plant dies, it provides an ideal habitat for many of the lower plant organisms.

Except for the tropical tree ferns, all trees are either gymnosperms or angiosperms, and thus superior plants. A tree can be defined as a perennial plant with a woody trunk supporting a *ramification*, or arrangement of branches, with thick clusters of leaves that form a *top*. The trees that lose their leaves and grow new ones every year are called *deciduous*; those that retain their leaves for several years are called *evergreens*. The latter are constantly losing and growing some new leaves, but so few at a time that we just don't notice them. Some trees last only a few years; others live for thousands of years. The size of trees varies tremendously – from the modest little rose tree in

Top left: section of angiosperm tree trunk showing centre pith, annual growth rings formed of xylem (spring wood light, summer wood dark), phloem (dark brown), and bark. Below: section of gymnosperm tree trunk, greatly enlarged. Centre: bamboo, a grassy plant, has stems that sometimes reach 20 inches (50 centimetres) in diameter. Top right: strawberry plant growing from runner stems. Below: rhizome, or underground stem, swollen with stored food.

Left: climbing liana with woody trunk wrapped around a tree trunk. Centre: liana stems are used by some people in tropical regions in place of rope. Right: climbing and hanging lianas intertwined, typical of tropical forests.

our gardens to the mighty sequoias, which soar 300 feet (90 metres) in height, or the baobabs, which may reach a circumference of 95 feet (nearly 30 metres). Tree trunks tend to be more or less cylindrical and straight, but there are also prismatic trunks, flattened trunks, trunks swollen like a ball, and creeping and climbing trunks.

About three hundred species of cacti (native only to the Americas) offer a wide variety of rounded trunks or stems. Some plants, such as the violet and the strawberry, have stems that creep horizontally along the surface of the soil and are known as *runners*. Climbing stems come in many different forms. The *twiners*, such as the morning-glory and sweet potato, have stems capable of winding around their support in the form of a spiral. The grape and Boston ivy plants have a rigid stem with twigs modified to form attachment organs, called *tendrils*. Some of these plants climb by holding on to their support with their down-reaching branches, which look like small hooks; some take hold with hair-like outgrowths from the stem or with outer roots. Highly robust climbing plants are the lianas, typical of the undergrowth in tropical climates.

Certain plants have evolved stems of distinctive shapes that are specialized for food storage and reproduction. One of these adapted stems is the *bulb*, the

underground stem of a plant such as the onion, lily, narcissus, or hyacinth. The bulb is a large, globular bud with a small stem, roots at the bottom, and many fleshy leaves growing from the top of the stem. The leaves store food and often sprout new buds that can be separated and used to raise new plants. A *rhizome* is a stem that grows new plants horizontally. Most rhizomes are perennial, extending themselves and sending up new plants year after year. Some, like those of the iris, are enlarged with stores of food. A *tuber*, of which the potato is the most common example, is the swollen, food-packed, growing tip of a rhizome. Left in the soil over the winter, such rhizomes die, but their tubers produce new shoots from their buds in the spring.

When a seed soaks up water and begins to *germinate*, or sprout, the first element to spring from the seed is the *radicle*, which responds to the pull of gravity by growing downwards. The radicle forms the primary root, which branches out into secondary roots, which in turn branch out, and so on. The root has four functions: to anchor the plant, to absorb water with nutrient minerals dissolved in it from the soil, to transport this material to the stem, and, in some cases, to store food. *Diffuse* root systems have many slender roots, the main ones all about equal in size; in *tap-root* systems, the primary root grows

Different types of roots. Top left: respiratory roots of mangrove grow above water. Below: primary and secondary roots of dandelion plant. Centre: aerial roots of tropical plants, called alien roots because they form on trunks and absorb humidity from the atmosphere. Top right: taproot of sugar-beet plant. Below: large; diffuse roots, called bundled, or faggot, roots.

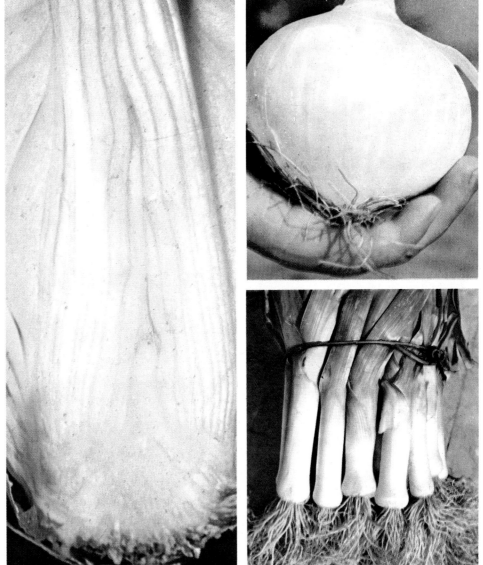

Different types of bulbs – storage buds developing underground from stems. Top left: cross-section of tulip bulb. The pre-formed inflorescence in the centre requires a cold spell before developing normally, so bulbs are usually planted in the late autumn. Below: hyacinth bulb with abnormal flower produced by planting in a hothouse. Centre and top right: section and exterior of onion bulb. Below: leek bulbs.

fastest and remains the largest root. At the tip of a root is the *root-cap*, which protects the dividing cells above it as they push the cap through the xylem cells, and so on. Many of the outside, or *epidermal*, cells in this region extend themselves horizontally, forming the fine *root-hairs* that snake their way between soil particles and perform much of the absorption.

The third essential elements of a superior plant are its leaves, which usually form a handsome cover for the plant. Each leaf originates in a *leaf bud*; as it blossoms the leaf emerges, still partly convoluted, moist, and light green in colour. Within a few days it has opened and assumed its definitive appearance. The *blade*, or body, of the leaf is usually attached to the plant stem or branch by a *petiole*, or leaf stalk, that supports the blade and enables it to orientate its upper surface toward the light. Blades that are attached directly to the stem, with no petiole, are called *sessile* (seated). The leaves of different plants vary immensely in shape (see page 92), but are almost invariably green because of the presence of chlorophyll. The upper surface of the blade is covered by a single layer of epidermis cells with thick outer walls and a waxy coating of *cutin*, which makes the leaf's top surface nearly waterproof. The thinner epidermis of the blade's underside is perforated with *stomata*, through which

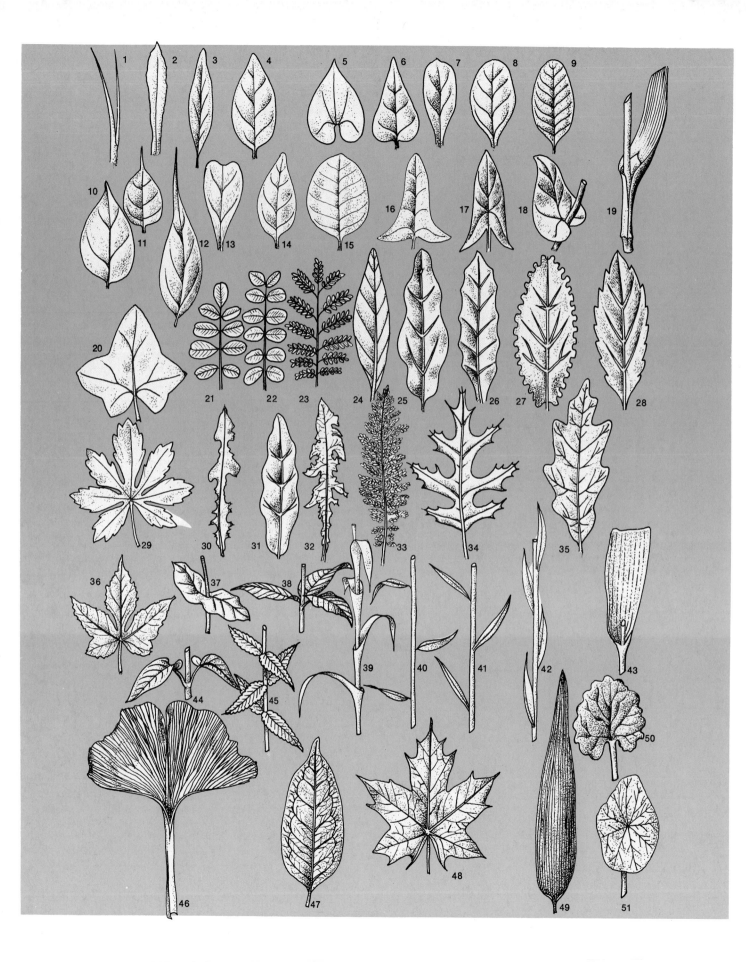

Opposite: types of leaf, differentiated by blade shape (1–36), position on stem (37–45), vein pattern (46–49), seating (40–43, 50, 51). (1) needle-shaped; (2) linear; (3) lanceolate; (4) oval; (5) cordate; (6) truncate; (7) cuneate; (8) obovate; (9) elliptic; (10) acute; (11) mucronate; (12) acuminate; (13) emarginate; (14) obtuse; (15) round; (16) hastata; (17) sagittate; (18 amplexicant; (19) sheath; (20) palmate-lobate; (21) odd-pinnate; (22) even-pinnate; (23) bipinnate; (24) smooth-edged; (25) sinuate; (26) sinuate-denate; (27) dentate; (28) serrate; (29) palmate-division; (30) sectile; (31) crenate; (32) runcinate; (33) bipinnate-sectile; (34) cleft; (35) lobate; (36) palmate; (37) connate leaves; (38) verticillate; (39) folded at tip; (40) patulous-sessile; (41) upright-sessile; (42) close-sessile; (43) sheath-sessile; (44) opposite; (45) decussate; (46) fan-shaped; (47) reticulate; (48) palmate; (49) parallel; (50) tip-to-base; (51) peltate.

Right: leaves of alfalfa plant (Medicago sativa), once planted widely as forage for grazing cattle, now supplanted in many places by other forage plants.

the leaf breathes and transpires. Within the blade, a network of highly ramified *veins*, or vascular bands, distributes water and minerals from the stem to all parts of the leaf and conveys food manufactured in the leaf to the stem. A *simple* leaf consists of a single blade; a *compound* leaf is composed of a number of leaflets. Leaves have evolved into organs of different shapes and colours for specialized functions such as food storage, reproduction (sepals, petals, carpels), protection (thorns or needles, as in cacti), and food-gathering (insect traps).

The upper angle between a leaf's petiole and the plant stem or branch is called the *axil*, and the bud growing there is an *axillary bud*. Like the *terminal buds* found at the end of a stem or twig, the axillary buds are capable of producing new growth. Some plants have *stem* or *branch buds*, which produce only new lengths of twigs with leaves, and *flower buds*, which produce only flowers; others have *mixed buds*, which produce twigs bearing both leaves and flowers. Axillary buds often remain dormant and sprout only when the leaf or the terminal bud on their twig has broken off. In temperate regions, buds on woody stems are covered by overlapping *bud scales*, which protect them against dryness, cold, and injury.

Flowers are among the most beautiful things in nature. Their amazingly varied colours and fragrances have evolved because they facilitate reproduction by luring animals – especially insects – to serve as pollinating agents and thus promote fertilization. The flower is basically an extension of the stem or twig, bearing highly modified leaves that are able to bear spores. In some plants these specialized leaves are not as striking in appearance as those of the angiosperms, or flowering plants. In the gymnosperms, for example, the ovules are borne uncovered on the base of a *scale*, and the scales are usually grouped together in the familiar *cone*. The smaller, male cones produce spores that differentiate into young male gametophytes in the form of pollen grains that are spread by the wind to the larger, female cones; such a system needs no animal lures to ensure fertilization.

Flowers develop from the *receptacle*, or tip, of a flower-bearing twig called the *peduncle*. Many angiosperms have *complete* flowers – flowers with all of these four floral organs: *sepals*, *petals*, *stamens*, and *carpels*. (Flowers lacking any of these organs are called *incomplete*.) The sepals are usually small, green, leaf-like blades, which enclose and protect the other floral organs during their development; together, the sepals form the *calyx*. From the calyx emerge the

Some different types of leaves. Left: convolvulus plant with amplexicant leaf. Top centre: leaves that have evolved into thorns. Below: sundew plant leaves serve as insect traps for the carnivorous plant. Right: rose leaf is compounded of leaflets in pinnate arrangement.

Top left: flower of the water-lily, an aquatic plant common to still waters. The flower of an angiosperm is a specialized branch system composed of the plant's reproductive elements and accessory parts. Below: section of camellia flower, revealing stamens and pistils, the male and female sex organs. Right: close-up of pollen-bearing anthers at tip of camellia stamens. Note pollen grains in partly opened anther.

petals, which together form the alluringly colourful, and usually fragrant, *corolla*. Each stamen, or male reproductive organ, is a slender filament topped by an *anther*, wherein microspores develop into gametophytes that clump together as pollen grains. Carpels are the seed-bearing organs, one or several of which form a *pistil*. A plant may have one or several pistils, each with an enlarged, somewhat spherical *ovary* containing the *ovule*, in which the egg-cells are formed. A long, slender *style*, or neck, leads up from the ovary to a slightly enlarged *stigma* at the top of the pistil, where pollen grains must come to rest in order to fertilize an egg-cell.

Perfect flowers, such as roses, tulips, sweet peas, and orchids, have both stamens and pistils on the same flower. *Imperfect flowers* are either *staminate* (having stamens only) or *pistillate*. Plants such as willows, cottonwoods, and aspens bear either staminate or pistillate flowers – never both types on the same plant – and are described as *dioecious*. *Monoecious* species, such as walnut, oak, and corn, bear both staminate and pistillate flowers on the same plant. Flowering plants are commonly *cross-pollinated*, with pollen from the anther of a flower on one plant being carried to the stigma of a flower on another plant by the wind, an insect, or sometimes by water. However, perfect plants, being

bisexual, tend to pollinate themselves before the corolla opens and exposes the stigma to the agents of cross-pollination. This happens frequently in species such as peas, tobacco, wheat, barley, oats and other flowers. Flowers that can produce large amounts of seeds after self-pollination are called *self-fertile*; orchids and other flowers that produce seeds only after cross-pollination are called *self-sterile*.

Flowers are not always isolated on a flower-bearing axis; more often they form *inflorescences,* or groups of flowers. Each species has its own characteristic inflorescence. In one basic type, called *indeterminate*, the tip of the main stem keeps growing and new flowers continue to grow from the stem. *Simple indeterminate* inflorescences have a single main stalk bearing flowers in one of these forms: *raceme* – flowers on pedicels, or individual stems, of equal length and equally spaced on the main stalk (snapdragon, hyacinth); *spike* – flowers arranged like raceme but *sessile*, or seated directly on the main stalk (cat-tail, plantain); *simple umbel* – flowers on pedicels of equal length growing from the apex of the spherical or hemispherical inflorescence (ginseng, ivy); *corymb* – like raceme but lower flowers have longer pedicels, making the cluster

*Top left: gentian flowers (*Gentiana asclepiadea*) are complete (calyx and corolla), perfect (stamens and pistils), and gamopetalous (petals united at base). Below: *Aristolochia, *with its long yellow perianth. Top centre: cornflower, a composite with fertile flowers in the centre and widely spaced ray flowers. Below: Unisexual flowers of the courgette plant. Top right: *Linaria *flower, with bilateral symmetry. Below: snapdragon, another bilaterally symmetrical flower.*

Inflorescences – complexes formed by a number of flowers and their stems – are characteristic for each botanical species. The first five shown here are all indeterminate inflorescences in which the main stems continue to grow. The three top flowers are of the raceme type: (left to right) digitalis (foxglove), evening primrose, and wisteria. Below left: flowering rush, umbel ('umbrella') inflorescences; centre: aster, with inflorescent 'head'; right: valerian, a compound umbel, or 'domed cluster'.

flat-topped or convex (cherry, wild hydrangea); *head* – small flowers without pedicels crowded together on the surface of the disc-shaped inflorescence (daisy, chrysanthemum). *Compound indeterminate* inflorescences have several main branches, each bearing flowers (carrot and wild carrot, or Queen Anne's lace, have compound umbel inflorescences). *Determinate* or *cymose* inflorescences are less common; they have the first flower at the tip of the main stalk and others growing successively at lower levels (forget-me-not, phlox).

After pollinization and fertilization, the egg-cells develop into embryo plants, each well stocked with food and enclosed in a tough seed coat formed of hardened ovule tissue. In many plants a *fruit* then develops around the seeds, in a process called *carpogenesis* (*carpo* means 'fruit'). The fruit develops from the wall of the ovary – known as the *pericarp* – which may thicken and becomes differentiated into three different layers of tissue, the *exocarp* (outer layer), *mesocarp* (middle layer), and *endocarp* (inside layer). These layers develop differently according to the species. Dry fruits are those in which the pericarp loses its moisture as it ripens and becomes a cellulose membrane connected with other lignified elements. In some cases the ripened pericarp *dehisces*, or

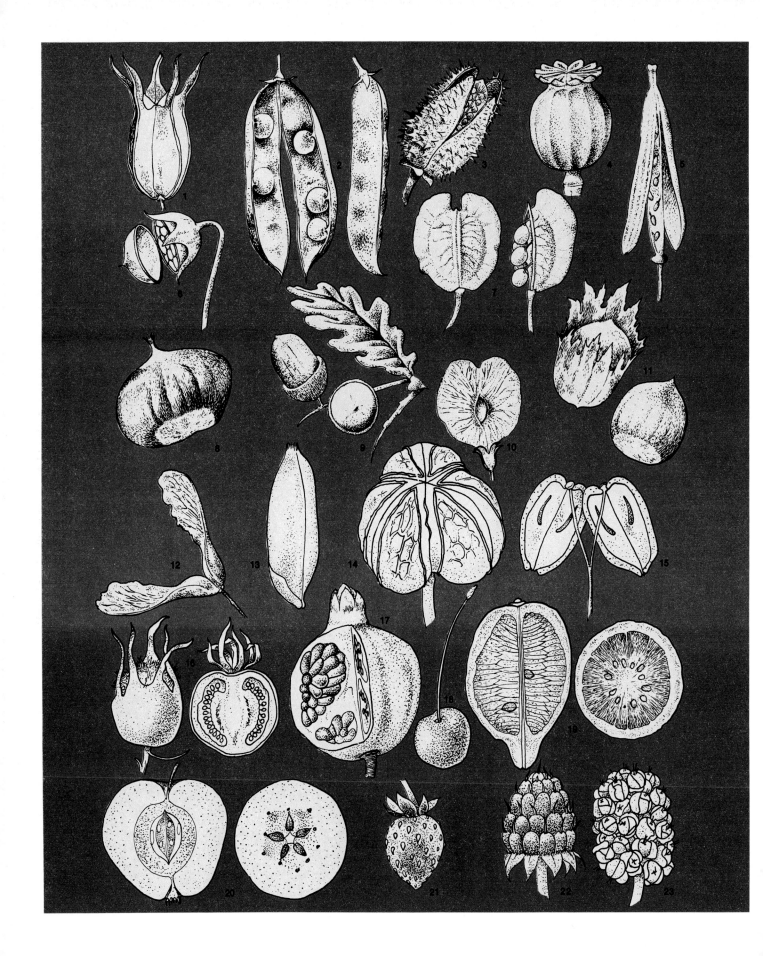

The chart on the opposite page shows different types of fruit.

Dehiscent dried fruits: (1) follicle (delphinium); (2) legume (pea, open and closed); (3) capsule; (4) porous capsule (poppy); (5) silique (mustard family); (6) lidded capsule; (7) siliquous.

Indehiscent dried fruits: (8) chestnut; (9) acorn (oak); (10) samara, or 'winged' (elm); (11) hazelnut, or filbert; (12) disamara (maple); (13) caryopsis, or grain (wheat); (14) tricoccus (spurge family); (15) diachene (parsley family).

Fleshy fruits: (16) nuculanium (medlar); (17) balausta (pomegranate); (18) drupe (cherry); (19) hesperidium (citrus fruits).

Accessory fruits: (20) pome (apple).

Aggregate fruits: (21) strawberry; (22) blackberry (raspberry); (23) sorosis (mulberry).

The diagram (right) illustrates various ways in which the ovarium is changed into fruit. (1) pistil; (2) seed; (3) specific ovarium tissue. In transformation, the ovarium may become a fruit dry and tough, like a capsule (4); woody, like a nut (5); fleshy, like a berry (6); or woody within and fleshy outside, like an olive (7). In these diagrams the fleshy parts are red, woody parts green.

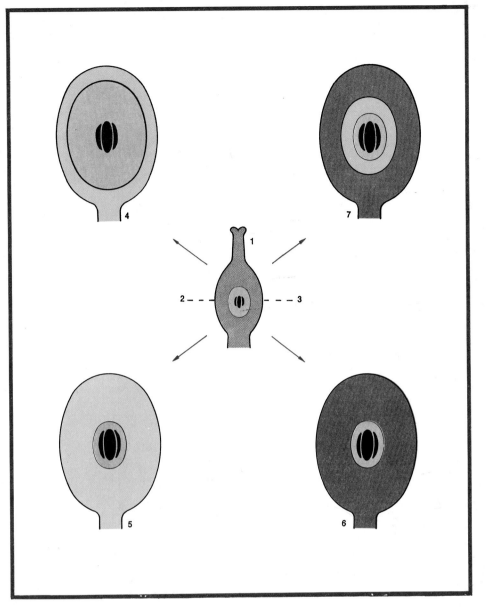

opens and expels the seeds; *indehiscent* fruit, such as acorns, must be opened by an agent, such as a squirrel, before the seed can germinate. Fleshy fruits are those in which the mesocarp cells accumulate a supply of nutritive substances, rich mainly in sugar, alcohol, ether, and essential oils. Frequently the endocarp lignifies and forms a *stone*, as in a *drupe* (cherry, apricot). To a botanist, a fruit is any kind of ripened ovary in which seeds are formed.

THE GYMNOSPERMS

The gymnosperms, or naked-seed plants, are believed from fossil evidence to have evolved about four hundred million years ago from fern-like plants. Today they include only about seven hundred species, as compared with the more than three million species of angiosperms, or enclosed-seed plants. However, the oldest and the largest living plants are both gymnosperms. Forests of gymnosperm trees cover immense stretches of terrain, mainly in the northern hemisphere, and represent tremendously valuable natural resources.

Gymnosperms have woody trunks, often of considerable size, though some species never grow higher than a foot and a half (half a metre). As a rule, gymnosperms are rich in branches; some, like the famous Cedars of Lebanon, have ramifications of unrivalled beauty and majesty. Yet the *cycads*, a group of tropical gymnosperms found only as far north as Florida, have no branches at all, only a simple tuft of leaves at the top that makes the plants resemble palms. Since the pine and fir trees are the best-known gymnosperms, we tend to think of them all as having needle-shaped leaves. However, the shape of gymnosperm leaves varies considerably; they may be fan-shaped, pinnate, lanceolate, scale-shaped, or needle-shaped. Except for a few species that are very similar to angiosperms, the gymnosperms have leaves with parallel veins.

The gymnosperms have no floral structure, in the precise meaning of that term. However, they have inflorescences of fertile leaves, or *sporophylls*, that are scale-like in structure and usually are joined in a spiral arrangement called the *cone*, or *strobile*. Each scale of the male cone produces pollen, and two ovules grow on the base of each scale of the female cone (usually larger than the male cone). Pollination occurs when the wind shakes loose clouds of yellow pollen grains, some of which reach the female cones. It may take nearly a year for the pollen to send tubes into an ovule and produce the sperm that fertilizes one of the two eggs there. As the embryo develops, the ovule changes into a seed coat, so the mature seeds, though not enclosed in fruit, are at least not completely 'naked' when they drop from the cone and perhaps develop into seedlings. The gymnosperms have thus evolved a structure and reproductive process better adapted to survival on land, under varying conditions of dryness, temperature, and so on, than have the ferns, from which they may be derived.

Studies of gymnosperm fossils indicate that they first made their appearance on earth during the Devonian period, which began four hundred million years ago, and were most widely scattered during the Carboniferous period, three hundred and fifty million years ago. The immense masses of these plants and their immediate forbears that then were deposited in swamps covering much of what is now Europe and North America account for much of the earth's supply of coal.

Of seven types of gymnosperms, three are extinct, known only by fossils. The four living groups are the *cycads*, the *ginkgos*, the *gnetales*, and the *conifers*. The conifers constitute the largest group, so we will discuss them last. The cycads date back to the Triassic period, and during the Cretaceous period represented two-fifths of all the existing vegetation. Today they include about a hundred tropical species. The trunk, up to 60 feet (18 metres) high in some species, is unbranched and bears a crown of large, evergreen leaves, pinnate and fern-like. The leaves range from a few centimetres to almost 9 feet (3 metres) in length and are often thorny or sharp-edged. However, the species of

Some elements of gymnosperms, or bare-seed plants. Left: the leaves of pine trees are needles; these are ice-coated. Top centre: juniper 'berries' are fleshy cones with seeds hidden within. Below: inflorescence of unisexual cones in the shape of an ament, or catkin (Scotch pine). Right: scale-like foliage of Araucaria *tree.*

the genus *Zamia* that are found in Florida have a short, tuberous stem and grow only about 3 feet (1 metre) high.

The ginkgos originated in the Permian period and were characteristic of the natural landscape roamed by the colossal prehistoric reptiles. Today they are represented by only one species, the *Ginkgo biloba*, which is regarded as a living fossil. The great trunk is often as tall as 80 feet (25 metres) and is very ramified. The leaves are fan-shaped and divided into two lobes; their forked veins resemble the foliage of the maidenhair fern, so the tree is often known as the maidenhair tree.

The gnetales are mostly shrubs or vines, although a few are small trees. Certain characteristics of various species of the three genera have led some botanists to suggest that these plants might be the missing evolutionary link between the gymnosperms and the angiosperms. However, this seems unlikely because gnetales do not appear in the fossil record until after the angiosperms had already become abundant. One species, *Welwitschia mirabilis*, is a strange plant found in only a few arid parts of South-west Africa. It has a large, squat, woody, tuberous stem with a long root and two long, strap-like leaves that keep growing throughout the plant's life, which could be a hundred years or more.

The conifers, or 'cone-bearers', are the most widely known and the most numerous among gymnosperms. They originated in the Carboniferous and Permian periods, but the more ancient species are all extinct. Nearly all conifers are trees and they tend to grow tall, some species reaching more than 300 feet (over 90 metres) in height. Most of the conifers are *monoecious*, producing both male and female cones on the same tree. Their leaves are simple and either needle-like or scale-like. Known familiarly as 'the evergreens', nearly all of the conifers are continually shedding leaves, but never all at one time. Conifers are found all over the earth, though predominantly in the northern hemisphere, and they cover one-third of all the existing forest area.

With about five hundred and fifty species, the conifers can be divided into six families – Pinaceae, Taxodiaceae, Cupressaceae, Taxaceae, Podocarpaceae, and Araucariaceae.

The Pinaceae, or pine family, includes the pines, larches, spruces, hemlocks, Douglas firs, and true firs. Nearly all are evergreen trees. The pines (genus *Pinus*) are distinguished by needles that grow in bundles of two to five and range from 1 to 4 inches (2 to 35 centimetres) long, depending on the species. Pines of different species are native to all parts of the United States except the

A variety of conifers. Left: redwoods of California include trees up to 3,500 years old and the world's tallest known tree, 368 feet (over 112 metres) high. Top right: pines grow squat, rather than tall, on a sea coast. Below: stately Italian cypress trees dwarf olive trees in Tuscany.

Remains of the vast Cedar of Lebanon forest that once covered the mountains of that country. Some are thousands of years old, but most were cut down to build ships, temples, and palaces for ancient civilization.

Midwest. The world's oldest living thing is a bristlecone pine (*Pinus aristata*) that has been growing at an altitude of 10,000 feet (3,000 metres) on a California mountain for about 4,600 years; its gnarled, twisted, and stunted trunk testifies to the environmental stresses it has survived there. The larches, or tamaracks (*Larix*) grow needles in clusters at the end of short shoots but shed them all each autumn, unlike most conifers. The cones grow upright on larches and the true firs, but are pendulous on most conifers. True cedars, such as the Cedars of Lebanon (*Cedrus libani*), are native to the Middle East, and they have also been planted throughout Europe and North America as ornamental trees; their foliage and cones are similar to those of the larches, but they are evergreens. The spruces (*Picea*) are distinguished by their short, sessile needles growing singly from woody 'pegs' on the twig. Hemlocks (*Tsuga*) have flat needles of varying length with a short, slender stem growing singly from a soft, woody bulge on the twig. Douglas firs (*Pseudotsuga*) grow to heights of 250 feet (75 metres) and diameters of 8 feet (2.5 metres); their flat, linear needles are narrowed at the base into a short petiole growing directly from the branch. The true firs (*Abies*) have needles similar to those of the Douglas fir, but without petioles, and upright cones, like the larches.

These gymnosperms are important factors in the food chain, as well as being valuable commercial products. Many animals feed on the seeds contained in the white pine. Chipmunks and mice also eat the seeds. The wood of these trees has long been used for furniture and house-building, and early man used the evergreen branches for shelters.

The Taxodiaceae, or redwood family, includes the world's largest living things. The giant sequoias (*Sequoiadendron giganteum*), found only on the west side of the Sierra Nevada, grow to about 280 feet (85 metres) with a trunk-base diameter up to 30 feet (9 metres); the oldest one standing is about 3,800 years old. The redwoods (*Sequoia sempervirens*), in groves along the California coast, have narrower trunks but reach well over 330 feet (100 metres) in height. Both are evergreens. A close relative, the deciduous bald cypress (*Taxodium distichum*), grows in the south-eastern states of America, mainly in swamps, and is distinguished by its fluted trunk and woody 'knees' protruding from its shallow but wide root system.

The Cupressaceae, or cedar or cypress family, includes true cypresses (genus *Cupressus*) and trees of various genera (*Thuja, Libocedrus, Chamaecyparis*) that are commonly called cedars. All have scale-like leaves growing on

Giant sequoia, with redwoods the last of forty sequoia species once flourishing in northern hemisphere forests, now grows only high on the west side of the Sierra Nevada. The tallest is 272 feet (almost 85 metres) high, with a base diameter of 30.7 feet (over 9 metres) and deeply cleft bark about 2 feet (61 centimetres) thick at the base. Top right: branches tend to droop downward, and thus foliage leaves the trunk bare for considerable lengths. Below: size of a fallen giant's trunk compared with a man.

Left: trees of the pine family thrive in the alpine zones of mountains the world over. Top right: a spruce-tree 'farm', where trees are grown for cellulose used in making paper. Below: a young red spruce; if it survives Christmas-tree hunters, it may grow to a height of 60 or 70 feet (18 or 21 metres) tall.

flattened or rounded branchlets, and bark that is very fibrous. The trees called cedars usually have aromatic wood. Members of the same family, the junipers (*Juniperus*) have scale-like leaves covering the branchlets, except for common juniper, which has pointed, concave, lanceolate needles in whorls of three. The tiny, semi-fleshy, blue, or reddish cones are commonly called 'juniper berries'.

The Taxaceae, or yew family, includes *Taxus baccata*, the European yew, and *Taxus brevifolia*, the Pacific Coast yew, both evergreens with flat, linear needles and a green seed protruding from a red aril like a stuffed olive in reverse colour. The Cephalotaxaceae or plum yews are hardy evergreen shrubs and small trees, growing mainly in South-east Asia. They have upright branches and, as the name suggests, fleshy fruit. The trees of the Araucariacea family are native to the southern hemisphere, but two species, *Araucaria araucana*, the monkey puzzle tree, and *Araucaria excelsa*, the Norfolk Island pine, are grown extensively in Europe and North America as ornamental trees. Finally the Podocarpaceae must be mentioned. The Podocarpaceae include sixty highly diverse species, among them the yellow-wood (*Podocarpus latifolius*) of South Africa, the Japanese *Podocarpus macrophyllus* and the New Zealand white pine, *Podocarpus dacrydioides*.

ANGIOSPERMS: Dicotyledons

The most highly evolved and specialized plants are the angiosperms – the second, and by far the larger, group of seed plants. Angiosperm, you may recall, means 'seed-in-vessel', or 'fruit'. It is this fruit and the ovary from which it forms that distinguish the angiosperms from the gymnosperms, for the ovules of the latter develop on the female cone scales, bare of any tissue that could enclose them in fruit. But the ovary is only one of the organs making up the fascinating complex of primary and secondary reproductive elements that we call the *flower*, and only the angiosperms have flowers. From the exotic orchid to the common dandelion, from the delicately fragrant lily of the valley to the bitterly aromatic rhododendron, the angiosperms display an amazing variety of forms, colours, and scents. These characteristics delight most humans and lead many to help flowering plants reproduce for aesthetic or economic purposes.

But the angiosperms evolved these ways of advertising their wares long before humans existed. Their 'customers' were the animals – mainly insects and birds – that were lured to a flower by its appearance or scent, found its nectar or pollen tasty and nourishing, and, while satisfying their own needs, helped perpetuate the plant species by accidentally carrying pollen from flower to flower and from plant to plant. Unlike the gymnosperms, very few angiosperms are pollinated by the wind; those that are, such as the grasses, sedges, and a number of trees, have evolved flowers with little or no 'advertising' power, since none was needed to perpetuate the species. At the same time, entomophilous pollination has evolved to such a point that quite often a species of plant can only be pollinated by one species of insect. Birds – especially humming-birds – pollinate certain flowers, and others are pollinated by bats or snails.

We might say a further word or two here about the pollinating agents and the flowers on which they feed. We have already mentioned that birds' eyes are red-sensitive, so they are attracted to red berries. For that reason, too, humming-birds are seen to prefer red flowers to blue ones that secrete the same amount of nectar. On the other hand, most pollinating insects, with the exception of butterflies, are red-blind. Bees' eyes are sensitive to yellow, blue, and ultraviolet.

The intricate and subtle evolutionary patterns have provided that wind-fertilized plants have little or no nectar – a sugary fluid – in their blossoms. Plants that must depend on living agents such as bats, birds, or insects (in Australia there is even a nectar-eating mouse) for fertilization do secrete nectar. And that nectar provides food for the pollinators. Such plants also are usually characterized by coloured petals to attract the creatures that become, in effect, plant breeders. In addition, spots or lines of colour often act as 'nectar guides' that help insects locate the source of their food more easily. These lines or spots are usually visible to the human eye, but some can be seen only by insects.

The pollinators, too, have adaptations for feeding. Insects such as bees and butterflies have long tongues that can probe into the depths of flowers. Humming-birds have long beaks and pointed tongues as well.

This, then, is a classic case of the mutual dependency that characterizes so much of nature. Neither plant nor insect (nor bird nor bat nor mouse) could exist without the other. The plants give food to certain creatures; those creatures, by eating, pollinate the plants and ensure their continuation.

Top left: oak woods. A member of the genus Quercus, *the oak often has a mighty trunk, and some reach 120 feet (over 36 metres) in height. Below: male flowers of the oak are aments, or catkins. Centre: many oaks are deciduous, some evergreen, and some have leaves that wither in autumn and fall the following spring. Top right: an oak in the open usually develops a stately head. Below: acorns, the fruit of the oak, usually mature in two years; most are too bitter for humans, but not for many small animals and birds.*

Another difference between the two groups of seed plants lies in the trunk or stem. In the gymnosperms, it is always woody; in many species of angiosperms it is herbaceous, and sometimes distinguished by subterranean components such as rhizomes, tubers, or bulbs. The very structure of the stem is different. In the gymnosperms, it consists only of *tracheids*, or perforated conduits sealed at top and bottom, which provide support; in the angiosperms, it consists of fibres, as well as trachea, which provide added strength.

Botanists divide the angiosperms into two classes, Monocotyledoneae and Dicotyledoneae, or monocots and dicots for short. *Cotyledons*, or seed leaves, are the one or more tiny leaves attached to the embryo plant in a seed. Their function is to absorb and digest nutritive substances stored in the seed and feed them to the embryo as the seed is germinating; they may also serve as food reservoirs, in which case the cotyledon is fleshy, like those of garden beans. In some cases the cotyledons are the first leaves to appear on the newly sprouted stem; in other cases they remain underground. Cotyledons vary in number: a gymnosperm has two, as a rule, but some species have as many as twenty; an angiosperm of the class Dicotyledoneae (dicot) has two cotyledons and one of the class Monocotyledoneae (monocot) has only one.

107

About three-quarters of the existing angiosperms are dicots, and they are found everywhere in the world except in marine environments. They are plants of arboreal, herbaceous, or liana structure, and many are hermaphroditic, having flowers with both male and female sex organs. The flower parts, such as petals or sepals, usually appear in twos, threes, fours, fives, or multiples of those numbers. The dicots are subdivided into two sub-classes according to the structure of their flowers. The sub-class Archichlamydeae, which we shall examine first, includes plants with flowers that are *apetalous* or *polypetalous*. Apetalous flowers lack petals because they have either no corolla or no *perianth* (calyx and corolla); polypetalous flowers have separate petals. Of the twenty-four orders comprising this large sub-class, let us first look at the Fagales, mentioned in the chapter on botany.

The Fagales order consists of two big families, the Fagaceae and the Corylaceae. The Fagaceae, or beech family, has about six hundred species. The arboreal form prevails, although there are shrubs, too. Leaves are mostly deciduous, except for certain evergreen species found in the south. These plants are also called *cupoliform*, because the fruit is often covered or encased in a 'cupola', or dome, formed of fleshy or woody leaves and covered with scales or

Specimens from the hazel or birch family. Top left: hazel fruit, an indehiscent nut (one that does not open when ripe to eject the seed). Below left and right: flower and fruit of the hornbeam tree, shown at top right shading a street. Centre: European white birch.

*Left: beech forest. such forests are widespread in central Europe. Top right: beech fruit, two or three angled nuts in burr with four vavles. The nuts are edible and rich in carbohydrate, oil, and protein. Below: the European beech (*Fagus sylvatica*).*

prickles. Flowers are *ament* (catkins) or clustered in *heads*. The American beech (*Fagus grandifolia*), with smooth blue-grey bark and two or three nuts in a spiny husk, grows to 100 feet (over 30 metres) tall and 3 feet (1 metre) in diameter. The European beech (*Fagus sylvatica*), is smaller and has darker bark.

The chestnuts (*Castanea*), deciduous trees with serrated, lanceolate leaves, two or three flattened nuts in a spiny husk, and fairly smooth dark brown bark, include about ten species native to southern Europe, the eastern United States, northern Africa and eastern Asia. The Spanish chestnut (*Castanea sativa*) is widespread throughout the Mediterranean countries, where it is cultivated for its nuts. The American chestnut (*Castanea dentata*) was once a valuable hardwood, but it has been almost wiped out by the chestnut blight, a bark fungus introduced from the Far East at the beginning of this century.

The oaks (*Quercus*), with over four hundred species, are certainly the most important of the Fagaceae. The plants of this genus are mainly arboreal, some deciduous and some evergreen. Save for a few exceptions, they are found only in the northern hemisphere. There are two species of oak widespread in northern Europe. *Quercus robur* and *Quercus petraea*, both renowned for their

strong and durable timber. *Quercus robur*, the English oak, is a very broad, not especially tall tree. The tallest European oaks are all sessile or durmast oaks, *Quercus petraea*. The evergreen holm oak (*Quercus ilex*) and cork oak (*Q. suber*) are native in southern Europe. The cork oak provides much of the world's cork, from its thick bark, which is carefully stripped off without damaging the living trunk. Each tree yields a fresh crop about every seven years. About sixty species grow in North America, and they are divided into two groups. Most of those in the 'white oak' group have leaves with rounded lobes or marginal teeth without sharp tips; their acorns mature in one year and have sweet meat. Trees of the 'red oak' group have leaves with sharp-tipped lobes or teeth, or with smooth edges; their bitter-tasting acorns usually take two years to mature. The Corylaceae, or hazel family, has two sub-divisions; the Coryleae 'tribe' includes the hazel (*Corylus*) and the hop-hornbeam, or ironwood (*Ostrya*); the Betuleae tribe includes the birch (*Betula*) and alder (*Alnus*). Most familiar of these is probably the birch, whose smooth bark is marked by horizontal *lenticils*, or corky lines, and can be pulled from some species like sheets of paper.

The Juglandales order has about sixty species, all members of the

Left: chestnut trees, also members of the beech family. The chestnuts have wide-spreading, dense foliage. They flourish on hill slopes, at an altitude of between 900 and 3,000 feet (300 and 1,000 metres). Top and below right: fruit, leaves and forest of Spanish chestnut, which grows to a height of about 70 feet (21 metres). Two or three flattened nuts, highly edible, are contained in the prickly burr.

Left: tree-size saguaro cactus with flowers at branch tips. These giant cacti with spine-reinforced fleshy trunks reach heights of 25 to 50 feet (7 to 15 metres). Their red fruit is sweet and edible. Top centre: Echinocactus, with long, very strong needles. Below: flower of the column-like Cereus. Nectar of cactus flowers feeds birds, insects, and bats. Right: cactus of the Opuntia genus, which includes prickly pear cactus, provides a meal for a desert rodent.

Juglandaceae, or walnut family. This includes the walnuts and butternuts (*Juglans*), and the hickories and pecans (*Carya*). The walnuts are highly prized for their wood, particularly in furniture manufacture. Perhaps best known, the English walnut (*Juglans regia*) is native to south-eastern Europe, India, and China, and is also grown extensively in the United States for commercial nut production and as an ornamental tree. The Salicales order has only the Salicaceae, or willow family, with two genera and some 325 species that grow in the temperate regions of the northern hemisphere. The willows (*Salix*) include many shrubs and some trees, all of them deciduous. The most familiar species is the weeping willow (*Salix babylonica*), which is native to China but is widely grown as an ornamental tree in Europe and the United States because of its 'sadly' drooping branches. The other genus, *Populus*, consists of the poplars, or cottonwoods, and the aspens, fast-growing species that are cultivated as shade trees, windbreaks, and for cellulose used in paper manufacture.

Of the four families in the order Urticales, the most varied is the Ulmaceae, or elm family, with more than a hundred and fifty species of trees and shrubs, found worldwide but mainly in temperate regions. Sadly, in recent years the elm has been severely reduced in numbers by the Dutch elm disease, a fungus

111

disease spread by a bark beetle. The hackberry and related berry trees are also members of the elm family. The Cannabinaceae, or hemp family, has just two genera: *Cannabis*, or hemp, provides tough inner-bark fibre that is made into rope, and leaves and flowers that constitute marijuana; *Humulus*, or hop, is an essential in brewing beer. The Urticaceae, or nettle family, is known best for its stinging species, *Urtica urens*.

Three orders of predominantly tropical plants offer some interesting representatives for consideration. For example, there is the familiarly spicy member of the Piperaceae family (order Piperales) known to botanists as *Piper nigrum*. Its dried berries and seeds are ground up into black pepper for your table. Some parasitic and semi-parasitic herbs and shrubs of the Santalales order are found in North America and south-eastern Europe. Members of the Santalaceae, or sandalwood family, they have chlorophyll but suck sap from the roots of neighbouring trees and shrubs. One such parasite is the East Indian sandalwood tree (*Santalum album*), whose fragrant, yellowish heartwood is much used for ornamental carvings and woodwork. Mistletoe (*Viscum album* of the Loranthaceae, or mistletoe family) is a semi-parasitic shrub, drawing only water from its host plant. True mistletoe is native to Europe; an American

Top left: flower of the Hottentot fig, cactus-like plant of Africa. Below: stone flowers (Lithops) resemble certain types of pebbles. Right: unusual tangle of saguaro cactus limbs. With water stored in trunk and limbs from only a single day's rainfall, some of these cacti can live for more than a year.

Left: inflorescence of Echeveria, *which thrives in arid and semi-arid parts of Mexico. Top right: a myriad tiny needles distinguish many cacti in the* Opuntia *genus. Below: red flower of* Cleistocactus, *a cactaceous plant with column stem.*

species, *Phorandendron flavescens*, is known as false mistletoe.

The Cactacea, or cactus family, is the only one in the Opuntiales order. Most of its more than a thousand species grow in arid regions of North America, especially the south-western states and Mexico. Adapted to survive in extremely dry areas, the cacti have wide, shallow root systems to absorb any available ground water, and fleshy stems that swell up and store water for as long as a year. Photosynthesis is carried on in the green stems because the leaves, which would increase loss of water through transpiration, have been reduced to simple spines that give the plant some protection against animals. Nevertheless, many desert animals get food and moisture from the stems and fruit of small cacti; and in the trunks of the saguaros (*Cereus giganteus*), which reach 50 feet (more than 15 metres) in height woodpeckers dig holes that owls and other birds also nest in. Large, colourful cactus flowers (in one case bigger than the plant) paint the desert landscape briefly in spring, and the bats, birds and insects visiting them for nectar carry pollen from plant to plant.

Plants of the Centrospermae order are mainly herbaceous, and some are cultivated for food. For example, the Chenopodiaceae family's five hundred or so species include spinach (*Spinacia oleracea*), and the garden and sugar beets

varieties of *Beta vulgaris*. The Aizoceae family includes two interesting species shown on page 112. The Caryophyllaceae, or pink family, has about eighty genera and two thousand species, including many wild flowers and the handsome, fragrant, and familiar carnation (*Dianthus caryophyllus*), a native of Eurasia that has been extensively cross-bred and exported to all parts of the world.

The insect-eating pitcher-plant (*Sarracenia purpurea*) and sundew (*Drosera*) are members of the two small families of the Sarraceniales order. Tea and violets are notable contributions from the Parietales order. Its Theaceae family are trees and shrubs with showy flowers of the genera *Camellia* and *Thea*. (Tea leaves for our popular beverage come from the Asian *Thea sinensis*.) The Violaceae family, with more than a hundred species of violets, includes the garden pansy (*Viola tricolor*), which was hybridized from wild pansy, and other wild violet species.

In the order Geraniales, the flax family, Linaceae, provides the valuable fibres used for more than four thousand years to make linen, and seeds from which oil is obtained. The Geraniaceae provide many species of *Geranium*, both wild and cultivated. The Rutaceae, or rue family, with more than a

Top left: castor-bean plant. Its large leaves are peltate and palmately cleft. Right: its raceme inflorescence has reddish pistillate flowers at top, stamenate flowers below. Below left: the soft-spined tricoccus fruit has one seed in each of its three compartments. Right: castor-bean exterior and cross-section showing the hypercotyl at the bottom of one of the dicotyledons.

Flowers of the buttercup family. Top left: buttercups of many species grow in meadows and humid woodlands; most are yellow, a few white. Below: marsh-marigolds, found mainly in swamps and beside brooks. Top centre: anemones have violet, red, or white sepals, but no petals. Below: purple clematis, a climbing vine found in rocky woods and slopes. Top right: a white-flowered clematis native to Europe. Below: Adonis vernalis contains an alkaloid used as a medical substitute for digitalis.

thousand species of trees, herbs, and shrubs, mainly in South Africa and Australia, includes the many species and varieties of orange, lemon, grapefruit, tangerine, and lime trees (*Citrus*) introduced and cultivated in warmer parts of the United States and Europe. Besides food, these fruits yield an oily essence used in making perfumes. The Meliaceae, or mahogany family, provides the highly prized hardwood obtained from trees in tropical and sub-tropical areas. Members of the family Euphorbicaceae are very diverse. European spurges (*Euphorbia*) are lowly herbs, but many desert species resemble large cacti. The genus also includes the familiar poinsettia (*E. pulcherrima*) native in southern America. Another species in the Euphorbicaceae is the castor-bean plant (*Ricinus communis*), found in warm parts of North America, but native to Africa, where it is cultivated for its pharmaceutical oil. (The seed is poisonous to eat.) Other valuable members of this family are the Brazilian rubber tree (*Hevea brasiliensis*) and the tapioca plant (*Manihot esculenta*).

The order Ranales embraces nine families with many familiar wild flowers and trees. Water-lilies make up the Nymphaeaceae family. The crowfoot family, Ranunculaceae, includes the buttercup (*Ranunculus*), the *Hepatica*, *Anemone*, *Clematis*, and larkspur (*Delphinium*). In the Magnoliaceae family we

Some species of poppy. Top left: the California poppy, now found cultivated and wild in Europe. Below: Papaver somniferum, the opium poppy. Centre: corn poppy, common in Europe, less common in the United States. Top right: Papaver radicatum survives in Arctic cold. Below: section of opium-poppy bud.

find the much admired magnolia trees, native to the warm regions of North America and Asia but grown also in Europe and Africa. The evergreen southern magnolia (*Magnolia grandiflora*) reaches a height of 90 feet (nearly 30 metres). Some species are deciduous, and one that sheds its leaves each autumn in the south retains them in the north until the new foliage forms. In the same family the stately tulip poplar or tulip tree (*Liriodendron tulipifera*) has unusual two-lobed leaves and tulip-like flowers. In North America, where it is native, this tree reaches heights of 150 feet (45 metres) and diameters of 6 feet (almost 2 metres). The Annonaceae, or custard-apple family, includes the pawpaw (*Asimina triloba*), a tree 10 to 35 feet (3 to 10 metres) high with deciduous leaves up to 12 inches (30 centimetres) long and 4 inches (10 centimetres) wide, purple-petalled flowers, and a long, irregular fruit that is edible when ripe.

The best-known family of the mainly herbaceous Rhoeadales order is probably the Papaveraceae, or poppy family. The common poppy (*Papaver somniferum*), with a mostly white or purple corolla, the bright scarlet corn-poppy (*P. rhoeases*), and the yellow celandine-poppy (*Stylophorum diphyllum*) all grow wild in Europe and now also in the United States, having been introduced from Europe. In return the orange-gold California poppy

Plants of the rose family. Top left: cherry trees in blossom. Below: pyracantha. Centre: inflorescences of the apricot tree. Top right: flowering almond tree. Cherries, plums, peaches, apricots, and almonds all belong to the genus Prunus. *Below: creeping cinquefoil, named for its five-petalled flower.*

(*Eschscholtzia californica*) has been introduced into Europe. Bleeding heart (*Dicentra eximia*) and Dutchman's breeches (*D. cucullaria*) belong to the subfamily Fumarioideae. The Capparidaceae are woody plants including the lowly caper (*Capparis spinosa*), which grows wild in Europe; its floral buds, picked or salted, are used in sauces. There are many food plants in the Cruciferae, or mustard family, including the radish (*Raphanus sativus*); white cabbage, kale, and mustard greens (genus *Brassica*); and watercress (*Nasturtium officinale*).

The record for species most dissimilar in appearance should probably go to the Rosales order, although the various species retain enough common elements to justify their classification in a single order. Succulent herbs of the Crassulaceae, or orpine family, are sometimes confused with cacti because spongy tissues in the stem and thick leaves store abundant water, enabling them to survive in dryish soils or on rocks and walls. The Saxifragaceae are a large family of herbs and shrubs named for the saxifrage, or 'stone-breaking' plant, an early-blooming wild flower. (The name was given to European species bearing granular bulblets that were supposed to dissolve kidney stones.) One member of the family is the familiar hydrangea. Another is the

Species of the Rosa *genus have been cross-bred to produce thousands of rose varieties. Top left: 'Belle blonde', a variety with blondish-yellow tint. Below: 'Prelude'. Centre: the dog-rose, Rosa caninae, is a common wild species. Top right: 'Message'. Below: intensely red 'Suspense'.*

currant (*Ribes*), whose juicy berries make excellent jams and jellies. The Hamamelidaceae include witch-hazel shrubs (*Hamamelis*) and the sweetgum tree (*Liquidambar styraciflua*), ornamental and a source of furniture wood. The Platanaceae are the massive sycamore (*Platanus occidentalis*), and the smaller Oriental plane and London plane, all distinguished by their blotchy trunks caused by peeling bark.

Plants of the Rosaceae family – trees, woody shrubs, and herbs – are widely cultivated for both ornament and nourishment. Their fruits take the form of drupe, pome, achene, berry, or capsule (see page 98). Many trees of the *Prunus* genus are cultivated for fruit: the almond (*Prunus amygdalus*), of which the oval drupe is not edible, though the seed is; and the various species of peach, apricot, plum, and cherry. The pear, the quince, and – perhaps the most common fruit of all – the apple, are all of the genus *Pyrus*. Other exquisite fruits from plants in the rose family are the raspberry and blackberry (*Rubus*) and strawberry (*Frageria*). The genus *Rosa* comprises about a hundred species of ornamental rose plants, but varieties are counted by the thousands, because hybrid specimens are constantly being induced.

Still in the Rosales order, we have the Leguminosae family, composed of

Some plants of the legume family. Top left: pods of the lupin plant, cultivated for green manure, fodder, and its edible seed. Below: the Scotch broom plant, so-called because of its stiff branches, has flowers with five sepals and a papillonaceous, or butterfly-like, corolla. Centre: one of the few leguminous lianas, Strongylodon macrobotrys. Top right: open pea pod. Below: lentil seeds.

trees, shrubs, and herbs bearing a fruit known as *legume*. According to the structure of their flowers, they are divided into three sub-families. The first, Mimosoideae, includes mainly wooden and thorny species with tiny flowers that form such thick inflorescences they give the impression of being a single flower. The bright yellow-flowered catclaw acacia (*Acacia farnesiana*) is often mistakenly identified as the white or pink-flowered sensitive mimosa (*Mimosa pudica*). In the sub-family Caesalpinioideae, the Kentucky coffee tree (*Gymnocladus dioica*), with purplish-brown legume pods 4 to 10 inches (10 to 25 centimetres) long, grows through the east-central part of the United States. The sub-family Papilionoideae includes many useful plants: Dyer's greenweed (*Genista tinctoria*) is used for dyeing. The clovers (*Trifolium*), lupins (*Lupinus*), and alfalfa (*Medicago sativa*) are raised for fodder; they also serve as 'green manure' when ploughed into the soil, because of the nitrogen fixed in their roots by bacteria living there. The common peanut (*Arachis hypogaea*), garden pea (*Pisum sativum*), lentil (*Lens culinaris*), kidney bean (*Phaseolus vulgaris*), and lima bean (*Phaseolus limensis*) are all useful foods. The decorative, and sometimes tree-smothering twining shrub wisteria is a legume of this sub-family, and so is the liquorice plant (*Glycyrrhiza lepidota*).

Few familiar plants are found in the order Myrtales. One in the Onagraceae family is the yellow evening-primrose (*Oenothera*), whose light-yellow species grow wild in many parts of the country. The most important member of the Myrtaceae family is the eucalyptus, with some five hundred species native to Australia and many cultivated in other warm areas of the world. In particular, many species have been introduced in California, including the blue gum (*Eucalyptus globus*), a tree that rivals the sequoias in height, sometimes reaching over 300 feet (90 metres). Various species supply timber, tannin from the bark and medicinal oil from the leaves.

The Malvales order has a number of interesting and valuable species. The Tiliaceae, or linden family, has more than forty genera, mostly trees, shrubs, and herbs found in the tropics or south of the Equator. However, the American basswood (*Tilia americana*) occurs throughout the north-eastern quarter of the United States and is valued for its strong wood, as a shade tree, and as an ornamental tree. Many species of the Malvaceae, or mallow family, provide pharmaceutical substances. The white-woolly seeds of *Gossypium herbaceum* supply most of the world's cotton. *Althaea rosea* is the garden hollyhock. Herbs and shrubs of the rose-mallow genus, better known as *Hibiscus*, display

Plants of the mallow family. Left: dried leaves and flowers of Malva sylvestris, *from which decongestant and expectorant medicines are extracted. Top centre:* Malva sylvestris *growing. Below: ornamental species of rose mallow (hibiscus). Other species provide textile fibres or digestive and refreshing brews. Right: fruit of* Gossypium herbaceum, *the cotton plant. The long fibres used to make textiles are attached to the seeds, which are separated by ginning.*

*Top left: horse-chestnut tree (*Aesculus hippocastanum*) in full blossom. This member of the Hippocastanaceae family is widely grown in parks and streets as an ornamental tree. Below: the compound leaf has five to seven leaflets and yellows early in the autumn (right). Buckeye trees have similar flowers, but most are yellow. The Ohio buckeye seed is enclosed in a capsule with short spines; others have smooth capsules, and the horse-chestnut capsule has longer spines.*

a variety of large, showy flowers; *Hibiscus esculentus*, native to Africa, is better known as okra. The chocolate tree (*Theobroma cacao*) belongs to the Sterculiaceae family and is native to South and Central America. And last, but definitely not least, there is the Bombacaceae family's colossal baobab tree (*Adansonia digitata*), which grows in Africa to a circumference of 95 feet (nearly 30 metres) at the base.

In the Sapindales order, we find poison ivy (*Rhus radicans*) and poison oak (*Rhus toxicodendron*) in the Anacardiaceae family, along with the sumacs (*Rhus*) and the cashew (*Anacardium occidentale*), grown in tropical America for its nuts and gum; various species of holly (*Ilex*) in the Aquifoliaceae family; a panorama of maple trees (*Acer*) in the Aceraceae family; and the buckeye trees (*Aesculus*) in the Hippocastanaceae, or horse-chestnut family.

The Rhamnales order has two principal families consisting of woody plants, some of which are climbing. The Rhamnaceae family's five hundred or so species of trees and shrubs are found mainly in temperate and tropical regions. *Rhamnus frangula*, whose wood is suitable for making wicker furniture and other objects, grows all over Europe and has become naturalized in the United States, where it is spreading rapidly and may become obnoxious. More

important is the Vitaceae, or vine family with about ten genera and five hundred species; the grape is the fruit of genus *Vitis*. Wines are made chiefly from varieties of two species, the European *Vitis vinifera* and the American *Vitis labruscana*. While there are more than six thousand varieties of these two species, very few have proved successful for making fine wines. About a hundred years ago the sub-species *Vitis vinifera sativa*, the most widely cultivated in Europe, was affected by the aphid *Phylloxera vastatrix* which threatened to wipe out the species. Fortunately it was discovered that the roots of the native American vines were immune to phylloxera, so American vine stocks were imported and the remaining cuttings of the old European vines were grafted on to them.

The arrangement of the flowers in umbrella-like clusters distinguishes plants of the order Umbelliflorae, which provides delights for the tongue and eyes, plus one to be wary of. The Araliaceae family is not impressive, though it includes ginseng (*Panax quinquefolium*), prized in Asia for its medicinal root, and the ornamental English ivy (*Hedera helix*). But the Umbelliferae family, with some 2,500 herbaceous species found mainly in the northern hemisphere, includes these food and flavour plants: chervil (*Anthriscus*), coriander

Some umbelliferous plants, members of the parsley family, which have umbrella-like inflorescences. Top left: ornamental flower of the Astrantia *genus. Below: poison hemlock. Centre: edible umbelliflores include fennel, celery, carrot, and parsley. Top right: eryngo (*Eryngium campestre*), a prickly plant with a turnip-like root. Below: chervil, an aromatic herb used for flavouring foods.*

Flowers of the primrose family. Top left:
Primula acaulis, *an early-spring mountain
primrose. Below:* Primula elatior, *or
greater primrose, a meadow flower. Top
centre:* Androsace *species grow in dry
sands and gravels or rocky woods. Below:*
Cyclamen europaeum, *a European flower
of sub-mountainous woodlands. Top
right: head of the* Androsace helvetica.
*Below: loosestrife, or golden club
(*Lysimachia vulgaris*).*

(*Coriandrum sativum*), celery (*Apium graveolens*), parsley (*Petroselinum crispum*), caraway seeds (*Carum carui*), fennel (*Foeniculum vulgare*), dill (*Anethum graveolens*), parsnip (*Pastinaca sativa*), and carrot (*Daucus carota*). Seed from the Asian anise (*Pimpinella anisum*) is used to flavour biscuits, sweets, and liqueurs. The plant to beware of is *Conium maculatum*, better known as poison hemlock – which was used to put Socrates to death. (Hemlock-parsley is called *Conioselinum* by botanists, for reasons you can probably work out.) And last – a delight for the eyes – are the plants of the Cornaceae, or dogwood family. This includes more than a hundred species, of enormous variety – creepers, shrubs and trees of every size. There are almost as marked variations between the flowers. The leaf, always oval with curving veins, is the only immediately obvious link between them. They grow mainly in temperate regions and the species include the European cornelian cherry (*Cornus mas*) and the American flowering dogwood (*Cornus florida*) and red-osier dogwood (*Cornus stolonifera*).

The five remaining orders are quite small. They are the Aristochiales, with wild ginger (*Asarum*) and Dutchman's pipe (*Aristolochia durior*), a vine whose blossom strongly resembles a brownish, curved pipe; the Polygonales, whose

123

sole family is Polygonaceae, the buckwheat family, with many species; the Leitneriales, whose corkwood tree (*Leitneria floridana*) has wood lighter than cork; the Podostemonales; and the Myricales.

All the dicot angiosperms we have discussed so far are members of the subclass Archichlamydeae, with flowers either without petals or with two or more separate petals. The other sub-class, Metachlamydeae, consists of plants with *gamopetalous* flowers, having the petals more or less united. Characteristically, such flowers have at the base of the corolla a single petal, which rises in a tubular or funnel-shaped *blade,* then opens out either in a circular fashion or in some irregular way. There are twelve orders of Metachlamydeae.

The Ericales order consists of plants whose characteristics are most like those of the polypetalous Archichlamydeae. In the Pyrolaceae family we find the herbaceous pipsissewa (*Chimaphila umbellata*), which originated in Eurasia and has developed eastern and western variations, and the small, waxy-white Indian pipe (*Monotropa uniflora*), a saprophyte growing on other plants' roots or on rotting vegetation. The Ericaceae family includes many species of rhododendron, azalea, and pinxter-flower – all of the genus *Rhododendron* – which are found in woodland areas but also are planted widely for their showy, ornamental flowers and lush, evergreen foliage. In the same family are heather (*Calluna vulgaris*), various heaths (genus Erica); also bilberry and cranberry, both of genus *Vaccinium*. The elegant arbutus or strawberry tree (*Arbutus unedo*) grows wild in the extreme west of Europe and is planted elsewhere.

The order of Primulales has but one family, the herbaceous Primulaceae, whose best known genus is the primrose or cowslip (*Primula*). The common primrose (*Primula vulgaris*) flowers between February and May, the cowslip (*Primula veris*) in April and May. Several other species, mostly with red or purple corollas, are grown in gardens.

The order Plumbaginales again has just one family, Plumbaginaceae. Its species include the delicate sea lavender (genus *Limonium*), or marsh rosemary (*Limonium carolinianum* and *Limonium nashii*), with its tiny lavender flowers and the pink-flowered thrift (*Armeria maritima*) and the blue-flowered *Plumbago europa*, which grows wild in Europe and is also cultivated as an ornamental plant.

Many woody plants of sizeable proportions, found in tropical and subtropical regions, belong to the Ebenales order. Among some three hundred species in the Ebenaceae family are the persimmon tree (*Diospyros virginiana*), which yields an orange to reddish-purple fruit that is deliciously sweet when ripe; the Texas persimmon (*Diospyros texana*) has smaller fruit, black in colour. The ebony tree (*Diospyros ebenum*), a native of South-east Asia, yields the hard, black wood prized for piano keys and furniture. In the Sapotaceae family, the sapodilla tree (*Achras zapota*) of South America produces latex gum called chicle, used to make chewing gum.

There is only one family in the Oleales order. Trees and shrubs of the Oleaceae family, which has twenty genera and about three hundred and fifty species, are widespread in warm and temperate climates, mainly in the northern hemisphere. The best known and economically most important are the olive trees (*Olea europa*), native to the Mediterranean area but now cultivated widely. The olive varies from bush size to about 30 feet (9 metres) in height. The *sativa*, or sowable, variety produces fleshy stone-fruits from which the valuable olive oil is pressed. A wild olive (*Osmanthus americanus*), with cream-white flowers and blue fruit ranges from shrubs to 75-foot (23-metre) trees. The ash trees (*Fraxinus*) are also members of the olive family. *Fraxinus excelsior* is one of the tallest European broadleaves. It has been known to reach a height of 148 feet (45 metres). The white ash (*Fraxinus americanus*) grows to 79 feet (24 metres). The manna ash (*Fraxinus ornus*), grown in Italy, yields a laxative substance called mannite. The European lilac (*Syringa vulgaris*), the Asian *Syringa persica* and the *Forsythia* species are other members of the family. Common privet (*Ligustrum vulgare*) and other species, often grown as ornamental plants in Europe and Asia, now grow wild in woods and swamps in the eastern and southern United States.

Top: green olive drupe, elliptic-ovoid in shape. The colour changes to purple-black in ripening. Below: olive press at work. The fruit of the olive is 15 to 20 per cent oil.

Plants of the olive family. Top left: inflorescence of the lilac, Syringa vulgaris, *a flower with a strong perfume. Below: flowers of the winter jasmine,* Jasminum nudiflorum. *Centre: olive trees, whose leaves are different shades of green on top and bottom, producing a chiaroscuro effect. Top right: ash trees, whose wood is valued for its elasticity. Below: Manna ash trees, source of a laxative substance called mannite.*

The Gentianales order includes four families of herbs and shrubs. In the Loganiacea family, the fragrant yellow jessamine or evening trumpet flower (*Gelsemium sempervirens*) grows on a twining or trailing woody shrub in the south. The Gentianaceae family's most noteworthy member is the gentian (*Gentiana*), whose vase-shaped, usually blue or purple, flowers, modified in the many species, appear in late summer and autumn in a variety of habitats. Perhaps the most beautiful of European gentians is the alpine *Gentiana aculis*. The foliage is inconspicuous, but the large flower is a deep and brilliant blue. From the foothills of the Alps comes the yellow-flowered *Gentiana lutea*, used to flavour a liqueur called *Enzianwasser* – from the German name for the genus. The fringed gentian (*Gentiana crinita*) has a misty, violet-blue corolla held in a four-sided calyx, usually a bronzy yellow-green. The bottle gentian (*Gentiana andrewsii*) has intense violet-blue flowers in cluster of two to five, with corollas so tightly closed that self-fertilization would seem a necessity of survival. Nevertheless, bumble-bees often manage to force their way in, perpetuating the benefits of cross-fertilization. The Apocynaceae are chiefly tropical plants, most of them with milky, acrid, and often poisonous juice. One familiar member, however, is the periwinkle, or common myrtle (*Vinca minor*),

whose creeping stems bear blue-violet flowers along roadsides and forest edges. The Asclepiadaceae family is named after the milkweed (*Asclepia*). This plant is particularly valuable to the famous monarch butterflies, which lay their eggs on the milkweed. The larvae feed on the leaves and acquire from the plant's acrid juice a bitter flavour that persists through their transformation into butterflies, discouraging would-be predators from eating the insect at any stage in its life.

The Tubiflorae order has an abundance of species and is economically important because of certain cooking vegetables and herbs. The Convolvulaceae family, most of whose members have twining stems or trunks, includes the common morning-glory (*Ipomoea purpurea*) and the wild morning-glory (*Convolvulus sepium*), the former native to tropical America. The Polemoniaceae has many species of the familiar genus *Phlox*. The memorable forget-me-not (*Myosotis scorpioides*) and its garden species (*Myosotis sylvatica*) belong to the Boraginaceae family. The taste-enhancing herbs are all in the large and cosmopolitan Labiatae family, so-called because the corolla of the flower has two lips – an inner upper lip and a lower lip with three lobes, which in some cases are joined together, in others divided. These

Plants of the nightshade family. Top left: black henbane, a slimy-leaved, poisonous plant widespread in northern Europe, Asia and Africa; it is a medicinal plant used as a sedative. Below: night-shade, a woody climbing plant naturalized from Europe. Right: fruit of the Chinese lantern, a species of ground-cherry. The red flower (right) contains the fruit, which is visible (left) in the cage formed of the flower's dried veins.

More members of the nightshade family. Left: hothouse tomato plants. Once shunned as food, the tomato is now widely relished; it is rich in mineral salts and vitamins. Top right: ornamental white flower of Datura arborea, *a plant traced back 53 million years; it is a member of the same genus as the poisonous and rank-smelling jimson weed, or thorn apple. Below: inflorescence of* Mandragora autumnalis, *a herb of the Mediterranean area used in medicine for its anti-asthmatic and sedative properties. The tap root is branched, but thick and fleshy, and some of the superstition associated with the plant may arise from a fanciful resemblance of the root to a human figure.*

plants are cultivated as garden ornaments and for medicinal purposes, as well as for their exquisite aromas. They include the peppermint (*Mentha piperita*), spearmint (*Mentha spicata*), sage (*Salvia officinalis*) – not the same as sagebrush – savory (*Satureja hortensis*), basil (*Satureja vulgaris*), oregano or wild marjoram (*Origanum vulgare*), thyme (*Thymus serpyllum*), catnip (*Nepeta cataria*) and rosemary (*Rosmarinus officinalis*). The *Lavandula* genus has aromatic species from which perfume essences are extracted. In the Solanaceae family we find these valuable plants: the cultivated potato (*Solanum tuberosum*), which originated in South America; the melon-like eggplant or aubergine (*Solanum melongia*); the cultivated tomato (*Lycopersicum esculentium*); and the tobacco plant (*Nicotiana tabacum*), which originated in tropical America. The Scrophulariaceae family provides the common toadflax (*Linaria vulgaris*), the even more familiar snapdragons (*Antirrhinum majus* being larger than *Antirrhinum orontium*), and the purple foxglove (*Digitalis purpurea*), whose dried leaves yield the powerful heart stimulant named for the genus.

The Plantaginales order has only the Plantaginaceae, or plantain family, whose members are homely herbs that grow in empty lots and unkempt gardens. In contrast, the Rubiales order – mainly woody or herbaceous plants

growing in tropical or sub-tropical regions – has both attractive and economically valuable species. In the Rubiaceae family, a plant we seldom see, but whose fruit is widely savoured, belongs to the *Coffea* genus. Coffee originated in Abyssinia (Ethiopia) and was introduced in Europe by the Venetians late in the fifteenth century; today it is cultivated on a vast scale, an important item in the economy of countries such as Brazil and Colombia. Some other members of the Rubiaceae family are the chinchona tree (*Chinchona officinalis*), whose bark yields the anti-malaria medicine and bitter-tonic flavouring, quinine; the trees and shrubs of the genus *Gardenia*, originating in the tropics and cultivated widely for the fragrance and beauty of their white or yellow flowers; and the delicate bluet (*Houstonia caerula*) found wild in open fields all summer long. The Caprifoliaceae family's best-known member is the honeysuckle (*Lonicera*), whose showy flower exudes a sweet fragrance that contrasts sharply with the rank odour given off by the shrubs, herbs, and small trees of the elder genus (*Sambucus*) when they are bruised. Both honeysuckle and the elders are widely distributed in the northern hemisphere. One member of the Valerianaceae family is the garden heliotrope (*Valeriana officinalis*).

Some plants of the mint family. Top left: leaves of the sage plant, much used in cooking. Below: flower of wild marjoram, or oregano, whose aromatic leaves also aid the chef. Centre: flowers of wild mint. Top right: flowers of lavender mint, a Mediterranean area plant whose leaves give sachets and some perfumes their delicate aroma. Below: Prunella flower, with the family characteristic of two lower petals curving outwards to form the upper lip and three petals, also curved, forming the lower lip.

*The dandelion (*Taraxacum*) is one of the most common members of the composite family. Top left: horns tip the individual flowers composing the inflorescence of many species. Below: the light fibres of the pappus, or crown, of the dandelion achene (seed) ensures scattering by the wind. Right: close-up view of dandelion achene, which has no endosperm, or built-in food supply.*

The single family in the Cucurbitales order, Cucurbitaceae, is small but includes some well-known fruits: squash (*Cucurbita maxima*), pumpkin (*Cucurbita pepo*), muskmelon (*Cucumis melo*), cucumber (*Cucumis sativus*), and watermelon (*Citrullus vulgaris*).

The Campanulales order is unique in being composed of one of the smallest families in numbers of genera and species, together with the largest family of flowering plants. Plants of the order are mostly herbs, and widespread because they adapt easily to a wide range of climatic conditions. In the smaller Campanulaceae family, flowers with five-lobed, bell-shaped corollas are the rule. One interesting example is Venus's looking-glass (*Specularia perfoliata*), whose species name describes the leaf meeting around the stem. The garden bellflower (*Campanula rapunculoides*) is a perennial which often grows wild. Found in moist ground, the cardinal flower (*Lobelia cardinalis*) has a vermilion or deep red corolla (occasionally deep pink or white). Its pendant, three-lobed, lower lip offers no footing for insects, whose tongues are too short to penetrate the long flower tube anyway, so only the long-tongued humming-birds can cross-fertilize the plants. Meanwhile they multiply mainly by perennial offshoots.

The largest angiosperm family is that of the Compositae, or composites, with about a thousand genera and twenty thousand species. Their characteristic inflorescence is the *head*, composed of many tiny flowers tightly crowded together on a flat or convex disc-shaped receptacle. Often the flowers at the edge of the disc have large and brightly coloured corollas; these *rays*, as they are called, are commonly and incorrectly referred to as 'petals'. Thus the common sunflower (*Helianthus annuus*) is a composite of dark disc flowers surrounded by a margin of orange-yellow ray flowers. As in most composites, only the sunflower's disc flowers produce seeds. The *Chrysanthemum* genus has both garden and wild species, including the white daisy (*Chrysanthemum leucanthemum*). The yellow daisy, or black-eyed susan (*Rudbeckia serotina*), is widely known. The English daisy (*Bellis perennis*), a widespread weed of lawns, especially on poorly-drained clay soils, and the western daisy of the United States (*Astranthium integrifolium*) with its purplish rays, are closely related. The *Aster*, *Dahlia* and *Zinnia* genera are familiar garden plants, as in the fragrant chamomile (*Anthemis nobilis*). Some attractive but often pesty composites are the common dandelion (*Taraxacum officinalis*), whose young leaves make good salad greens; the goldenrod (*Solidago*), with many species,

*Left: bud of the globe artichoke, whole (below) and in section. This member of the composite family belongs to the same genus (*Cynara*) as some thistles (see opposite page). The tender white heart of the artichoke and the bases of its husk scales are delicacies when cooked. Below right: portion of the composite inflorescence of the artichoke. Top: portion of the gumweed inflorescence, showing disc and ray flowers.*

Left: fruit, spiny receptacle, and leaves of Cynara cardunculus, *one of the many thistle species. This species is cultivated for strips of the young plant, which becomes white, tender, and edible if kept out of light. Top right: stem and inflorescence of nodding, or musk thistle (*Cardus nutans*). Below:* Cirsium eriophorum's *small white or purple flower heads are covered by thick hair, earning it the common name of woolly thistle.*

usually having small heads of predominantly ray flowers, yellow like the disc flowers; the common ragweed (*Ambrosia artemisiifolia*), whose genus name may seem inappropriate to hay-fever victims of the plant's pollen; and the common thistle (*Cirsium vulgare*), whose head of honey-scented, crimson-to-magenta, tubular flowers is attractive but whose prickly stem and leaves catch the unwary walker. Composite food plants include the artichoke (*Cynara scolymus*), common chicory (*Chicorium intybus*), endive (*Chicorium endivia*) and best known of all, garden lettuce (*Lactuca sativa*).

131

ANGIOSPERMS: Monocotyledons

At the beginning of the preceding chapter we explained the main distinction between dicots and monocots; the first has two cotyledons in the embryos, the second has but one. Additional differences are found in the stems, leaves, flowers, and fruits of these two kinds of angiosperms. The dicot's stem always has its vascular bundles in a regular arrangement; the monocot stem has more vascular bundles, but they are scattered through the stem. Also, the monocots usually have no cambium – the layer of living cells that produces annual growth rings in the dicots with woody stems or trunks. Much oftener than the dicots, monocots develop underground elements of the trunk (rhizomes) and of the leaves (bulbs). Dicot leaves have veins that are netted, or branched; monocot leaves have parallel veins, without any main vein. In the monocot, flower parts, such as petals or sepals, occur in twos, threes, fours, or sixes, but not in fives, which are common in the dicot flowers. The dicots' fruits take all possible forms; the monocots' fruits are limited to the capsule form or modifications of it, such as the caryopsis of the wheat plant, containing but one seed.

There are about three times as many dicot species as monocot species in the great angiosperm sub-phylum, but certain monocots – rice and wheat in particular – provide the staple diet for some four-fifths of mankind. The monocots are unusually adaptable, with species that can thrive in climates that are extremely cold or hot, humid or dry. The monocots are divided into seven orders with twenty-six families.

The Farinosae order is so-called because of the starch reserve in each seed. The largest and most important family is the Bromeliaceae, composed mostly of herbaceous plants native to tropical America. Best known is the pineapple (*Ananas sativa*), widely cultivated today in Hawaii and the West Indies, especially Puerto Rico. The fruit of this short-stemmed plant is a berry, and the large, pine-cone-like part with delicious pulp that we eat is not the fruit, but an *infructescence*, so closely packed that it forms a single unit. Like many other members of this family, Spanish moss (*Tillandsia usneoides*) is an *epiphyte* – a plant that grows attached to another plant, but not parasitic on it; this moss hangs in long tufts from limbs of trees in the swamps of the Gulf Coast states.

Many plants of the Liliales order have subterranean bulbs or rhizomes, and some are *phylloclade*, having stems that resemble leaves and perform leaf functions. The Liliaceae family has more than four thousand species, including many in the genera *Lilium*, *Tulipa*, and *Hyacinthus* (the garden flower is *Hyacinthus orientalis*), plus the lily of the valley (*Convallaria majalis*), and the familiar wild flowers of the *Trillium* genus. Edible species in the lily family include garlic, onion, leek, and chives, all of the *Allium* genus, and *Asparagus officinalis* (which means 'of the stores'). Familiar ornamental flowers in the Amaryllidaceae family are the *Narcissus poeticus*, daffodil (*Narcissus pseudo-narcissus*), jonquil (*Narcissus jonquilla*), tuberose (*Polianthes tuberosa*), and the American aloe or century plant (*Agave americana*), which matures and flowers only once in many years, then dies. The Iridaceae family includes the many species of the genera *Iris*, *Crocus*, and *Gladiolus*.

The order Graminales comprises two large families – the Cyperaceae, or sedge family, and the Gramineae, or grass family. The former are grass-like or rush-like herbs, mostly with solid *culms*, or stems, and grow best in humid, marshy soils. Each seed is enclosed in an *achene*, or small, dry, hard-shelled fruit, that does not burst open. It would not be capable of germinating unless it

Plants with underground reservoirs for food in the form of bulbs are members of the lily family. Top and below left: flowers and bulbs of the garlic plant. Each bulb consists of a number of small bulbs, or cloves, each having a thin, rosy, inner skin and a white outer membrane. Centre: flowers of Lillium candidum, *a late spring lily. Top and below right: flower and bulbs of the onion plant (see also page 91).*

were first ingested by a bird, the fruit being corroded by the bird's gastric juices and then expelled. The *Carex* genus has about fifteen hundred species, almost all of them marsh plants widespread in cold and temperate regions. Some dangerous plants that usually grow together with rice belong to the *Cyperus* genus, whose *Cyperus papyrus* was used by the Egyptians more than four thousand years ago to make the paper for which the species is named.

The Gramineae, or grasses, are one of the most important groups of plants, not only because they include more than six thousand species extensively diffused over the earth, but because those species that produce the cereal grains are valuable nutritionally to humans and other animals, both wild and domesticated. The grasses are mainly herbaceous, with fibrous roots – all about the same thickness and length, in contrast with a large tap root with smaller branching roots – and they have culms that are either hollow or contain very little pith. The fruit, which we call 'grain', is a *caryopsis* capsule, having a thin pericarp attached to the single contained seed. The most important species are wheat (*Triticum aestivum*), cultivated today in innumerable varieties; rice (*Oryza sativa*), the staple food of about two-thirds of mankind; barley (*Hordeum vulgare*), oats (*Avena sativa*), rye (*Secale cereale*), corn (*Zea mays*),

and sugar cane (*Saccarum officinarum*). The stalks and foliage of some of these plants, as well as some of the grain, serve as fodder for both work animals and those from which our meat and dairy products are obtained. Another useful member of the grass family is bamboo – the woody and arborescent species of genera such as *Bambusa*, *Arundinaria*, and *Dendrocalamus*. These plants usually flower only at an advanced age, shortly before their demise; this occurs regularly, at intervals of fifty to a hundred years, and at the same time in any particular part of the world where bamboo grows. The culm dries, depriving the plant of its aerial vegetative part; however, the species survives by sprouting new shoots from the plants' rhizomes.

Perhaps unrivalled in the extensive variety of its colourful, irregularly shaped flowers, the Orchiaceae family of the order Orchidales is represented worldwide, with its broadest array in the tropics. The plants are perennial herbs; some have corms, or enlarged underground stems, while others are rootless saprophytes, living parasitically on other plants. The flowers are distinguished by three outer sepals of about the same texture as the three inner petals, one of which – called the *lip* – is usually directed downwards and often is spurred. Orchids are widely cultivated but most wild species depend entirely on

Plants of the grass family cover vast areas of the earth's land, wild and cultivated. Left: close-up view of a spike of wheat. This nourishing cereal grain contains vital protein and minerals as well as energy-producing carbohydrates. Top right: ripening spikes of wheat. Below: tall silos where wheat is stored for the market.

Flowers of the orchid family, a cosmopolitan group represented most strongly in the tropics. Top left: cultivated species of the Cypripedium *genus (lady's slipper or moccasin flower). Below:* Cataseum fibriatum, *an exotic orchid prized by connoisseurs. Top centre: extraordinary flower of* Odontoglossum grande. *Below: varicoloured flower of* Serapias vomercea. *Top right: star-shaped flower of* Dendrobium stretbloceras. *Below: another cultivated species of* Cypripidium.

insects with long tongues for pollination. Some wild species include the tiny-flowered green adder's mouth (*Microstylis unifolia*), the larger snake mouth (*Pogonia ophioglossoides*), and lady's slipper – white (*Cypripedium candidum*), yellow (*Cypripedium parviflorum*), and pink (*Cypripedium acaule*).

Two familiar plants of the order Scitamineae are natives of the tropics: arrowroot (*Maranta arundinacea*), whose tuberous roots are valuable sources of starch, and the banana, a perennial herb of the genus *Musa*.

Typical of plants of the Spathiflorae order are flowers crowded on a fleshy or spongy axis, or *spadix*, often with a coloured *spathe*, or shield, around the flower base. Members of the Palmaceae family grow mainly in tropical areas of the Americas and Asia and are valued for the fruits of species such as the date palm, with its sweet fruit, and the coconut and oil palms, both exploited for the oleiferous substances extracted from them. The family Araceae is mainly tropical, but the wild *Calla palustris* occurs in swamps in temperate regions, and lords-and-ladies (*Arum maculatum*) is a familiar plant of woods and shady hedgerows. Many Araceae have odours that seem foul to us, but are attractive to the flies which pollinate the flowers. The Lemnaceae, or duckweed family, is composed of small, stemless plants of one or more fronds that float freely on or

in water, produce flowers, and often have roots growing from the underside. The simplest and smallest known flowering plants belong to the genus *Wolffia*; *Wolffia arrhia* has leaf blades less than a millimetre in diameter.

Marsh and water plants make up most of the orders Pandanales and Helobiae. In the first, we find in the Typhaceae family the common and familiar cat-tail (*Typha latifolia*), whose cylindrical spike flower has stamens on the upper part, pistils on the lower, where they catch pollen shaken loose by the wind. In the Sparganiceae family, the bur-reeds (*Sparganum*) have globular heads, scattered on the stem. The Helobiae order includes pondweed, such as toad lettuce (*Potamogeton crispus*), arrowgrass (*Triglochin*), water plantain (*Alisma triviale*), and the flowering rush (*Butomus umbellatus*), with its sword-shaped leaves and rosy-coloured flower.

Left: fronds and fruit of the cocoanut palm, found in Florida and California and widespread throughout the tropics. It is valuable for its fleshy fruit and the latter's high oil content. Top right: characteristic palm trees with slender, graceful trunks topped by tufts of fronds. Below: flowers of the cat-tail family, common in marhses and shallow water. Only the dark brown pistillate part is left; the stamens, which grow out of the stem above, have already been dispersed by the wind.

PLANTS AND MAN

Living in cities or suburbs, as so many of us do today, we tend to regard plants primarily as amenities, rather than necessities of life. Yet without the food and oxygen that plants produce through photosynthesis, humans and other animals could not exist. While our culture seems determined mainly by science and technology, the basis of what we call civilization is our ability to grow and use plants effectively.

During their first million or so years of existence, humans and their immediate evolutionary predecessors got their food by hunting wild animals and gathering the fruits and roots of wild plants. Only a few individuals could subsist together this way, and when the food supply diminished they simply moved their camps. Natural processes eventually restored the plant and animal life in the areas they had abandoned.

Then, about ten thousand years ago, some humans began to exploit nature, instead of just living on its bounty. They lived in an arc-shaped region of the Middle East that was carpeted with wild seed grasses – the Fertile Crescent, historians later called it. Some years ago archaeologists digging there found the remains of the earliest village we know of – a group of small, mud-brick houses that were occupied about 8000 BC. In the rubble they found fossilized grains of wheat, stone tools for cutting seed grasses and grinding the grain into flour, and crude tools for breaking the soil to plant seeds. Fossil bones like those of our domestic sheep and goats indicated that these people also herded animals to supplement their meat supply. It seems clear that farming and herding had made it both possible and necessary for these people to settle in one place, build 'permanent' shelters, and live together in greater numbers than hunting and gathering alone would permit.

By 4000 BC people in the Fertile Crescent were herding cattle as well as sheep and goats, cultivating the soil with improved tools, irrigating their fields, and growing wheat and barley with fuller heads of grain than the wild varieties provided. The increasing food supply had far-reaching consequences. More children survived long enough to reproduce, quadrupling the population in four thousand years. New villages sprang up throughout the region. Freed from the need to hunt or gather their own food, some people now worked only at making pottery, or baskets, or cloth from plant fibres and sheep's wool. Some became property owners and traders of goods; communities organized and promulgated laws. The earliest writing we know of – pictograms carved into clay tablets and hardened by baking – records such things as laws, grain transactions, and tax payments in grain. Farming was spreading westward around the Mediterranean Sea and eastward to the Indus river valley. But farmers in the Fertile Crescent were already beginning to move northward as over-cultivation exhausted their soil, and irrigation speeded up the natural salination process, making the land too salty to farm. Thus, while farming was nurturing the first great civilizations, it was helping to change the once-Fertile Crescent into the desert it is today.

Meanwhile the vast blanket of forests surrounding the Mediterranean Sea was beginning to disappear as nomadic farmers and herders cut trees, burned the vegetation, farmed or grazed herds for two or three years until the thin topsoil was exhausted, then moved on to slash and burn elsewhere. Most of the famed cedar forests of Lebanon were cut down for firewood or beams to build ships, palaces, and temples for the early Middle East civilizations. Denuded of natural cover, the topsoil of these lands yielded to wind and rain that washed

much of it into the sea, leaving barren deserts on the African and Middle Eastern coasts and meagre soil on the European coasts. The need for fertile soil was an important motive for the earliest wars of conquest, and has been for most wars ever since.

Today a small and diminishing number of primitive people still live in remote areas by hunting and gathering or by slash-and-burn farming. But most of the earth's arable land is now under continuous cultivation. In crowded South-east Asia, most families subsist mainly on the rice and vegetables they produce by hand labour, working, irrigating, and carefully husbanding the soil of their small plots. Only in a few areas does this *intensive farming*, as it is called, produce a surplus of rice for export. *Extensive farming* is exemplified by the production of vast amounts of wheat, corn, soya beans, cotton, and other staple crops on farms of several thousand acres or more, as in the American Midwest. These large farms are worked by relatively few people operating oil- or gas-powered machinery that cultivates, sows, harvests, separates fruit from chaff, and even bales or bags the product for sale at home or abroad. Such farms are owned by families or corporations in places like the United States, Canada, and Argentina, or by farmers in organized 'collectives' in the Soviet

Left: rice paddy-field, patiently built and cultivated by intensive human labour in South-east Asia. Terraced hillsides exploit mountain stream water while protecting the soil from erosion. Right, top and below: fertile pasture land typical of the Mediterranean area.

Wheat harvest in an extensively cultivated region of Canada. The path left by the reaping machine, viewed from above, looks like an abstract painting.

Union. Between these two extremes are farms of all sizes, worked by hand, with the aid of horses or oxen, or with tractors and other machines. There are vegetable farms, fruit orchards, rubber plantations, dairy farms, and farms that just feed livestock for the market after the animals have been raised on natural-growing vegetation.

Even trees are being farmed, though not nearly fast enough to supply the wood needed for houses, furniture, and the myriad of cellulose-based products such as paper, cardboard, and certain synthetic plastics and fibres. Stands of virgin trees like the monumental California redwoods are endangered because their wood is so highly valued for houses and outdoor furniture. The sudden realization that our supplies of oil, natural gas, and coal are not endless, reminds us how dependent we are on the primeval plants that millions of years ago formed most of the fossil fuels we have been using up so profligately. They supply not only most of the energy to power our machines, but also the raw materials for many plastics and new substances our science and technology have 'created'. Once used, these fossil fuels can never be replaced.

Science has greatly increased our understanding of plants and how best to grow them. Scientists have produced new kinds of plants and new varieties that

are more sturdy, fruitful, faster-growing, and pest-resistant than the natural varieties. They have developed fertilizers to nourish the soil and poisons to kill plant and insect parasites on our crops. But pollution of the natural environment by some of these chemicals, degrading lakes and rivers and endangering many species of wildlife, reminds us forcefully that everything in nature is interrelated – that we can't get anything from nature without paying for it in some way.

The world produces more food than ever before, but the population is larger than ever and growing rapidly. About two thousand million people lack the right kinds of food to keep healthy; five hundred million more don't get enough food to sustain them. However we try to meet this problem, we must remember what happened to the Fertile Crescent and the Mediterranean lands when man began to raise plants for food – and make sure that we are working *with* nature, not against it.

Mechanical harvester cuts and separates the grain from the chaff. Mechanization substitutes the energy of fossil fuels for the energy of human farm labourers. Where this is practised, farm workers make up only about 4 to 10 per cent of the working population; in non-mechanized agricultural regions, farm labour often exceeds 50 per cent of the working population.

THE LIFE OF INVERTEBRATES

ZOOLOGY

Zoology is the study of animals. Because a vast number of different species of animals exists, and because each one of these species can be examined from various viewpoints, *zoology* is subdivided into a large number of headings, which are nevertheless closely allied with one another. Such headings are: *Morphology*, which ranges from the anatomical examination of the whole animal body, usually by means of dissection, to analysis of the single tissues and cells. *Physiology*, which deals with the functioning of the various organs and their reciprocal interactions. *Embryology*, which follows the development of the individual creature from the egg onwards, and illustrates the progressive appearance and differentiation of the various organs. *Ethology* studies the behaviour of animals with special regard to the patterns of locomotion and feeding and to their often highly complicated reproductive habits. Lastly, *ecology* studies the relations which exist between an organism and its environment, while *genetics* considers the laws which govern the hereditary transmission of characteristics from one generation to the next.

The first task of zoology, however, is still the *identification* and *description* of the various species and the interpretation of the differences existing between the various species. An organized knowledge of this world must, however, start with the introduction of a simple and universal method of classification.

The *modern classification* of the animals is organized on a hierarchic basis, following the method proposed by Linnaeus. At the lowest rung are the *species* grouped in categories which ascend gradually to the animal kingdom. Similar species are in turn grouped in a *genus*, similar genera in *families*, and similar families in *orders*; the orders are then grouped in *classes*, and the classes in *types* or *phyla*. The phyla known to us today – molluscs, annelids, chordates, etc., there being about twenty in all – represent considerably diverse modes of animal life; in other words each phylum has a particular constructive plan, the variations of which correspond to the various classes, families and species identified within the phylum.

According to Linnaeus, each species had been created independently of the other species and was essentially incapable of undergoing any modifications in the course of time. But it was the opinion of Lamarck and Darwin that each species is subject to continual modifications which lead to the emergence of new species after many generations. Two species which are similar to each other, therefore, evidently derive from a common progenitor. It is thus conceivable that all the species which we see around us, therefore, derive, just as evidently, from a single, very ancient founder or archetype; the process which has led from this progenitor to all the various present-day species is called *evolution*, and everything points to the certain fact that this process is still carrying on today.

The fairly perceptible differences that we observe between two species are, accordingly, a sign of the fairly remote age in which the two original progenitors, from which the two species are derived, were separated. The genus, family and all the other categories of zoological classification thus acquire a natural meaning, because the structural similarities correspond to a fairly close consanguinity or blood-relationship in the genealogical tree.

To construct the genealogical tree of an animal species one can begin to look through the archives where the remains of animals from the past are kept. These archives are in fact the *sedimentary rocks* in which bones, teeth, shells and any other of the harder parts of animal bodies are preserved. But these

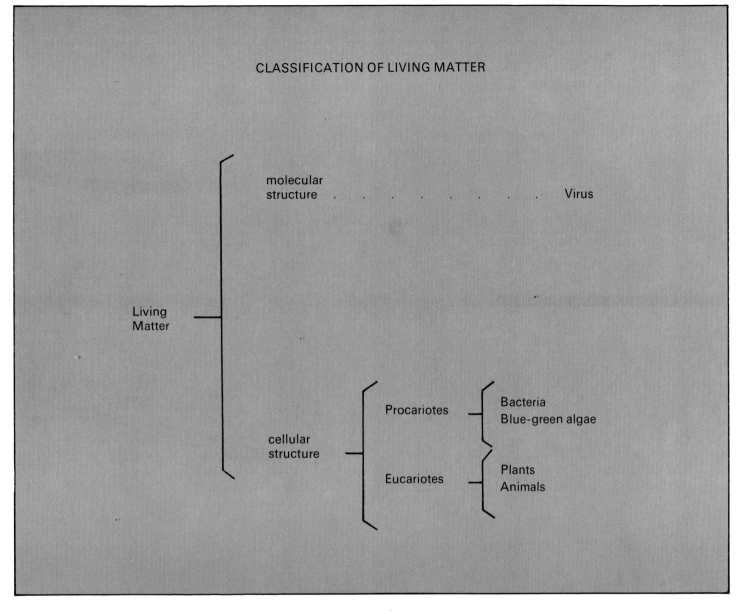

CLASSIFICATION OF LIVING MATTER

molecular
structure . . . . Virus

Living
Matter

cellular
structure

Procariotes — Bacteria
Blue-green algae

Eucariotes — Plants
Animals

vestiges of bygone life – called *fossils* – are not in themselves sufficient for our task of reconstruction.

Most of the useful elements at our disposal derive, once again, from comparisons between present-day species. In some cases it may be difficult to define the affinities of an animal. Take the dolphin, for example. This species has fins like those of a fish, but it suckles its young and breathes with lungs like a dog, horse, or any other mammal. Does the dolphin then have more affinity with fishes or mammals? In fact it only reminds one of fishes because it lives in the sea and has fins, but all its other characteristics link it with the animals which we classify as mammals: for this reason the dolphin is included in this group. In short it can be said that the dolphin is a mammal which is *adapted* to aquatic life, and that its feet and tail have been turned into fins for swimming. Its similarity to fishes is due to a simple process of convergence justified by the fact that both the dolphin and fishes live in water.

Now let us look at a bird. Its wings, which are covered with long feathers, are used for flying with, in the same way as the membranous wings of bats or the wings of a butterfly or fly are. The wings of these animals have a different origin and look different, but they serve the same purpose: flying. In this case we can

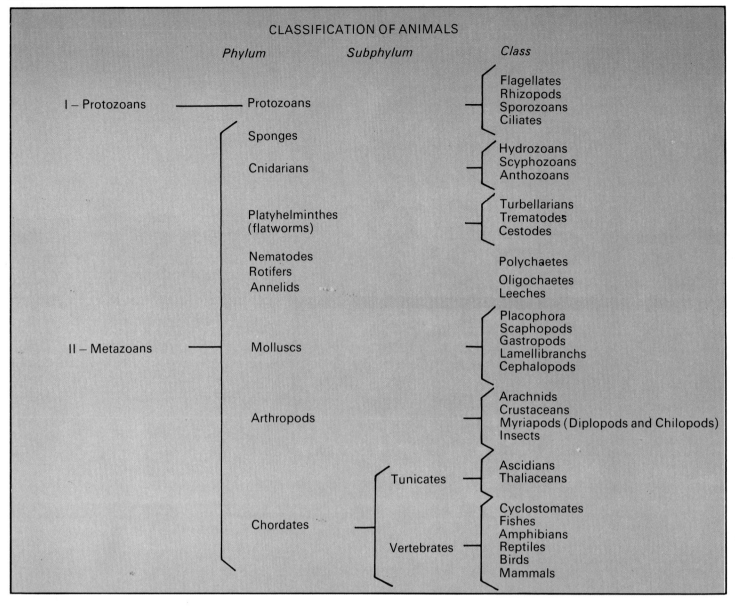

CLASSIFICATION OF ANIMALS

Phylum	Subphylum	Class
Protozoans		Flagellates / Rhizopods / Sporozoans / Ciliates

I – Protozoans

II – Metazoans
- Sponges
- Cnidarians — Hydrozoans / Scyphozoans / Anthozoans
- Platyhelminthes (flatworms) — Turbellarians / Trematodes / Cestodes
- Nematodes
- Rotifers
- Annelids — Polychaetes / Oligochaetes / Leeches
- Molluscs — Placophora / Scaphopods / Gastropods / Lamellibranchs / Cephalopods
- Arthropods — Arachnids / Crustaceans / Myriapods (Diplopods and Chilopods) / Insects
- Chordates
  - Tunicates — Ascidians / Thaliaceans
  - Vertebrates — Cyclostomates / Fishes / Amphibians / Reptiles / Birds / Mammals

say that these organs are *analogous*.

Now let us look at the front legs of a horse and the upper limbs of a man: these features are clearly not used for flying, but for walking or taking hold of and moving objects near at hand. The functions of the bird's wing, the horse's leg or a human arm are thus different.

For this reason these organs are not analogous; however, if we take a look at the skeleton, we find that it consists essentially of the same bones, differently developed, and that in all three cases, the wing, the leg and the arm are articulated in a similar manner on the spinal column; what is more, in the embryonic development, the *Anlage* of these appendices is similar. All this goes to suggest that the bird's wing, the front leg of the horse and the upper limb of a man are equivalent to one another and that the major differences displayed by them simply correspond to different adaptations to their environment.

In various animals the same appendices are used either for flying or walking, or taking hold of things. In zoological terminology these features are *homologous*, but obviously not analogous. Homologous organs, therefore, indicate kinship or relationship, but analogous ones do not. The latter are simply the result of modifications which appear independently in different

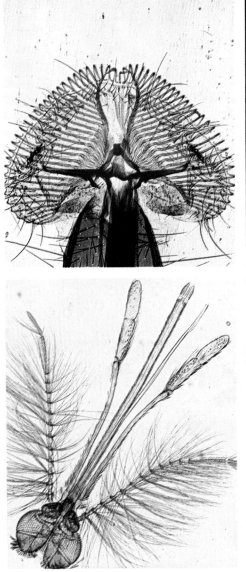

Left: examples of homologous organs; the buccal apparatus of the fly (top) and the mosquito (below). These have a different form and function, but they consist of the same differently modified parts. Right: examples of analogous organs. The wings of the bat (top) and those of birds (below) serve the same purpose, which is for flying, but show totally different structures.

animals living in one and the same environment.

And so in order to construct a natural classification of the animal kingdom, the zoologist has to be able to recognize the similar structures by pure analogy and identify the homologous organs despite their differences; these organs will in turn give an indication of common descent.

It is a general rule in fact that two animals resemble each other more closely when they are youngsters than they do when adults. The embryologist von Baer, a famous nineteenth-century scholar, reached the point when he had to admit that he could no longer successfully make a distinction, in his collection, between an embryo of a reptile and an embryo of a bird without the respective index-cards bearing the names of the species.

The study of animals of the past, which are being increasingly revealed in rocks in fossilized form, has strongly confirmed the hypothesis of an evolution of the animals species from the simplest to the most complex. It is true that the study of fossils tells us nothing whatsoever about the most primitive creatures, which must have corresponded to the most simple forms of present-day unicellular protozoans. It is extremely rare to come across soft remains of animals without skeletal parts. But our fossil documentation, however

incomplete it may be, does show regularities which can only be explained by the theory of evolution. It is a fact that in the oldest rocks there are no traces of vertebrates, and that the early fossils of this group – which date to four hundred and fifty million years ago – recall the most simple forms of present-day vertebrates, the jawless lampreys which have no paired appendices. The remains of real fishes are only found in more recent rocks, and the amphibians – which were the first vertebrates to appear on *terra firma* – did not appear until even later. Birds and mammals, the most complex and specialized of the vertebrates, are also the last to appear on the face of the earth. The same applies to many groups of invertebrates. Where insects are concerned, for example, the cockroaches and dragonflies appeared before the bees, flies and coleopterans, and their comparative primitiveness is related to this.

The problem of the origin of life on earth goes beyond the realms of zoology. But what can be said without any hesitation is that among the unicellular organisms it would not have been easy to distinguish between animal forms and vegetable forms. Furthermore, this problem is debatable in the case of present-day species as well. The safest criterion for distinguishing an animal from a plant resides in their different *manner of feeding*. The animal, be it

The presence of fossils in rocks, that is the evidence of quite profoundly modified organisms, has made it possible to reconstruct the plants and animals which covered the earth in bygone days. Top left: fossil nummulites, typical of the Tertiary period. Below: fossils of Turritella turris, a marine gastropod mollusc. Top right: a hymenopteran insect preserved in amber. Below: snail shells in a stage of not very advanced fossilization.

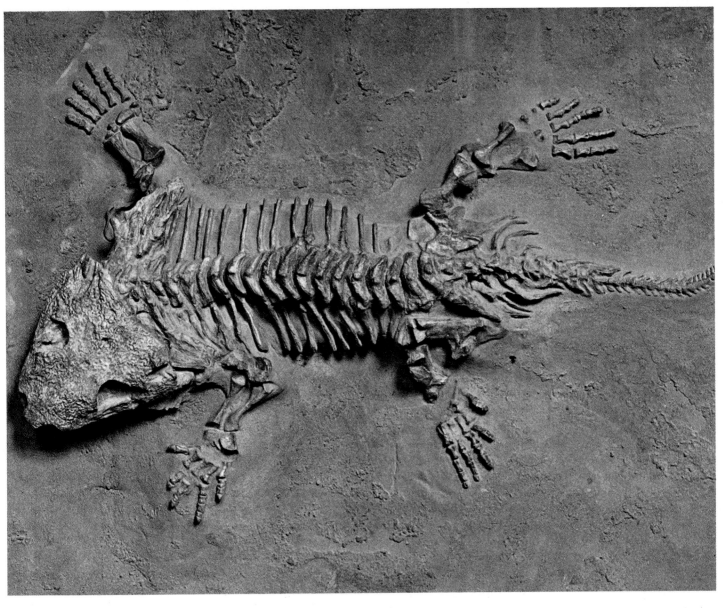

The fossil skeleton of Seymouria baylorensis, *a vertebrate which lived in the Palaeozoic period. The exceptional interest of this example resides in the fact that it is a form of life with features which lie somewhere between those of the amphibians and those of the reptiles.*

unicellular or pluricellular, needs complex organic substances already constructed and previously synthesized by plants (and then transformed by other animals). All the plant needs, on the other hand, is water, carbon dioxide and some mineral salts; the necessary organic substances are synthesized in the leaves, by means of the action of luminous energy which is caught by the chlorophyll. Now, in the case of unicellular forms, there are species which resemble one other in every respect except for the fact that the one has chlorophyll and the other does not. Are we then to call the former a plant and the latter an animal? It would be more apt to say that on the level of these simple organisms the traditional distinction between animal kingdom and vegetable kingdom cannot be given much credit. In fact, some species such as the Euglena behave as vegetables when they are in the light, but feed like animals when in the dark.

In any event, an animal needs food. A large number of species feed on vegetable tissues, using both living plants and their decomposing remains. Many eat just branches and leaves, like the ruminants and – on a more modest scale – the locusts or caterpillars: other creatures live actually in the vegetable tissue itself, where they burrow tunnels; still others feed by sucking the lymph

or sap through a long pricking rostrum. A whole other legion of species feeds on other animals; some are predators and hunt their prey, others suck blood or other vital fluids, and these are parasites. There are external parasites, such as the tick, which live on the blood of their hosts, and internal parasites, such as the fluke which lives in the liver, or the malarial plasmodium which destroys the red corpuscles. According to the different method of feeding, each species is differently organized. The cats have muscles for leaping, curved claws which are pointed, and sharp, fierce teeth for catching, killing and gnawing their prey. Cattle and the like have teeth with wide cusps, designed for grinding grass and a concamerate stomach in which the food can be swiftly stored, and then later re-chewed and digested at leisure. The leech or bloodsucker has suckers with which to cling to the victim's body, small teeth for tearing open the skin, glands producing a secretion which prevents the blood from coagulating before the creature has finished sucking, and lastly, intestinal sacs in which the blood sucked in is collected. The various organs of an animal are therefore designed to create a coherent whole, so as to solve the basic problems of the survival of the species. A mammal with predator's teeth will undoubtedly be a nimble animal, and this will be reflected in the muscular arrangement and the

The difference between animals and plants lies above all in the fact that the latter are capable of producing organic matter from inorganic substances (carbon dioxide, water, mineral salts) by using solar energy. Because of this characteristic they are called autotrophic organisms.

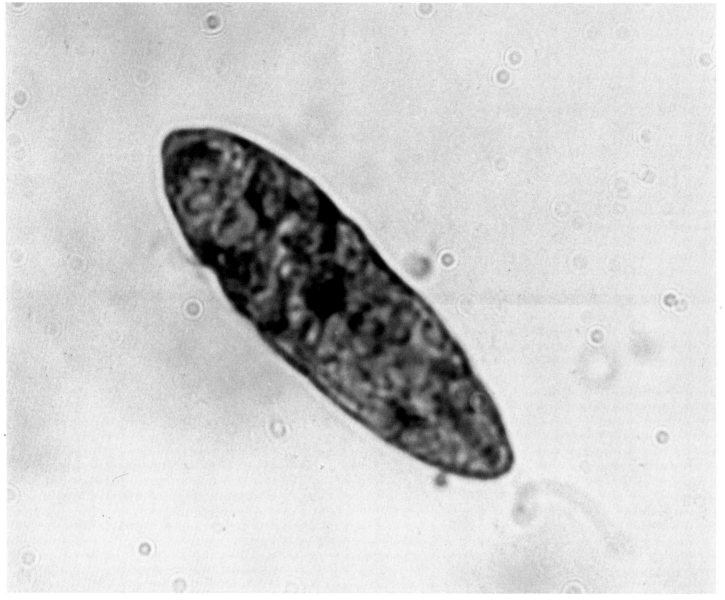

The illustration shows a Euglena, *a unicellular organism which has features which are intermediary between those of plants and those of animals. In fact it is equipped with chlorophyll like plants, but is capable of active movement by means of a flagellum. In addition, when there is no light at hand, it can feed like an animal.*

proportions of the bones in the trunk and limbs. George Cuvier, a French zoologist, living in the early nineteenth century, who set forth the bases of comparative anatomy, would maintain that in many instances just a few bones could suffice to reconstruct an entire skeleton and hence the overall appearance of the animal. In effect, in order to survive, every animal must respond to the demands of the environment in which it lives. The environment, on the other hand, does not remain constant as time goes by, but undergoes continual variations. Some of these variations affect the existence of the individual animal: day-to-day life is subject to the alteration of light and darkness, dry and wet conditions, tranquillity and the danger of being preyed upon – all problems which some individuals, in any social group, manage to cope with more effectively than their kin. Those individuals which from time to time turn out to be more suited to the environment will have a greater likelihood of surviving and leaving descendants behind them; in turn they will pass on to their offspring those characteristics which have made them more likely to survive than their peers. In every corner of the earth all the animal populations are in effect subordinate to the dictates of this natural selection, which was first perceived by Darwin as being the principal agent of evolution.

In addition to the continual day-to-day or seasonal variations, the environment undergoes long-term variations which sometimes affect large sections of the earth's surface. Today we know that in the last million years of the history of our planet very extensive regions of the northern hemisphere have on various occasions been covered by advancing ice-floes from the polar region, with alternating and recurrent returns to temperate conditions. These phenomena have obviously had repercussions on the fauna: many species became extinct during the periods of glaciation, while other species have in turn come into being as a result of single and fortunate groups being isolated in small areas encompassed by unadaptable tracts of land. It is in fact a general tendency of such isolated populations that they become more and more unlike kindred populations to the point when they turn into completely different species. This phenomenon can be observed, for example, in islands in the oceans and in caves and grottoes. The fauna of *islands* is extremely rich in specific species, which are all the more distinctive according to the age of the island which is host to them. The fauna of Madagascar is of particular interest in this respect: here there are whole families of animals which exist nowhere else in the world, such as the lemurs (prosimians). And the fauna of Australia is

Left: a koala eating eucalyptus leaves, the only food it really likes; top right: a cheetah's meal; the cheetah is a carnivore which hunts its prey using its attribute of great speed; below: an apodal amphibian eating an earthworm. Opposite: a diagram showing the alimentary chain which links organisms living in a lake.

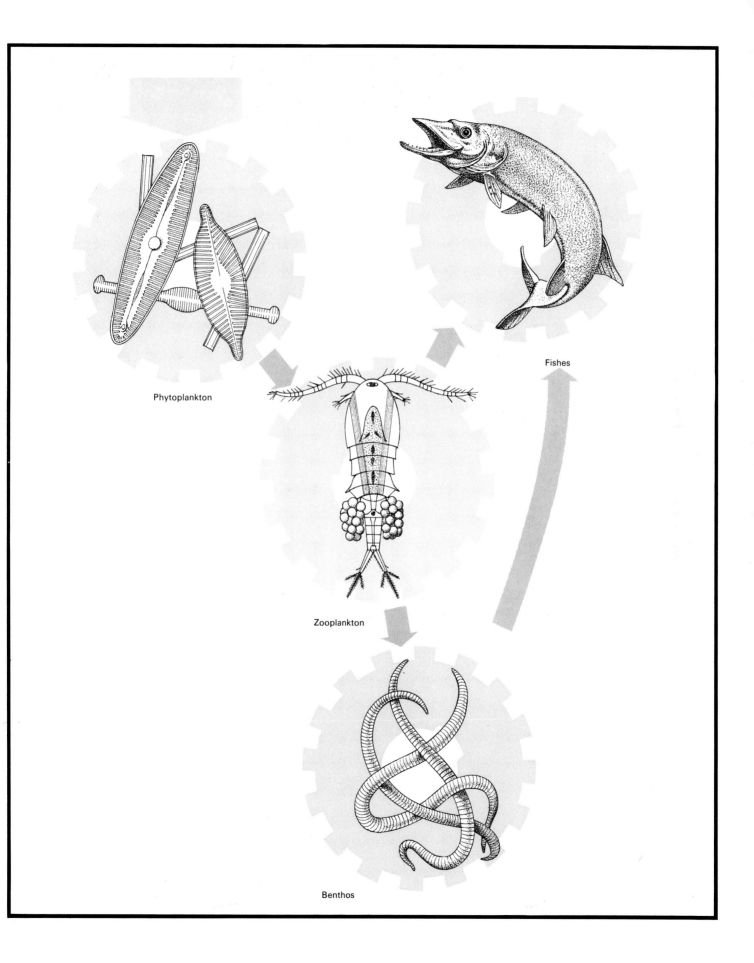

Phytoplankton

Fishes

Zooplankton

Benthos

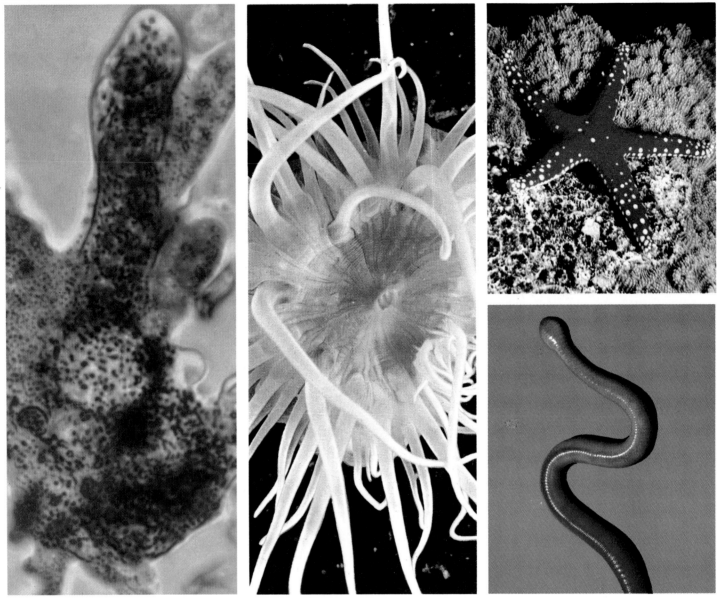

no less exceptional, with the duck-billed platypus and the echidna or spiny anteater, both primitive mammals which lay eggs, like birds, and which have no relations in the rest of the world. Other unique features of the Australian fauna are the kangaroos, the koala bear and the other marsupials, which do have distant relatives in the Americas, but belong to different families. Considerable interest attaches to the fauna of the islands, large and small, which are situated in mid-ocean, far from the coastlines of the various continents. The small archipelago of the Galapagos Islands, in the Pacific Ocean off the South American coast, houses a large number of thoroughly typical animals which often differ in every island in the archipelago. Charles Darwin landed on the Galapagos Islands during his early voyage around the world, and his study of the typical chaffinches, land and marine iguanas and the giant tortoises living in that remote spot provided him with many valuable ideas in support of his theory of evolution.

The fauna of *caves* is an extremely interesting one, too. In many parts of the earth there are animals which are adapted to live out their lives underground, in total darkness, and in conditions where the food available is almost invariably quite restricted. Such animals are blind, usually pale-coloured, or

Left: a protozoan (amoeba), a unicellular organism, that is, having just one cell. Centre: a coelenterate (jellyfish), a pluricellular animal with a radial symmetry. Top right: an echinoderm (starfish) with radial symmetry only in appearance and a far more complex internal structure. Below: an annelid (earthworm) in which one can see the division of the body into segments which are called metameres.

Left: an insect (death's-head moth); its body, divided into head, thorax and abdomen, with three pairs of feet and two pairs of wings. Right: examples of vertebrates (fishes); these are animals with an internal skeleton and a spinal column divided into vertebrae; they are metameric, or segmental, animals.

colourless, and illustrate how animal life is always prepared to colonize even the least hospitable environments. Grottoes shelter certain species of fishes and salamanders, but for the most part their inhabitants are coleopters, crustaceans and other invertebrates, which in some cases are considerably modified as compared with their kin who have remained above ground.

There are just as many and possibly more animals in the *deserts* of the world, where each brief rainy season causes an incredible burst of colour in the plant-life and the immediate return of the most varied forms of animal life: birds, reptiles, insects, molluscs and even creatures associated with water such as toads. And when it does not rain it is enough for the sun to sink low on the horizon to rouse small carnivores, mice and snakes to life. Obviously such creatures are equipped with special adapted features to enable them to survive in the desert. For example, there is no water to drink: thus the desert mouse has learnt to extract water from the seeds it eats. At night these animals can move about because of the low temperature (sometimes it can drop too low), but how do they manage to detect the presence of a prey in the dark? All the rattlesnake has to do is sense, from afar, the tiny amount of heat given off by a small mammal; nature has provided it with two appropriate organs – the sensory pits.

Left: an insect (death's-head moth); its body, divided into head, thorax and abdomen, with three pairs of feet and two pairs of wings. Right: examples of vertebrates (fishes); these are animals with an internal skeleton and a spinal column divided into vertebrae; they are metameric, or segmental, animals.

colourless, and illustrate how animal life is always prepared to colonize even the least hospitable environments. Grottoes shelter certain species of fishes and salamanders, but for the most part their inhabitants are coleopters, crustaceans and other invertebrates, which in some cases are considerably modified as compared with their kin who have remained above ground.

There are just as many and possibly more animals in the *deserts* of the world, where each brief rainy season causes an incredible burst of colour in the plant-life and the immediate return of the most varied forms of animal life: birds, reptiles, insects, molluscs and even creatures associated with water such as toads. And when it does not rain it is enough for the sun to sink low on the horizon to rouse small carnivores, mice and snakes to life. Obviously such creatures are equipped with special adapted features to enable them to survive in the desert. For example, there is no water to drink: thus the desert mouse has learnt to extract water from the seeds it eats. At night these animals can move about because of the low temperature (sometimes it can drop too low), but how do they manage to detect the presence of a prey in the dark? All the rattlesnake has to do is sense, from afar, the tiny amount of heat given off by a small mammal; nature has provided it with two appropriate organs – the sensory pits.

*This fossil animal (*Aysheaia pedunculata*) used to live in a marine environment; its present-day descendants are not very different, but have completely changed their environment; they are in fact found solely on the floor of tropical forests.*

Different but no less specific adaptations occur in animals living in the *abyssal regions* of the sea. No ocean deep is deep enough to discourage some form of animal life. Pale, blind shrimps with extremely long appendices, sea-lilies with flexible stalks, fishes with exaggeratedly dilated mouths, also blind in some cases, but sometimes, on the contrary, equipped with enormous eyes: these are just some of the creatures which have been brought to the surface by the deepest explorations. Many abyssal species, and particularly certain fishes, have photophores, that is, organs which produce a greenish luminescence which can pierce the fathomless gloom of the abysses. What do such animals live on? On the organic remains which reach the deepest places from shallower waters, first and foremost, and then, of course, on other abyssal fauna.

Another environmental condition of an extreme type which has not prevented colonization by many animal species is the *land area perennially covered by ice*; such areas are found in the polar regions and on mountain heights. The problems to be overcome here are determined by the extremely low temperature, and once more by the scarcity of food. And yet mosses, lichens, dwarf willow branches and scattered tufts of marsh-grasses are enough to nourish large mammals like the reindeer, caribou or musk-ox, while in

Top left: a nurse-shark, a marine predator. Below: a frog (Rana esculenta). The life of this amphibian, which in its adult life also lives away from water, is inextricably linked to water in as much as it can only develop in it. Top right: a colony of coelenterates (madrepores); these animals only thrive in very specific environmental conditions. Below: a carp, a fish which is capable of living in muddy water with a low level of oxygen.

alpine meadows with their more precipitous features there is living space for the ibex and chamois in Europe, and for the Rocky Mountain goat in America.

And when an underwater volcano reaches the sea's surface, forming a new island, the bare lava-strewn surface is quickly peopled with varied forms of vegetable and animal life, carried there by the swell of the sea, the wind and migratory birds.

Animals have thus colonized every available environment on the face of the earth, and the explanation of why there are such innumerable varieties of living species resides precisely in the need to respond to the diverse environmental demands made upon them. This variety can nonetheless be reduced to a small number of principal models, corresponding to the types or phyla of the zoological classification which we have outlined above.

The first phylum consists of the *protozoans*, organisms with just a single cell which must carry out all the vital functions: feeding, locomotion and reproduction. Examples of protozoans are the paramecium which is covered with constantly moving minute cilia, the amoeba which constantly changes its forms, and the terrible trypanosome which the tse-tse fly can inject into the human bloodstream, causing sleeping sickness.

155

Opposite: elephants in the savannah, an environment covered by herbaceous vegetation which can grow to a considerable height and includes scattered trees. Left: a tarsier, a nocturnal mammal living in tropical forests. Centre: a zebra. The herds of these equids are typical of the fauna of the African savannah. Right: a brown bear, the largest carnivore of the temperate forests.

Passing to the pluricellular organisms, we find a division of tasks into separate groups of cells within one and the same individual. This is the case with the *sponges*, which are in effect little more than colonies of protozoans. A sponge is like a small sac riddled with small holes, through which water is continually passing. Some cells manufacture the framework which gives the sponge its form and consistency; others, supplied with long moving filaments (the flagella) tirelessly exchange the water and extract from it any nutritive substances; and others are in charge of reproduction. Then there are the jellyfish and sea-anemones which, despite their somewhat different outward appearance, are based essentially on the same model: they are small sacs with an aperture, the mouth, which, in jellyfish, points downwards, and in sea-anemones, which are attached to the sea-bed, points upwards. A ring of tentacles draws food in towards the mouth, the food consisting of various sorts of animals. These animals are classified under the *coelenterates*, together with madrepores and coral, which are colonial forms with a skeleton.

Then there are the *flatworms*, which can crawl about either under water or in wet ground, such as planarians, or which live solely as parasites in the body of other animals, like tapeworms. This group has a real muscular system, and a

fairly complex nervous system, as well as other specialized organs.

The phylum of the *nematodes* includes worm-like animals with a cylindrical body which is non-segmental and covered with an extremely tough elastic cuticle which is virtually impermeable. There are thousands of species scattered throughout the sea, in fresh water and in the ground, but the best-known forms are those which live as parasites on plants and animals. Man, too, has many enemies among the nematodes, numbering some thirty different species in all, among which are the miner's worms, the filaria and the round worm.

The *annelids*, on the other hand, include the earthworms, the leeches and many other mainly marine species, the bodies of which are formed by a number of rings or segments in series, which are similar to one another and in which many internal organs and many external structures recur. Earthworms live mainly in the soil, where they feed on vegetable remains; leeches, however, live mainly in fresh water, although they may also be found in the sea, and some tropical species are even arboreal.

Snails, slugs, mussels, oysters and cuttlefish belong to the phylum *Mollusca*. The bodies of these creatures still consist of three main parts, which are differently modified in the various species: a foot, a mantle cavity and a mantle

Top left: a rattlesnake moving across sand. Desert animals have specific physiological mechanisms which enable them to tolerate extremely dry conditions. Top right: a school of penguins, typical Antarctic inhabitants. Below left: abyssal fishes (silver hatchet fish) with enormous eyes; other species which live at great depths have tentacles and luminous organs. Below right: the ibex, one of the Alpine mammals which are found at the highest altitudes.

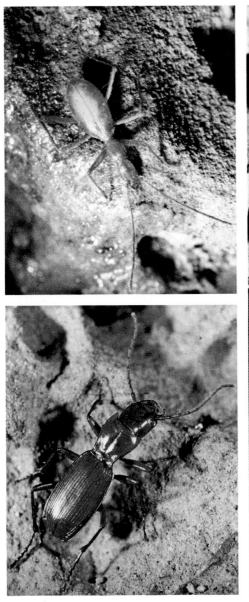

Animals living in caves have no eyes and are colourless, often with long, thin feet and antennae, which gives them a delicate, diaphanous appearance. Top left: an Aphaenops cerberus, *a typical cavernicolous insect, as is the* Speomolops sardous (*below*). *Right: a typical cave.*

which secretes a shell. The foot is conspicuous in snails, while in the case of cuttlefish it is transformed into a ring or crown of tentacles and a short funnel which it uses to blow out water with considerable force, which in turn enables the creature to move like a jet.

The shell is formed by a single piece in the snails, by two pieces in the mussel and oyster; in the cuttlefish it has become internal and in the case of the slug it has virtually disappeared altogether.

The *arthropods* are quite different. These include insects, spiders, scorpions, ticks, millipedes, crayfish and crab. Here, too, there are considerable differences between the various forms, but there are also important common features: the presence of an external skeleton which is fairly hard; the development of articulated appendices which are used for locomotion and other purposes. Insects have three pairs of legs and, generally, two pairs of wings; on the head they have a pair of movable antennae (constantly ready to detect mechanical or chemical external stimuli), eyes and a complicated buccal apparatus which is used both for chewing (cockroaches, locusts and grasshoppers, cockchafers) and for biting and stinging (mosquitoes, bed bugs, harvest flies), and for sucking in liquid food (flies and butterflies and moths). Spiders and scorpions

do not have any antennae and are equipped with a pair of short venomous pincers (the chelicerae), a pair of jaw-like feet of various types (these are the large appendices of scorpions which culminate in the chela) and four pairs of feet. Ticks are similar, but their buccal appendices are designed for biting and sucking blood. Millipedes, which are all land-animals, and crustaceans, which are mainly aquatic, are all masticatory creatures; they have different ways of moving (just consider the problems of an animal with three hundred feet!), feeding and mating. Because they are effectively specialized in very diverse directions, the arthropods are without doubt the most varied animal phylum.

In the case of the *echinoderms* we come across a phylum which is made up of purely aquatic animals: starfishes, sea-urchins, sea-cucumbers and sea-lilies. Here again the extremely varied outward appearance must not deceive the observer: in all these forms of life the zoologist can identify an aquiferous system, consisting of a system of ampullae and appendices which is used to take food and for hygienic purposes. It is thus incredible how similar these animals are when they leave the egg in the form of tiny larvae.

The last large phylum consists of the *Chordata* or *chordates*, namely, the vertebrates plus the ascidians, the salpa and the amphioux, all marine animals

The fauna of the ocean islands often has specific features, and as such is of great interest to scholars and students. The illustration shows the Conolophus subcristatus, *the land iguana of the Galapagos islands; the animals on these islands supplied Darwin with many ideas for his theories on the origins of the species.*

Care for the young (left), migrations (top right) and the relationships between members of a flock (or herd) show the deep resemblance between certain behavioural aspects in man and other animals.

of very different appearances but with characteristics which demonstrate their relationship with animals which are more familiar to us. On the other hand the presence of a brain and a spinal cord, the development of an internal skeleton composed of bones or cartilages, and the presence (if we exclude the lampreys) of two pairs of paired appendices – fins in the case of fishes and feet in the case of amphibians, reptiles, birds and animals – constitute three features which are more than enough to define a vertebrate.

However, zoology is not confined to a mere classification of the different species by detailed comparison of the structure or of the successive stages of development; nor does it just examine the pattern on the basis of which the environment influences the characteristics of the various animal populations.

One area of zoology is concerned with the study of *behaviour*, a discipline which is becoming a whole new and independent science, known as ethology.

And from this viewpoint it is possible to make a distinction between more primitive and more evolved animals: in the first case, the behaviour is limited to simple schematic responses to the action of external stimuli. A protozoan heading for the light, a flatworm crawling towards a shaded area, both are responding mechanically to the luminous stimulus on the basis of an innate

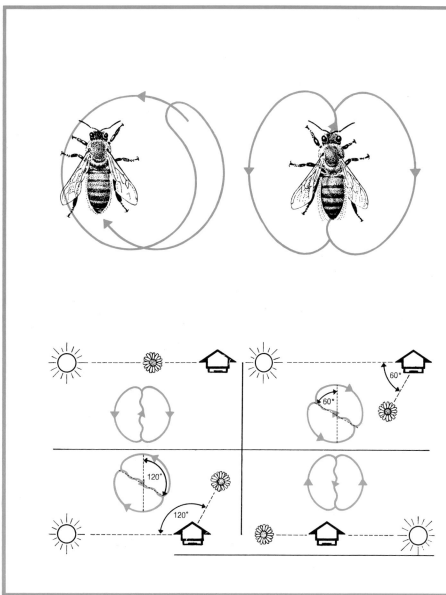

and essentially unmodifiable reaction. Of course external stimuli are also necessary to higher animal forms as well, in order for them to know what is occurring in the environment about them: a cat will place itself down wind in order to scent its prey without giving away its own smell; the male moth can identify the presence of a female of its own species, which emits molecules of a fragrant substance into the air, even at a distance of $1\frac{1}{4}$ miles (2 kilometres), by means of its feathered antennae. But in these creatures the perception of an external stimulus releases a very complex reaction; in other words the stimulus only supplies the animal with useful data, but the subsequent behaviour – catching a prey in the first instance, and finding and mating with a female in the second – unfolds according to a very articulate programme which is often (as in the case of the cat) subject to improvisation.

From the behavioural point of view, the *social* animals have made a fundamental qualitative leap forward. There are, of course, different types of society, and sociability. In the insect societies – termites, wasps, bees and, in particular, ants – there is a strict division of tasks among the members of the community, and this division sometimes entails extreme consequences. The reproductive activity, for example, turns out to be entrusted to one, or just a

Worker bees communicate with one another with special movements known as 'dances' which indicate to the recipient the angle between the direction of the sun and the direction required to find the flowers from which nectar and pollen can be extracted. Bees belong to the order Hymenoptera, the group which displays the greatest instinctive sense of all the invertebrates.

The ever greater alterations to the natural environment and indiscriminate hunting are threatening the survival of many species of animals. Top left: a young seal killed by fur hunters. Below: a wild duck, an animal which is threatened both by hunting and by the destruction of its natural environment, the marsh. Right: a herd of American bison, which were mercilessly hunted in the nineteenth century.

few, females, with the others remaining sterile and being earmarked as workers fulfilling the various needs of the common nest, including feeding their sisters. In the case of vertebrates, on the other hand, other forms of sociability emerge: some are temporary and not very binding, as in the case of birds which migrate in flocks, or antelopes which graze in large herds; others are more lasting and complex, as in the case of baboons and other monkeys who look after one another and defend themselves collectively against predators. A stable society is a sort of enlarged family in which the various members can exchange experiences, and in which a tradition can develop and a culture emerge.

Man has found himself continually confronted by the variety of the zoological world, in a series of events which have often had a dramatic outcome for the animals concerned. In fact, the list of animals totally destroyed by man within the span of history is quite a long one; suffice it to mention the moa, a huge bird living in New Zealand; the American passenger pigeon, whose numbers two centuries ago defied assessment; the dodo, a species of wingless pigeon from Mauritius. What is more there are many species which are now threatened with extinction: the American bison, and even more so the European bison; the sea otter; almost all the species of whales; and others.

THE INVERTEBRATES –
General Overview

An *invertebrate* animal means an animal without vertebrae, in other words, without an internal articulated skeleton to support the body and supply points to which muscles can attach themselves. But this is a purely negative definition, restricted to telling us what invertebrates are not. In reality, the variety of forms and habits and appearance presented by the invertebrates is far greater than that offered by the vertebrates which – from fishes right through to birds and mammals – also retain an unmistakable common family look.

Some invertebrates consist of just *a single cell*, like the amoeba, the paramecium or the malarial plasmodium. These forms are always small in size, in some cases microscopic: the malarial plasmodium, for example, lives out part of its life cycle within a red blood corpuscle, which in itself is a fairly small cell, resembling a small disc with a diameter of seven thousandths of a millimetre. But other invertebrates attain considerable dimensions, as is the case with certain tapeworms which may exceed 32 feet (10 metres) in length, or the giant crab found in Japan, where the male can spread its chelae or claws to a span of 9 feet (3 metres). The truly giant forms of invertebrates are all found among the molluscs: some species of giant squid may reach an overall length of some 62 feet (20 metres).

The invertebrates are found in every environment: sea, fresh water, and on land. Their normally average dimensions, habits and their at times not very conspicuous appearance means that they often go unnoticed by the unskilled observer, whereas even the most fleeting appearance of a bird or other vertebrate is usually enough to describe an environment.

And yet the presence of the invertebrates is certainly not limited to the flight of a many-coloured butterfly or the slow gait of a mollusc whose shell ends up high and dry on the seashore: because of their variety of species and the number of individuals, the invertebrates are the true, even if hidden, masters of every environment. A single ants' nest, in some species, may number a million individuals: and in a termites' nest the population may even exceed ten million. The floor of a forest may be host to thousands of specimens in every cubic decimetre, which may even belong to more than a hundred different species of invertebrates. The sea hare lays forty thousand eggs a minute and may lay as many as half a billion eggs before it dies. These are truly vast statistics, but possibly less staggering than those which illustrate the extraordinary variety of the invertebrates. In fact, we know of about a million different species, as opposed to not many more than three thousand species of mammals and about eight thousand species of birds. Every year the number of known species expands by several thousands, especially in the case of insects, which alone account for considerably more species than all the rest of the animal kingdom put together.

In the sea and in fresh-water environments we find many *sessile* invertebrates, which spend their entire life, or at least a great deal of it, anchored to submerged objects, like plants rooted to the bottom. Sponges really lead a sedentary life, often smothering submerged rocks, like crusts, as do madrepores, coral, mussels, oysters and other less obvious forms of life. For these animals the procuring of food is relatively simple: all they have to do is cause a continual flow of water and filter it, drawing in any suspended particles towards the mouth. In fact in the water there is the constant possibility of being able to fish enough to live on. On land things are somewhat different, and as a result a sedentary land animal would be unable to survive. It is true that plants

A diagram of the evolution of animals.

164

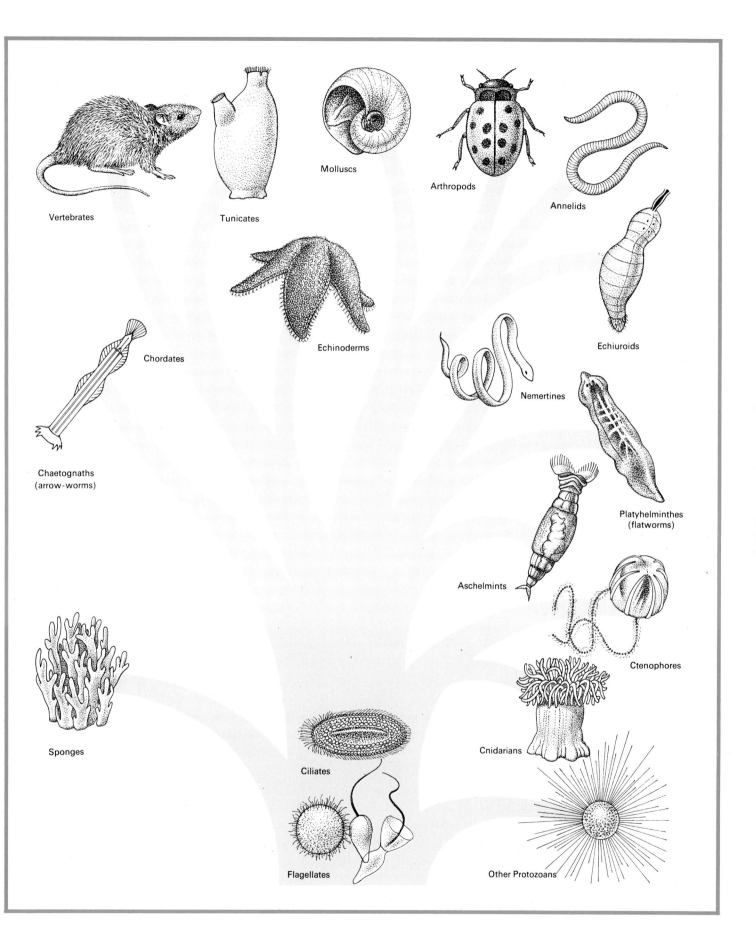

Vertebrates

Tunicates

Molluscs

Arthropods

Annelids

Echinoderms

Echiuroids

Chordates

Nemertines

Chaetognaths
(arrow-worms)

Platyhelminthes
(flatworms)

Aschelmints

Ctenophores

Sponges

Cnidarians

Ciliates

Flagellates

Other Protozoans

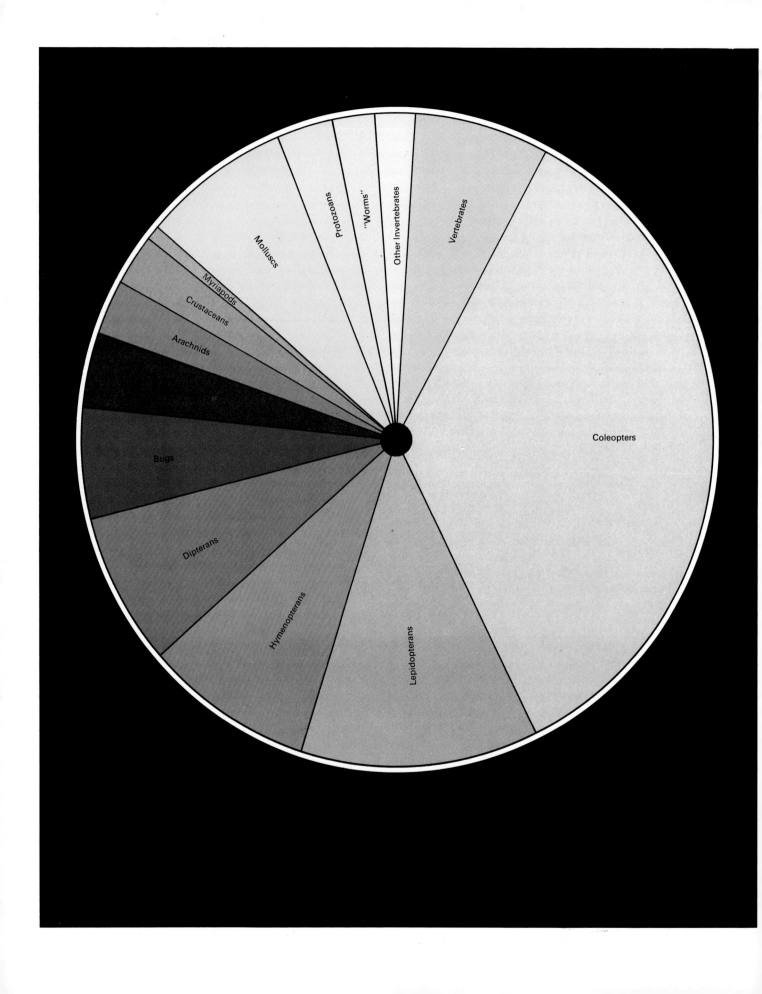

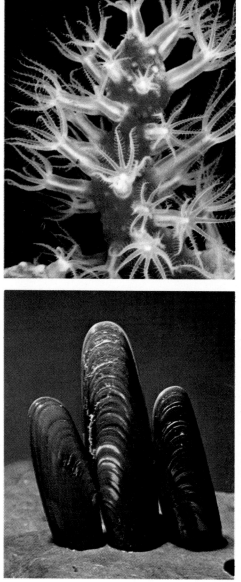

Examples of marine invertebrates which live attached to the sea-bed. Top left: red coral, a typical anthozoan coelenterate; anthozoan is from the Greek, and means flower-animal, indicating the extraordinary similarity of this organism to plants. Below: a group of date mussels, molluscs which bore holes in the rock by means of an acid mucus. Right: a marine worm belonging to the genus Spirographis. *Opposite: this diagram shows how the invertebrates account for the great majority of animal species known to date.*

growing on the land are rooted in and to the soil or ground, but all these plants require from the soil is water with some dissolved mineral salts: the rest of the job is done by the light which combines water and carbon dioxide in the leaves, and thus forms the sugar. For an animal, on the other hand, the ability to move is one of the conditions for its survival. Even the sedentary animals which we have mentioned are aware of this need, and it is a fact that, before they fasten themselves to submerged bodies or objects, they almost all pass through a brief period of active life, in the form of tiny moving larvae which move through the water by means of minute cilia with which they are covered. This method of propulsion, which also occurs in many small unicellular creatures, is not sufficient to move organisms which are longer than a few millimetres: it is, moreover, only effective in aquatic environments. In the larger invertebrates and in all land forms, locomotion is achieved in other ways. In all instances the *musculature* is of prime importance; by regularly contracting and releasing, this allows the animal to advance. The case of the earthworm is a good example of this. This creature is in fact equipped with bands of *longitudinal muscles* and *circular muscles* arranged like a ring. When the circular muscles contract, while the longitudinal muscles are released, the animal lengthens; when the

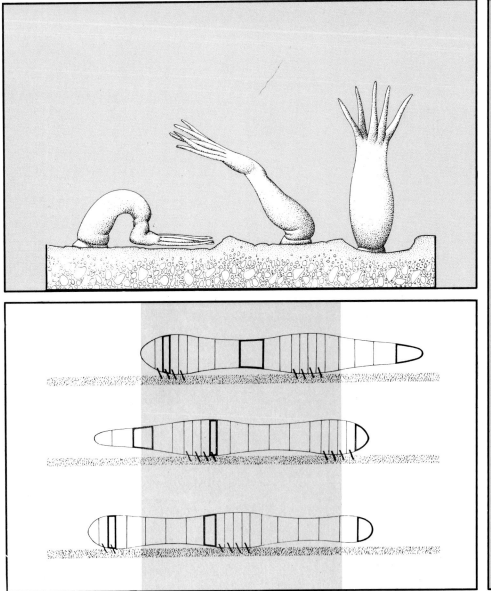

longitudinal muscles contract, while the circular muscles are released, the animal shortens. The curved bristles on the worm's ventral part stop it from slipping backwards and losing its hold; the alternate contractions of the circular and longitudinal muscles thus enable it to advance. But a far more efficient method of locomotion is supplied by means of moving and articulated appendices, such as the *legs* of the vertebrates and, in the invertebrates, the legs of insects, spiders, millipedes and crustaceans. These appendices are in fact used for very different purposes: in water they may be used as oars or fins, on land for walking or jumping (like the grasshopper); and it should be remembered that the antennae and mandibles of these animals are nothing other than transformed legs. Many invertebrates have also come to terms with the air, in which they compete in terms of speed and flight potential with other winged creatures, bats, and birds. The ranks of butterflies and moths include the hawkmoths, which can remain motionless in the air, in front of a flower, in the same way as humming-birds.

There are many ways in which the various species procure their food, and even more numerous types of diet. Among the unicellular forms there are even some, like the Euglena, which behave like plants when exposed to light, and

Top left: some of the movement positions of the polyclad Prostheceraeus vittatus. *Below: the movement mechanism of an oligochaete on a flat surface; note the successive contractions of the various segments which are equipped with bands of circular and longitudinal fibres. Right: the legs of a tarantula (*Lycosa tarentula*) move in groups of four co-ordinated diagonally (A–B). This co-ordination persists even if some limbs are missing (C).*

*The insects are the only invertebrates which can fly. Left: an orthopter (*Tettigonia viridissima*) which both flies and jumps. Centre: a vanessa, a butterfly which can survive long migratory flights. Top right: a crane fly, an insect which has just two wings (such creatures usually have four). Below: a dragonfly, one of the speediest and strongest insects in flight.*

like animals in conditions of darkness, absorbing from outside the organic substances which they need.

Another single-celled invertebrate is the amoeba, which changes its form constantly, emitting long protuberances (pseudopodia) which enable it to move slowly and swallow minute organisms which, once introduced into the cell, are demolished by digestive enzymes. We have already mentioned animals which have a *filtering process*: sponges, mussels and oysters take in all their food by pumping water and gathering the alimentary particles suspended in it. Then there are the *herbivores*, which move lazily across leaves or branches which they cut into with powerful mandibles (like grasshoppers, locusts and the grubs and caterpillars of a large number of moths and butterflies) or nibble at by rasping with their curved, sharp teeth (like snails). In practice there is not a single species or part of a plant which is not attacked by some form of invertebrate, whether it be the larva of a coleopter which burrows a tunnel in the stem or stalk or beneath the bark, or a female wasp which causes a growth to appear on a leaf (called a gall) within which the larvae develop, or a mite or a cimex which have a rostrum designed to prick and suck in the sweet, nutritious sap or lymph. On the whole herbivores have the appearance of rather slow and

169

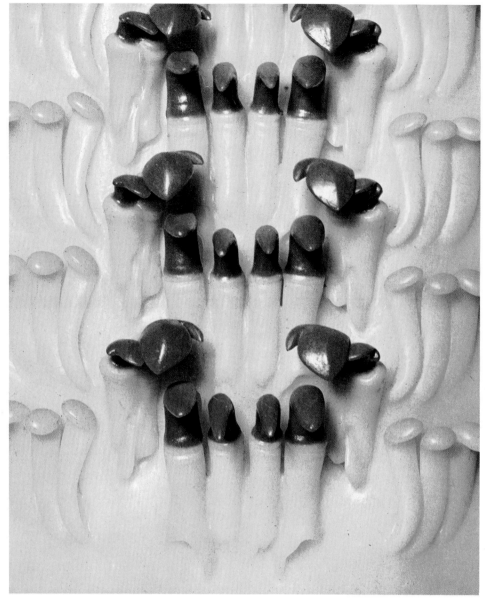

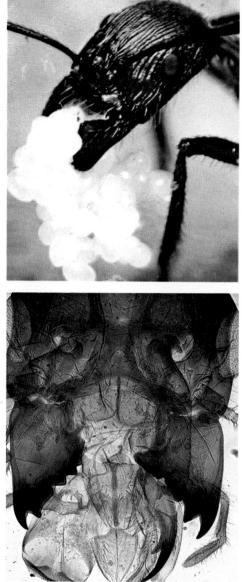

lazy creatures, like the snail and the caterpillar, something of a counterpart to ruminants among the mammals. The small invertebrate *predators* obviously correspond to the cats with their agile, slender bodies: these include the scolopendra, which are swift and fierce; the ground (or carab) beetles and the tiger beetles, which are constantly in search of prey. Just as each trade requires its appropriate tools, so it is to be expected that the mouth of a predator should be armed in a different way from the mouth of a peace-loving herbivore. In effect the long-toothed mandibles of a tiger beetle or the powerful, poisonous small pincers of a scolopendra are clearly in contrast with the broad cud-chewing surfaces in the mouths of very many vegetarian species. But not all predators hunt at will; some prefer to lay an ambush, like a large number of spiders which hide in the petals of a flower and wait for the victim to alight, or crouch on leaves and wait for the first hapless creature to end up in the threads of the sticky cobweb.

With this constant threat from predators, which naturally include many vertebrates as well, a lot of invertebrates have developed protective organs and defence strategies. One of the simplest responses to being preyed upon consists in an *increased fertility*. An individual which produces millions of eggs has

The invertebrates have the most varied diets, and are equipped with buccal apparati with structures which are often complex and sometimes bizarre. Left: the radula of a mollusc. This is a structure which functions like a rasp on the surfaces which the animal attacks. Right, top and below: two examples of buccal apparati in insects which masticate: in both cases we see the mandibles which are used for catching the food.

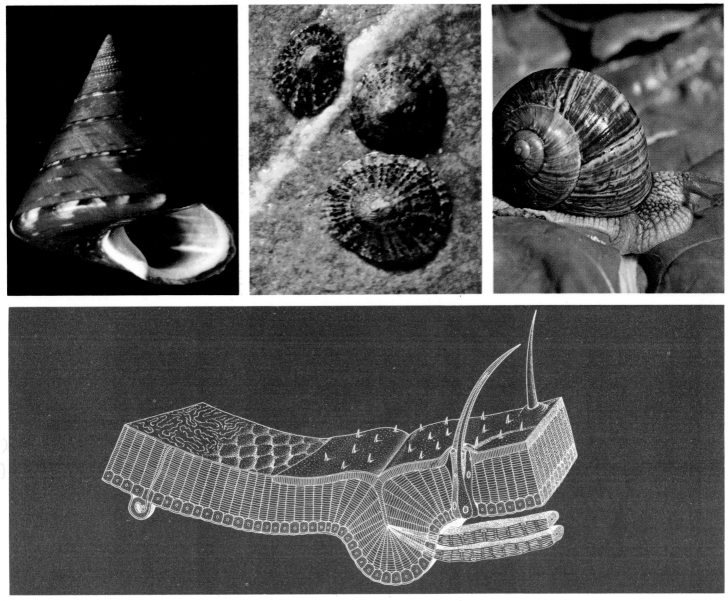

In the molluscs, animals which do not have a strong integument, the presence of a shell can be a good means of defence against predators. Top: the tough shells of two marine species, Calliostoma zizyphium *and* Patella vulgata *(the first spiralled, the second not spiralled), and the shell of* Helix pomatia, *the common land snail. In the insects, on the other hand, it is the tough quality of the integument which provides protection. Below: a schematic section of the cuticle in which one can see a stratified structure.*

quite a good chance of seeing some of its eggs developing completely and forming the next generation. But in some instances defensive tactics are more subtle. Many grasshoppers and locusts have *mimetic* (or camouflage) colours which make them hard for predators to see. In the case of the stick insect and the leaf insect, these perfectly imitate small twigs and dry leaves in both shape and vegetable consistency. Poisonous animals such as wasps obey an opposite principle: they are brightly coloured, with the precise function of alerting possible predators and inviting them not to come too close. Many species of flies, butterflies and moths and even coleopters have made use of this device as well, by having copied the pattern of wasps and passing themselves off as these, although they are harmless and attractive creatures. Another different form of defence is provided by thick, tough *tunics*, like the molluscs' shells or the hard carapace of a crab.

A protective tunic may also have another function. In fact land animals are in constant danger of losing too much water by transpiration: a shell which is waterproof or almost waterproof is thus a necessary requirement if activity is not to be confined to rainy days, as in the case of slugs.

The arthropods (insects, spiders, myriapods and crustaceans) owe much of

171

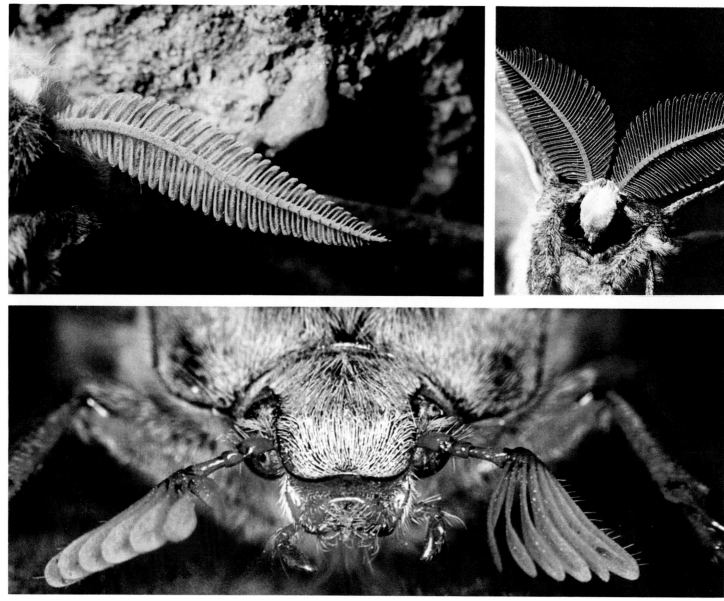

their successful survival in all types of environments to their external skeleton or *exoskeleton*, be it in water, on land or in the air. This skeleton, which is at once robust and elastic, enables them to meet three needs simultaneously: it supplies protection against predators, restricts transpiration, and gives them a wide range of movements, thanks to the development of articulated appendices, formed by mobile segments each of which is covered by a different skeletal section.

The individual world of each species depends on the characteristics of its *sensory organs*. In the invertebrates, and in particular in insects, we find a sensorial apparatus which is certainly in no way inferior to that of vertebrates.

The organs of *sight* are possibly the most widespread, starting with the unicellular protozoans, many species of which are equipped with a *stigma* which distinguishes between variations in the intensity of the light and even, to some extent, colours. Some molluscs (octopus, cuttle-fish) and the arthropods have very complex *eyes*. The octopus's eye is very like the vertebrate's eye and can identify forms. In crustaceans and insects, on the other hand, there is an unusual type of eye: this is the *faceted eye* or *compound eye* consisting of a large number of elementary eyes; these eyes look like a crystal with thousands of

The antennae are a typical sensory organ of the arthropods. Made up of a very variable number of joints, they can be of various types. Top: two examples of bipectinate antennae of moths. Below: flabellate antennae of a coleopter.

172

Eyes of invertebrates. Top left: the simple eyes of a pecten, a bivalve marine mollusc. Right: the eye of a crustacean. Below: compound eyes of insects. Left: a dragonfly. Centre: a wasp. Right: a fly.

facets, or a honeycomb with numerous small hexagonal cells. The eyes of many insects are so refined that they can detect the plane of polarization of light, which the human eye is unable to do.

Other sensory organs enable the invertebrates to detect movements of water or air around them, or to identify the presence or even the concentration of a chemical substance spreading round them. These sensory organs are usually situated at the front of the animal, near the mouth, and often on special appendices which are called *palpi* or *antennae*.

The fly, however, can taste with its front legs the drops of liquid which lie ahead of it: its taste-buds are in fact . . . on its feet. The eyes remain at the front even when the mouth is half-way down the ventral surface of the body, as in the case of planarians. In the pecten, however, the eyes are situated at the edge of the mantle, and in some marine worms they even appear on the last segment of the body, above the anus.

Reproductive patterns and the cycle of development are also amazing. The unicellular forms usually reproduce by a process of simple division: this gives rise to two individual cells which, after just a few hours, reach the size of the original cell, and are instantly ready to divide once again. In the malarial

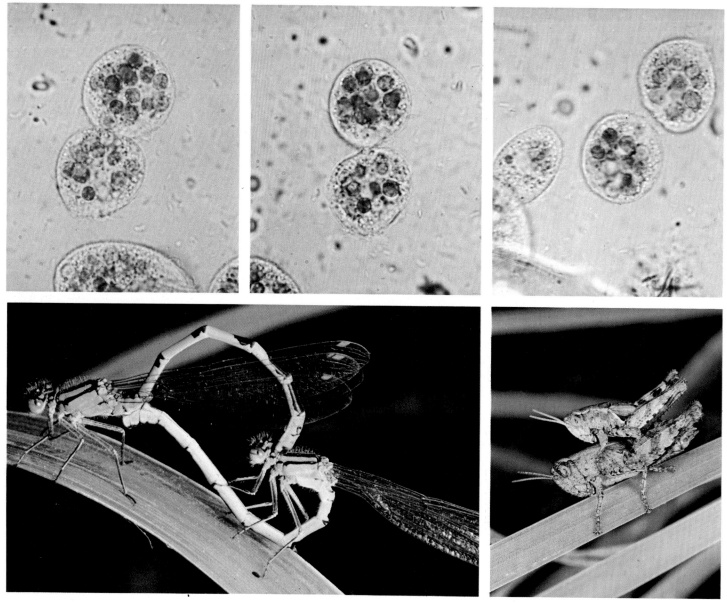

plasmodium and other parasitic protozoans, the cell undergoes multiple division, in other words it gives rise to several cells at any one time; this process is favourable to the parasite which can as a result spread more rapidly in the host. In the invertebrates, which consist of numbers of cells, reproduction is on the whole *sexual*, with *fecundation*, i.e. the merging of an *egg-cell*, produced by a female, with a *spermatozoon*, produced by a male. In some cases, however, even among the pluricellular forms, there is a form of *asexual* reproduction, thus vegetative. This applies to the fresh-water hydra, a tiny polyp which regularly produces buds or gemmae which detach and start their own life in the form of polyps like the 'parent'. This form of vegetative reproduction is very similar to the phenomena of *regeneration*, whereby starfishes redevelop their arms which have been amputated, or earthworms with their rear segments missing re-form the missing part.

Sexual reproduction also occurs in very different forms in other respects. In the first place it is not always possible to identify males and females in a given species: in the case of snails, earthworms, and other invertebrates, each individual produces both eggs and spermatozoa: these species are called *hermaphrodites*. In other cases, such as the aphids, the eggs develop without

Division (top) is the simplest type of reproduction occurring in animals: the division of an individual causes the formation of two equal daughter cells. Sexual reproduction is the most common method. Below: insects mating. Left: two dragonflies in the curious amorous position. Right: two grasshoppers.

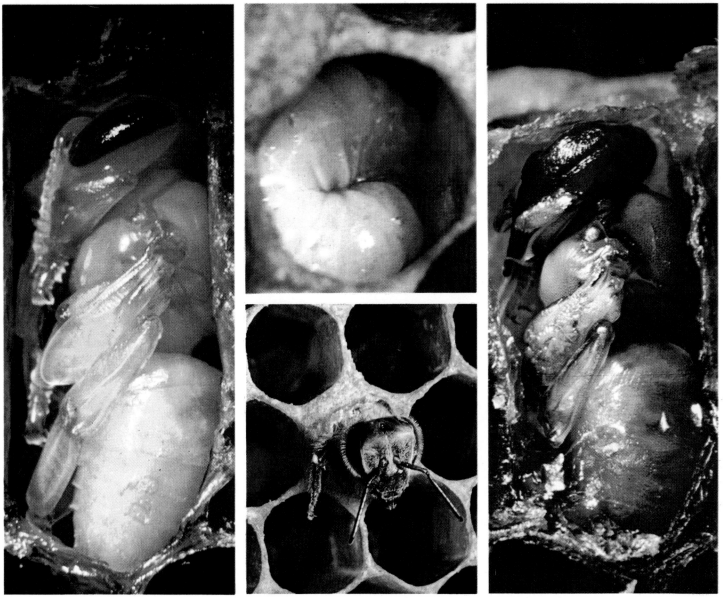

In the insects, the development of the various individuals passes through various stages in which the animal shows totally different structures and habits. Here we see various phases in the development of a bee. Top centre: the larva; left and right: two successive phases in the pupal stage; below centre: the adult insect emerging from the comb.

fertilization: here the term *parthenogenesis* is used.

Whatever the case the egg – whether fertilized or not – develops into an embryo, from which in turn a small individual like the adult (*direct development*) is born, or in other cases a small animal which differs considerably from the adult in both structure and behaviour (*indirect development*). Indirect development applies to butterflies and moths, the larva of which is the caterpillar, or to the fly, the larva of which is a small white worm or maggot with no appendices. The larvae of many marine invertebrates are very curious indeed: they are tiny transparent creatures which move through the water by means of their cilia and subsequently turn into small sedentary creatures. The transition from the larva to the adult is the process of metamorphosis which sometimes occurs abruptly in the course of a few minutes, as in many marine worms, and at others takes place slowly, as in butterflies and bees, the pupae of which turn into a cocoon or small enclosed cell. Life expectancy varies: in insects it is unevenly divided between a long larval life and a short adult one: the mayfly spends a year in the water as a larva and only a few hours as an adult, while a species of cicada spends seventeen years underground before acquiring wings and progressing to trees to sing.

THE MARINE INVERTEBRATES

The first marine invertebrates that we come across are on the seashore. Wherever the coastline is low-lying and sandy we find shells with elegant or curious shapes at every step, often damaged by water and sun, but also often still brightly coloured. Shells have almost become a symbol of the animal life of the sea, and people often put shells to their ears to hear that 'rumble of the sea'.

Indeed life probably started in the sea. Very many primitive organisms, animal and vegetable alike, are associated with this environment, and the oldest traces of life from past millenia have been found in sedimentary rocks from very old marine depths. There can be no doubt that the primordial seas must have been quite different from the present-day oceans. The water, for example, was not salty as it is now: the dissolved salts in the oceans originate in fact to a large extent from the slow erosion and disintegration of ancient rocks. But the marine environment is, as a whole, subject to less dramatic changes than those to which inland waters and land environments are subject, and this comparative stability has managed to preserve, in the sea, organisms which seem to have remained unchanged for many millions of years.

The coelacanth, an extraordinary fish living in the Indian Ocean which was not discovered until 1938, is extremely similar to some fossil species dating to the Cretaceous period, when the earth was still inhabited by dinosaurs, and the appearance of man was still almost a hundred million years off. Even more ancient is the family of the king-crab, an animal akin to a large crab with a long pointed tail, which is in fact related (even if somewhat distantly) to the scorpions, as well as to the trilobites, a species of creature which died out two hundred and fifty million years ago. The larva of the king-crab, with its flat, leaf-like body, shows, when compared to this extinct group, a great resemblance which clearly indicates a relationship between the two.

For every group of animals living in fresh water or on land, it is almost always an easy matter to find a marine family from which it is derived.

The only major group which is absent in the sea is the insects, if one excludes the gerridae (*Halobates*), a small cimex which skates on the surface of the water in the Sargasso Sea, in the same way as water gnats (or marsh treaders) and water striders do in ponds. The abundance of the organisms living in the sea allows many species to live a type of sedentary life which would be completely out of the question on land: sponges, oysters, bryozoans and other creatures spend their life filtering the sea-water which supplies them not only with oxygen with which to breathe, but also with various types of alimentary particles: bacteria, microscopic algae, protozoans etc.

Countless organisms, both animal and vegetable, in fact live suspended in the water; often they are unable to move actively; these organisms are collectively called the *plankton*.

If you lower a very finely meshed net into the water you can easily collect a cross-section of this world in all its different shapes and colours. First, we find the algae; these are usually very small, unicellular, and often vividly coloured: in fact they have a large amount of chlorophyll and other special substances which are designed to absorb solar energy and use it to manufacture organic substances. The water of the sea absorbs many rays, as a result of which the light which reaches the various depths progressively changes colour, as well as being reduced in intensity of course. For every type of ray there is a pigment which can absorb it, and for this reason the algae living at different depths are differently coloured. This vegetable plankton, or *phytoplankton*, supplies the

Anyone can take stock of the incredible variety of the marine invertebrates on the basis of the huge quantity of remains thrown up on the seashore by the ocean swell (left). One often comes across unusually beautiful forms, like the pearly nautilus shell (right, in section).

food for the animal plankton, or *zooplankton*. This latter category includes extremely different creatures, some of which even attain a considerable size. We find eggs and larvae of crustaceans, worms, fishes, the graceful jellyfish and the delicate but fearsome siphonophores, the young stages of the sea-urchins and starfishes, many species of shrimps and many more animals still.

Many larvae which we find in the plankton do not take any food throughout the period which they spend in the upper layers of water. Emerging from eggs which contain a great deal of yolk, they carry with them a food reserve sufficient for several days. In the meantime, carried hither and thither by the motion of the water, they are dispersed and colonize new territories; their development continues all the while, until they are ready for metamorphosis. When this occurs, they descend towards the sea-bed and in the space of a few hours turn into sedentary creatures like sponges and sea-anemones.

Many other larvae, on the other hand, take in food which is made up of other planktonic organisms, mainly algae. A swift and ordered movement of the cilia is enough to carry to the tiny mouth the even tinier alimentary particles. But in this case as well, the life as a free agent does not last more than a few days, or a few weeks at the outside; metamorphosis then takes place.

But not all the planktonic organisms show this alternation between a mobile phase, in other words a planktonic phase, and a sedentary phase attached to the sea-bed. Many species, in fact, have become completely planktonic. These include a large number of copepods, tiny crustaceans whose long, feathered anténnae form organs for floating and at the same time oars which enable them to make certain active movements. Then there are the pteropods, small molluscs related to the snails, but equipped with a light shell, or even shell-less, which can stay afloat and even move through water by means of characteristic fins. Copepods measure only one millimetre, while pteropods are ten times this size. Sometimes they appear in enormous masses, like the euphausiaceans, small shrimps which form the 'clouds' known as the krill with which whales sate themselves. Jellyfish are also typical planktonic organisms. The transparent body of a jellyfish, shaped like an umbrella or hemisphere with the mouth opening downwards, consists to a large extent of water, up to 98 per cent in some instances. Some species of jellyfish spend their entire cycle of development in the upper layers of the sea, but most of the species alternate a sedentary phase at the bottom with the mobile planktonic phase. An upturned jellyfish, attached to a submerged object, is called a polyp. In effect a polyp

Stages in the collection of marine plankton. Top left: a plankton net before being lowered. Right: the net is hauled in after a drop. Below left: the plankton is poured into a glass container and comes to rest on the bottom (right).

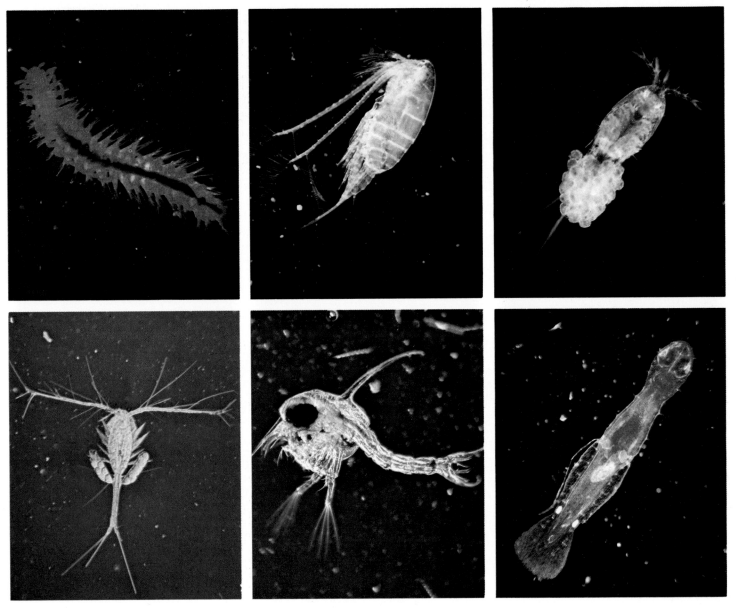

A cross-section of planktonic animals.
Top left: a polychaet annelid. Centre:
a young specimen of the crustacean
Calanus gracilis. *Right: a female copepod*
crustacean. Below left: a copepod with
long appendices which enable it to float.
Centre: larva of decapod crustacean.
Right: a chaetognath (or arrow-worm).

usually gives birth, as it were, to a jellyfish by gemmation, the jellyfish then detaching itself and leading an independent life. Jellyfish produce eggs later on; these are fertilized and develop into small larvae, the planulae, which fasten themselves to the sea-bed and produce new polyps. The cycle thus starts all over again. In many species the polyp phase is not very conspicuous and is short-lived, although the general pattern is retained. Despite their harmless appearance, which might even be called inconsistent, the jellyfish are voracious carnivorous animals which also eat fishes which they stun or daze by means of the stinging cells with which they are provided.

The small sea-snake has a strange appearance, and often occurs in very large shoals, which is their means of dealing with predators. This creature has a totally transparent body, elongated like that of a fish, with a tail and two fins which make it look deceptively like a small aquatic vertebrate. In reality it has neither skeleton nor paired fins, and instead of having a jaw its mouth has a simple crown of spines which serve to seize hold of the prey. Sea-snakes, though diaphanous and inconsistent, are also carnivores and even cannibalistic.

Among the smallest creatures present in the marine plankton, we find the radiolarians, the acantharians and some foraminifera, creatures consisting of a single cell (quite a large one, incidentally) with an often elegant and complex mineral skeleton. Each of these three groups has adopted a different material to construct its skeleton: the radiolarians in fact use *silica*, as do the diatomeae (although these are classified among the algae); the acantharians use a somewhat rare compound, *strontium sulphate*, while the foraminifera resort to an inorganic compound well known to many animals, *calcium carbonate*, which is also used to form the skeletons of corals, madrepores and certain sponges, and – to a large extent – the shells of molluscs.

The skeletons of radiolarians show a wide variety of forms, in some cases resembling delicate lace, in others comparable to honeycombs, often with needles, points, anchors and other subtle filigree-like features. In the acantharians, on the other hand, linear structures are dominant, with stars of subtle rods issuing from the centre of the cell, sometimes ramified, sometimes at right angles. For the most part the foraminifera are bottom species, but they also include planktonic forms, the skeleton of which is equipped with small empty chambers for floating, as in the case of the globigerinae, the remains of which have been accumulating on the ocean bed for millions of years, giving rise to characteristic ooze which is slowly transformed into rock.

Although they are equipped with cilia or fins, the planktonic animals are unable to make any considerable movement. On the contrary, they are at the mercy of the motion of the waves and currents, which carry them along without encountering any resistance. But all the same, the plankton does make typical and regular active movements. Every day, in fact, these tiny organisms tend to head towards the upper layers of the water during the daylight hours, and return to deeper parts during the night. Lighted water is a necessary and vital condition here, because algae can produce an organic substance by means of chlorophyll and other pigments: many flagellate algae thus move continually towards areas where the light is stronger, and take advantage of the hours of darkness to divide, descending by gravity to deeper waters. The small crustaceans and other animals which feed on plankton naturally follow these movements as a guarantee of food.

The plankton also shows impressive quantitative variations which are associated with the seasonal round, and thus determine the migration of many species of fishes whose food consists of it. Likewise the composition of the plankton varies from sea to sea: as a general rule the cold seas have a particular abundance of plankton, while warmer seas do not.

Many planktonic animals are luminescent; at night, especially in summer, the water often twinkles with the presence of countless protozoans; jellyfish, larvae and worms which give off a cold red, green or bluish glow.

In progressing from the surface waters to the bottom of the sea, following the cycle of development of so many animal species, it is as well to consider for a moment the likelihood of survival for a tiny oyster or sea-urchin larva. Here, as in many other forms, the eggs are produced in hundreds of thousands, and sometimes even in terms of millions by each individual.

It is evident that this fertility must serve to counterbalance the effect of innumerable causes of death. A small ciliate larva, which at the point of metamorphosis is often no longer than one millimetre, has no hope of being able to flee in an active sense from its predators. Its only weapon is its transparency, which makes it hard to see, as well as the fact that its size is too small to interest many voracious predators. But other animals live by gathering everything that comes their way and thus countless larvae end up as victims of a wide variety of enemies. But the most delicate moment occurs usually at *metamorphosis*, when the small larva has to find an available nook on the sea-bed where it will turn into a small creature now resembling the adult form, except in terms of size. On the sea-bed the larva finds itself pitted against not only predators but also against all the other sedentary or encrustant organisms, among which the most dreaded competitors are none other than the members of the same species. Once the moment of metamorphosis has been achieved, the animal may encounter other problems before reaching maturity, which is why the end product of such vast numbers of eggs and larvae will be just one or two

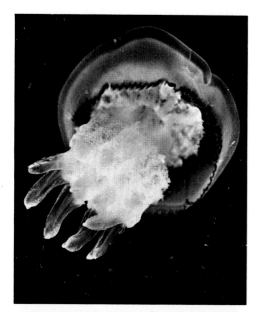

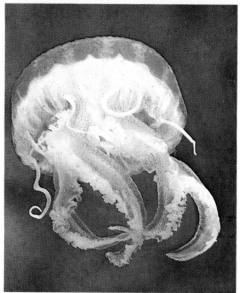

Although the medusae can move quite actively, much of their movement is caused by currents: these animals can therefore be considered as an integral part of the plankton family.

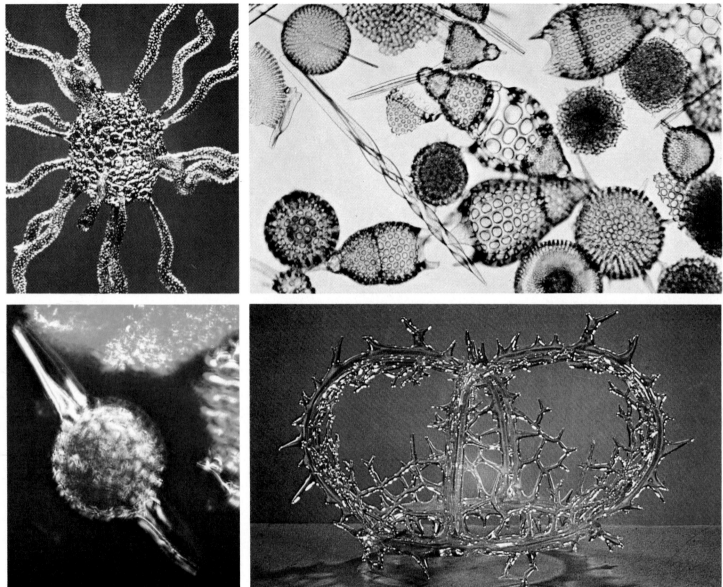

The radiolarians are unicellular marine creatures, ranging in size from fifty microns to a few centimetres. They are usually planktonic in form. Top left: Orosphaera serpentina. Below: a Stylosphaera specimen. Top right: radiolarian skeletons. Below: the extremely graceful shape of the skeleton of Acanthodesmia corona.

oysters or sea-urchins in the adult stage.

On the sea-bed we find the most evolved intervebrates from the behavioural viewpoint, the octopuses and their relatives. The octopus is a large mollusc which can reach a weight of 55 pounds (25 kilos); its body is principally composed of a voluminous sac of entrails, covered by a cutaneous duplicature (the *mantle*) which is open in the ventral section; a short and massive head is attached to the entrail sac; this head has two large, slightly protruding eyes and a mouth provided with a pair of horny jaws – a sort of parrot's mouth – and surrounded by a ring of eight tentacles. These, which are flexible and equipped with rows of suckers, enable the animal to move gracefully on the sea-bed and in and out of the rocks where it sets up home and which it leaves only in order to hunt for food. Its favourite victim is the crab, to which it is associated by a curious common characteristic: both, in fact, have blood which is blue in colour, because of the presence of a coloured substance containing copper (*haemocyanin*) which is used to fix oxygen in the same way as *haemoglobin* (which contains iron) which gives the colour of the red corpuscles in the blood of human beings and the other vertebrates.

The octopus develops directly, that is, from the eggs – which are fairly large —

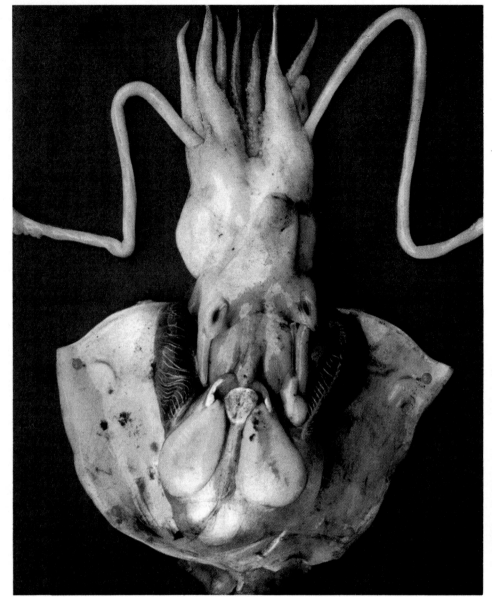

and a miniature octopus emerges, not a larva. The same occurs in the cuttlefish, the eggs of which, wrapped in a blackish-violet involucrum and left in clusters on submerged objects, are commonly known as sea-grapes. In addition to the eight tentacles boasted by the octopus, the cuttlefish has two more longer tentacles which, when not active, are drawn back into special pouches situated besides the mouth. The cuttlefish, which is much less tied to the sea-bed than the octopus, moves typically by jet propulsion, that is, by expelling the water initially collected in the mantle cavity with great force through a small funnel.

Hydrodynamic forms come into play in order to carry out movements of this type: the body of the cuttlefish is in fact quite elongated and the mantle has a long, narrow fin running all the way along both sides of its body; an internal shell (the so-called *cuttlebone*) gives the animal its necessary rigidity. The squid is even more slender in shape; this creature nearly always lives in open water, some way from the sea-bed; it has a brightly coloured triangular fin at the apex of the body, which is supported by a long feather-shaped shell. Octopus, cuttlefish and squid have, to all appearances, brilliantly solved the mechanical problems of locomotion. But an octopus is also capable of solving quite different problems. Equipped with eyes which are comparable to those of the

Cuttlefish (top right) are marine molluscs with no external shell; they have tentacles around the mouth. They are equipped with glands (in the illustration at left they look like two dark masses with white markings) – which enable them to expel a dark liquid which clouds the water and conceals them from the enemy in question. The females lay their eggs (below right) by fastening them to submerged objects.

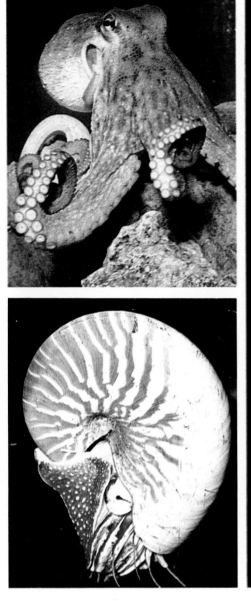

*Three more marine molluscs with
tentacles (cephalopods). Top left: an
octopus. Below: a pearly nautilus. This
animal, with a concamerate shell, has been
living in the sea since the far-off days of
the Palaeozoic era. Right: a female
argonaut (or paper-nautilus) with shell;
the male, on the other hand, is shell-less.*

vertebrates, and with a complex and highly organized nervous system, the
octopus can distinguish shapes, recall past experiences, and associate ideas in
the form of conditioned reflexes. From the behavioural viewpoint these
invertebrates – the cephalopod molluscs – are one of the three best-devised
zoological groups, together with the hymenopteran insects (wasps, bees, ants
etc) and the mammals (which include the species *Homo sapiens*).

When they are disturbed, however, these creatures get out of trouble by
means of a reaction which does not require a great deal of intelligence or
ability: they squirt a cloud of black ink from a special pouch, and having so
disconcerted the possible predator, make their getaway at great speed.

Some of the animals related to the octopus and the cuttlefish have an outer
shell. One such is the nautilus, whose light, mother-of-pearl shell is subdivided
by numerous partitions into a series of chambers, the last of which, being the
largest, is occupied by the animal itself. But each one of the partitions of the
shell is pierced by a hole, through which passes the *siphon*, a sort of long tube
into which the mollusc's body extends. The internal chambers, which are fairly
full of gas, help the animal to float. The (pearly) nautilus, which inhabited
every sea to some degree in past geological eras, is now confined to the Indian

Ocean and nearby waters.

The argonaut (or paper-nautilus) is another cephalopod which has a shell. In this case, however, the shell is a soft one formed by two modified tentacles and not by the mantle, as with other molluscs. Only the female has a shell and she uses it as a nest to incubate the eggs during the reproductive period.

But let us return to the sea-bed. Depending on the nature of the *substratum* (muddy, sandy, or rocky), the development of the algae and the depth, the fauna encountered will vary accordingly. The pastures made up of algae and those plants with thread-like leaves offer shelter to a large number of herbivores, which graze there like the large ruminants of the prairies and savannah. Apart from the numerous types of fishes, the most voracious vegetarians are undoubtedly found among the gastropods, i.e. among molluscs with a one-piece shell which is usually in the shape of a coil or spiral, and also including some species which are shell-less or almost shell-less, like the land slug and the large sea hare. The mouth of these molluscs is armed with numerous rows of fine teeth, which constantly grind up the tender vegetable tissues. But many gastropods are carnivorous: by means of an acid secretion, some make a hole in the shell of another mollusc which they then eat; others

The fantastic and almost unreal nature of the ocean depths is quite often caused by the presence of curious marine invertebrates which take on the form of plants. The illustration shows tubicular polychaetes and coral growth in the Gulf of Siam.

Animals with similar internal structures may often have quite different outward appearances; this is the case with the echinoderms, a group to which the sea-cucumber, starfish and sea-urchin, depicted on this page, all belong.

kill their prey (which may include small fishes) with the poison which they inject with a kind of long tooth. Among this latter type we find the cones, whose elegant shells are much prized by collectors.

However, the most fearsome predators on the sea-bed, among the invertebrates, are the starfishes. When they arrive at an oyster-bed, it can mean the speedy death-knell of the precious molluscs. Efforts to combat these predators by pulling them to pieces are futile, of course, because of their remarkable regenerative capacity; indeed, a complete new individual may sometimes develop from an arm which includes a small piece of the central disc. When it attacks an oyster or mussel, the starfish demonstrates its strength and patience. Two arms are attached to one of the mollusc's valves, three arms to the other valve, all clinging tightly by means of hundreds of tube-feet; the reaction of the mollusc is suddenly to lock the valves; the starfish tries hard to prise them apart, and the victim resists. But in the end the mollusc has to open them just a crack in order to breathe: the starfish leaps at the opportunity, keeps the valves of the mollusc open and inserts into it its own stomach, turned inside out outside the mouth. The digestive enzymes start to eat into the mollusc which gradually releases its grip. At a later stage the starfish 'recovers'

its own stomach with the prey inside it already half-digested.

Despite their different outward appearance, sea-urchins are related to the starfishes; they have a skeleton which is bristling with spines. There are in fact two different types of sea-urchin. The first, the so-called ordinary or common sea-urchins, have a more or less spherical form or, if not quite spherical, show an almost perfect radial symmetry. Their bodies are armed with conspicuous spines, which are often long and sharp; the mouth, which invariably opens downwards, has five sharp teeth articulated on brightly coloured calcareous plates moved by strong muscles; this complex buccal apparatus is known as the *Aristotle's lantern* – the great Greek philosopher was also a keen observer and recorder of animal life – and enables the sea-urchin to gather food by rasping the rocks on which it lives: algae and animals thus end up in its long intestine which opens upwards with the anus.

The other type of sea-urchin looks quite different when compared with this one. The body is flattened, the spines are shorter, the Aristotle's lantern is non-existent, and the anus is situated at the rear. Such dramatic modifications clearly correspond to an essential change in habits. These latter sea-urchins in fact live in the sand, where they burrow constantly not far from the surface,

The sponges are marine invertebrates which live on the sea-bed, with a rather simple internal structure; sometimes brightly coloured (centre), their shape is often only vaguely defined, and adapts itself to that of the surface to which it fixes itself. Top left: Axinella damicornis. Below: Oscarella lobularis. Top right: Clionia celata. Below: Euspongia officinalis.

186

*This benthonic creature (*Cynthia papillosa*), with its extremely simple sac-like form, has internal structures which suggest that it is distantly related to the vertebrates.*

swallowing sand and mud containing alimentary particles, in the same way as earthworms. The strangest forms of these urchins are the sand-dollars found along the American coast; the body is flat like a coin, usually with a slightly eccentric hole in it which seems intentionally designed for making necklaces.

The sea-urchins move slowly. The tiny tube-feet which emerge from ten series of plates are equipped with suckers which enable the animal to cling firmly to rocks; these tube-feet alternately attach and detach themselves, thus allowing the creature to move; during such manoeuvres the spines act as levers. The sea-cucumbers or holothurians also move by means of tube-feet. These animals have a fleshy body, which in fact resembles a cucumber, and live on the sea-bed, feeding on debris and small creatures which they scoop up usually with the tentacles around the mouth. They are not very conspicuous or attractive creatures, although the Chinese eat them – dried sea-cucumbers are marketed under the name of *trepang* – and one of their unusual habits recalls the starfish. Just as the starfish ejects its stomach outside the mouth to start its external digestion of a mollusc, so the holothurian literally abandons its stomach and other entrails when it is seriously disturbed. As the lizard's tail, these entrails then re-form, thus allowing the animal to resume its normal life.

*The segmentation of the body of this marine worm (*Eupolymnia nebulosa*) with tentacles suggests its affinity with common land animals like the earthworm; both in fact are annelids.*

There are also some invertebrates on the sea-bed which are thoroughly sedentary, such as the sea-lilies, corals, sea-anemones, sponges, ascidians and many molluscs.

There are two different types of sea-lily. Some have a genuinely vegetable appearance, with a long peduncle anchored to the bottom, at the top of which there is a calyx with ten arms that look like so many petals. Each arm in turn has a series of pinnules which, by moving constantly, draw food towards the mouth, which is situated at the centre of the calyx. The other type of sea-lily has no peduncle: the calyx is anchored to the bottom by means of a tuft of small roots; this type could uproot itself and transplant itself elsewhere.

The sponges are among the most widespread and variously formed creatures of all those inhabiting the ocean deep. Some are tiny, encrustant on rocks, shells or even on the dorsal carapace of crabs; some are ramified, like bushes; and others are enormous, like great goblets, such as the gigantic Neptune's cup sponge, which grows to a height of 5 feet (1.5 metres), and is found in deep waters. The colours of the sponges also vary greatly and are often brilliant with reds, yellows, blacks and browns. Like animals, sponges are simply organized and, to use a bureaucratic expression, fairly decentralized. Essentially they are

Left: a group of date mussels, marine molluscs which live inside rocks which they burrow into by means of special secretions. Top right: a murex, *used in olden times for the purple substance which acted as a dye. Below: a* Cardium *with its strange prickly shell.*

formed by quite a number of small chambers, the inner walls of which are lined with cells – the *choanocytes* – which have a flagellum which constantly produces a water current; the water enters and exits through pores and tubes which are quite elaborately arranged, and differ from one sponge to the next. Commercial sponges are actually nothing more than the skeletal framework of certain marine sponges, from which all the cells have been eliminated by steeping and other such treatments. This skeleton consists of a somewhat uneven network of *spongin*, a substance akin to the keratin of hair and fur. But not all sponges boast this type of skeleton; many, on the contrary, have a number of small needles with one, two or more points which more or less lock together and form a solid but fairly elastic structure. These needles may consist of silica or even calcium carbonate, depending on the case in point.

The actiniae, better known as sea-anemones, are likewise vividly coloured. These creatures are related to the medusae or jellyfish and coral, and are sedentary like these, but not colonial. A wide adhesive disc keeps the animal firmly anchored to the bottom; the body resembles a short cylinder; the mouth opens in the centre of the upper basal section, surrounded by vast numbers of long, fleshy tentacles, and one should beware of their attractive appearance;

these animals are in fact equipped with stinging cells which we have already come across elsewhere. An animal which has such efficient weaponry can catch fairly large prey, even if it is in no way mobile.

However, some small sea-dwellers run no risk whatsoever by venturing into the sea-anemone's tentacles. Some fishes – and they differ in different parts of the world – have won the confidence of this fearsome animal-flower and take much delight in spending long and fearless periods among the stinging tentacles.

Of course relationships are not always this ideal on the sea-bed. Animals which live at the expense of others, in some cases sucking their blood literally to the point of death, are certainly not rare. Among these parasites we find the sacculina, a bizarre creature which, in the adult stage, resembles a small sac attached to the abdomen of a crab, mouthless, eyeless and legless. In fact, to live as a parasite all that are needed are organs which suck, and the sacculina is generously endowed with such organs: the small sac beneath the crab is connected to a cluster of ramified tubes which thrust their way inside the victim and suck from it the vital fluids. Given the 'anonymous' appearance of the sacculina, no one would conceive of its being a crustacean, in other words, a

Top left: a chiton, a mollusc with its shell formed by a number of plates. Centre: a group of mussels, attached by the byssus. Below: dentalium or tooth-shell. Right: a group of lamellibranchs belonging to the genus Chlamis.

The presence of two pairs of antennae, pedunculate eyes, five pairs of legs with strong chelae (or nippers) on the front pair and a tail fin which enables them to move rapidly backwards (top left) are all features of the group of macruran crustaceans which includes many kinds of crayfish and prawn, and the lobsters. Below left: Alpheus glaber, *with asymmetrical nippers. Right:* Homarus americanus.

relative – though none too close a relative – of its victim. In effect when the parasite emerges from its egg it looks quite different from its later appearance; during the larval state it is hard to tell it apart from many other crustaceans, particularly acorn barnacles (or balanids) and goose barnacles (or lepadids).

Acorn and goose barnacles are also strange crustaceans: after a brief free-living period as larvae they attach themselves to submerged objects to which they will remain attached for the rest of their lives. Once the larva has selected a spot for its metamorphosis, it attaches itself by means of an antenna which, in the case of the goose barnacle, becomes a long fleshy peduncle, whereas, in the case of the acorn barnacle, its size reduces. These latter eventually take on the shape of a small hollow tooth or tiny volcano-like cone, often ivory-white and as hard as stone: from the top a tuft of cirri or tendrils issues forth which, once again, is used to swirl the water roundabout and gather the suspended alimentary particles; when threatened by dryness or danger, the cirri are quickly drawn in beneath the four small, mobile calcareous plates. Acorn barnacles are commonly found on rocks, mussels and other submerged objects and are among the marine invertebrates which best cope with the problem of dryness which confronts them every time the tide goes out. Some species of

Examples from four classes of echinoderms. Top left: an Echinaster *(class Asteroidea). Below:* Ophioderma *(class Ophiuroidea). Top right:* Anthedon mediterranea *(class Crinoidea). Below:* Cucumaria *(class Holothuroidea).*

extremely brightly coloured acorn barnacles live attached to whales or giant turtles, which they use as means of transport.

While acorn barnacles usually install themselves on static submerged objects, goose barnacles prefer to anchor themselves to pieces of floating wood. In appearance a goose barnacle resembles an acorn barnacle supported by a peduncle.

The forms and habits of the marine crustaceans are truly amazingly varied. The isopods form a large group, the prototype of which is the woodlouse, which is commonly found in gardens and damp cellars. Among the marine isopods we find many forms which happily eat debris, and which closely resemble the better-known land isopods: these are found among the algae, on floating pieces of timber and in the tunnels dug by shipworms. But some species are parasitic, probably because this is a less taxing way of life. Thus we find such forms in shoals of fishes to which they fasten themselves with their curved, hooked legs, or inside other crustaceans, in which case they are modified to such a degree that their real affinities are unidentifiable. Likewise among the typically planktonic copepods we find curious parasitic forms which live in the digestive tubes of molluscs, ascidians and holothurians, or hitch lifts – with

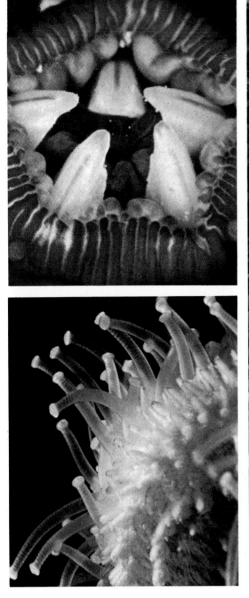

Details of a sea urchin. Top left: the buccal apparatus, with a complex structure, known among zoologists as Aristotle's lantern. Below: a detail of the ambulacral apparatus. Right: the outward appearance of the animal, bristling with spines.

their enormous sacs of eggs – from certain fishes. The best-known and most typical crustaceans are nevertheless the crabs, prawns and crayfish, and lobsters. There are countless species of crabs, some just a few millimetres in length – like the tiny pinnotheres or pea crab which lives rather like a mess-mate or commensal inside the shell of certain bivalve molluscs – others brightly coloured, like the splendid giant crab in Japan. Most crabs have a planktonic larval life, in the form of small, large-eyed creatures, with a strange appearance, and labelled with odd names by zoologists, such as zoea and megalopa; later a more critical change transforms them into animals which live on the sea-bed, covered by a tough dorsal carapace and armed with a pair of efficient pincers, one of which is often more colourful than the other. But some crabs continue to swim even after they have reached the adult state: the last pair of legs, which are flattened like blades, guarantees propulsion in such cases. Some species peculiar to rocky depths let algae and sponges grow on their back for camouflage; the dromioids manage to transplant to their backs those organisms around them which seem most suited to the purpose.

Other fascinating animals live in submerged rocks and may go unnoticed. The bonellia (*Bonellia viridis*), for example, is a curious bottle-green worm,

the body of which is formed by two parts: the lower, sac-like section stays concealed in a rocky crack, while the upper section – when not retracted – extends as far as a metre outside the crevice. This upper half of the bonellia is simply a proboscis which is dilated at the tip to form two fleshy lobes through which food is drawn in. It is an unusual looking worm, undoubtedly, but it would merit little attention were it not for the fact that all the individuals of the type described are females; there are males, but these are minute animals just a couple of millimetres in length which live as commensals in the female's proboscis. When the bonellia emerges from its egg, its sex is still undefined; the choice takes place during its development. The larva in fact produces a male if, at a certain critical moment in the development, it finds itself in contact with a female of the species; otherwise it will become a female. These are not in fact unusual behavioural patterns in the reproductive biology of the marine invertebrates. The example of the polychaetes is a good illustration of this. This is a group of worms which resemble the earthworm quite closely, but differ principally because they have colourful tufts of bristles planted on a series of protuberances which recur on the various segments of the body.

There are two main types of polychaetes, the errant and the sedentary. The

Above: two curious bottom-dwelling creatures. Left: a sedentary polychaet. Right: a flatworm belonging to the order Polycladida. Opposite: a sea-anemone, an animal which, despite its appearance, is quite akin to the medusae, or jellyfish. Small fishes of the genus Amphiprion, commonly known as clown-fish, swim in and out of the tentacles, thanks to the establishment of a symbiotic relationship with the anemone.

former, as the word suggests, spend their lives wandering about on the sea-bed among the sponges, in the sand, or even letting themselves be transported here and there by movement of the water like planktonic organisms: in such cases the body becomes flat and diaphanous and small fins appear on the sides of the body to enable the creature to float. The sedentary polychaetes, on the other hand, are generally tubicular, that is, they manufacture calcareous or membranous tubes from which their front extremity emerges, often provided with feathered, brightly coloured cirri. In many polychaetes living on the sea-bed it is possible to distinguish a forward portion of the body with a normal structure, and a rear portion, somewhat like a necklace, the segments of which are filled with eggs and spermatozoa. Once mature, these 'tails' are detached and scatter their contents in the water. Other polychaetes, on the contrary, resort to vegetative reproduction: the body constructs in a crosswise sense in several places, thus giving rise to separate segments which divide spontaneously and grow to form complete new individuals. Gemmation occurs in the family of syllids: at the rear end of the body a few protuberances form which swiftly develop into small polychaetes which once more detach themselves and embark on an independent life. There are numerous

It is hard to be convinced that the living beings depicted here are in fact animals, because they are so like flowers; in fact these coelenterates, which sometimes have long tentacles (left), are often fearsome predators and even manage to catch quite large prey. Top right: a detail of red coral. Below: a colony of Epizoanthus.

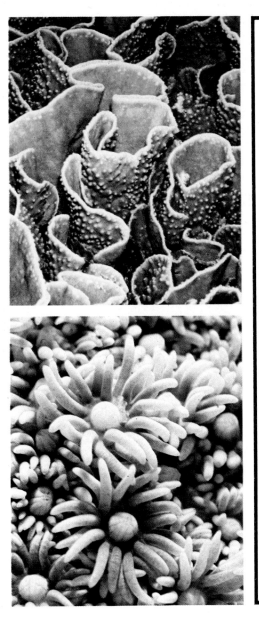

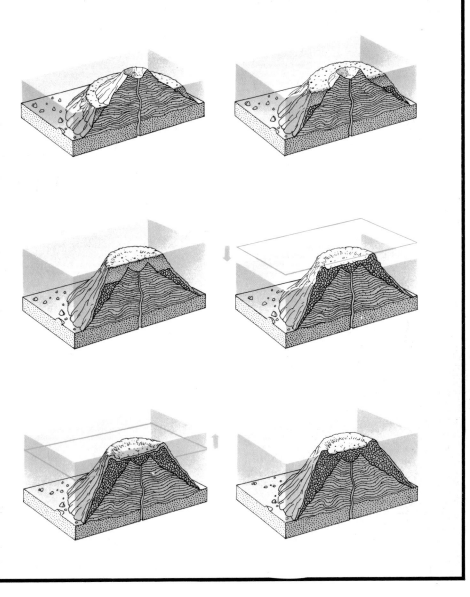

Left: two views of colonies of coelenterates. The constructive activities of animals like these, combined with variations in the sea-level, have given rise to typical formations known as atolls in the tropical seas. Right: various stages in the formation of an atoll on top of an old volcano.

phosphorescent species among the polychaetes, including the *Chaetopterus*.

The reproductive behaviour of molluscs also varies; here, hermaphrodite forms are frequent, as in the case of the oyster and other bivalve molluscs, as well as many gastropods. In this case the mollusc behaves initially as a male, and later as a female. In the cephalopods, i.e. cuttlefish, squids and the like, the sexes are separate and the male usually has a special tentacle with which to fertilize the female. This tentacle, known as the hectocotylus, is sometimes detached from the male's body: this is why its presence in the female's body was at one time mistakenly taken for a case of parasitism.

Among the most interesting molluscs mention should be made of the murex, from which the ancient Mediterranean peoples used to extract the purple substance and dye cloth with it. The technique required to obtain the precious dye was in no way a simple one, and attempts to repeat it in more modern times have not been widely successful. Today the murex (or dog-whelk) is only of interest as an edible food, and because it has an elegant shell, decorated with tubercles or spines.

Where molluscs are concerned, the major practical interest in them revolves around the bivalves, which are important from a gastronomic point of view,

such as oysters and mussels, not forgetting the famous pearl oysters, of course.

Leaving aside its splendid mother-of-pearl effect (and its commercial value) the pearl is in many cases nothing more than the elegant coffin of a worm. It is in fact not a rare occurrence for a tiny creature or some kind of tiny body to wedge itself in between the shell of the mollusc and the mantle which secretes it: the presence of the extraneous object irritates the mantle, which increases its secretory activity around the spot, and, in a word, surrounds the intruder with one or more layers of mother-of-pearl. Thus the pearl is born, often irregular or incomplete and only in certain cases presenting the perfect shape which renders it so beautiful and sought-after.

There are also species notorious for the havoc they wreak, with their constant tunnelling.

The most notorious of all are the shipworms: these long-bodied, worm-like bivalve molluscs are only minimally protected by their thin shells. They burrow tunnels in submerged wood and the planking of boats and ships. As they burrow, the shipworms line their tunnels with a calcareous secretion, and this makes them appear to be living in long white tubes that have been driven into the wood. The shipworms often riddle wood to such a point that it is reduced to

A coral atoll in the Great Barrier Reef of Australia. The sweeping circle around the central island forms a lagoon in which, at low tide, it is possible to observe a world seething with marine animals and plants.

A small coral island belonging to the Tuamotu Archipelago, one of many such groups which go to form Oceania.

a fragile sponge – not unlike the consequences of the activities of termites.

Other bivalve molluscs even dig their way into rock: this is the case with the rock piddocks (*Pholas*) and date mussels (*Lithodomus lithophagus*), both, incidentally, highly sought-after by gourmets. Not infrequently, these molluscs, which use both mechanical and chemical means when excavating, remain imprisoned in the niches which have been formed during their growth; in fact they continue to erode the rock still deeper, without greatly widening the aperture with which the excavation commenced: what is more, all they need is to guarantee themselves a constant exchange of water in order to breathe and feed. Most bivalve molluscs are nevertheless sedentary. The species which burrow tunnels in wood or rock never abandon their dens; mussels remain anchored to the sea-bed by the byssus, a bunch of fibres like coarse silk which people in bygone days used for sewing and weaving; as they grow, oysters as it were weld their lower valve to the rock; clams and cockles remain deep in the sand and move very little. But other genera are more active: some types of clam, for example, jump by using their large, fleshy foot, and pectens move quickly along by jet propulsion, beating their valves.

Despite their highly varied forms, the shells of molluscs all have an

199

essentially identical structure. They reveal various superimposed layers of calcium carbonate crystals, sometimes in the form of calcite, in other cases in the form of aragonite; the mineral substance is combined with an organic compound known as conchyolin. The shape and appearance of the shells are the result of a regular growth process, which naturally reflects the variousness of the environmental conditions.

It is possible, in the shell of a mollusc, to read several years of history, almost like the rings of a tree trunk. In warm seas, where the vital processes are more accelerated and where the water contains a high level of carbonates, the shells are more massive and heavy. The giant clam in particular is peculiar to tropical waters and being heavy it thus remains resting on the sea-bed with its valves ajar, feasting on the organic substances produced by the microscopic algae which lodge in its tissues. Because these tiny vegetable organisms need light for their vital chemical reactions, the giant clam houses them beneath special transparent epidermic cells with a lens-like function which concentrates the available light-rays. The importance of these unicellular algae is so great for many marine animals that they have made special cunning arrangements to encourage the growth of the algae. Another bivalve, the corculum, has even

The seashore, where there is an abundance of organic matter thrown up by the ocean swell, is host to numerous forms of highly specialized invertebrates, which may often live alternately in and out of water depending on the movement of the tides, and which can tolerate the extremely high concentration of salt present in the pools formed by the spray from the waves as they break over the rocks.

Of all the creatures which frequent the coast, the crabs are among the most familiar and feared because of the strength of the powerful pincers on their front legs (below). Top left: a fiddler crab, so-called because of the considerably developed right pincer which to some extent resembles a fiddle being held in the arms of the animal. Right: a running crab beside a limpet, a very common gastropod mollusc on rocky coasts.

incorporated a series of transparent peepholes on its shell, in such a way that the light reaches the algae even when the valves are closed.

Other marine animals which live in association with the microscopic algae with one cell are the various types of coral, and the madrepores. These constructive organisms are classified as coelenterates, along with medusae (jellyfish) and sea-anemones, because they are organized in the form of colonies of polyps, each one of which corresponds to the model described above. Coral and madrepores are demanding creatures; in fact they need warm, clear and well-lit water: these requirements – especially the last – are dictated precisely by the algae living in their tissues. The result is that the vast majority of the species live in tropical waters and only the odd one finds its way to temperate seas. Wherever the environmental conditions respond best to their needs, the madrepores flourish in a variety of forms and in great abundance. Some are ramified, others lobate, digitate and convolute in the form of brain convolutions.

The places where the activity of the madreporaria is most conspicuous are the Great Barrier Reef of Australia and the countless atolls in the Pacific Ocean. The former is separated from the continent by a wide and shallow

strait, in the waters of which – being warm and fairly calm – there is extremely intense animal life. The movement of the sea is continually snapping off the coral skeletons, the fragments of which join together to form a genuine rock-bank on which generations of polyps settle.

The atolls have a fairly typical and unusual form: they are small ring-like islands, in the centre of which is an enclosed, shallow lagoon. At first glance it is not easy to explain this appearance. In fact these are underwater volcanic cones, colonized by coral when the crater thrust its way above the surface of the water. Around the crater these constructive animals have continued to live their lives for thousands of years; when they die they deposit successive layers of carbonates. In the last few thousand years the general raising of the water-level due to the melting of huge quantities of polar ice has in turn submerged many of these volcanic cones, but the coral has soldiered on: it is thus possible for a very high wall, built entirely by polyps, to rise up above the cone of lava.

Madrepore and coral alike have been busily at work since remote geological periods. Many mountains are to a great extent formed by coral limestone, deposited in the course of millions of years. And such mountains contain the indelible traces of the superhuman activities of the marine invertebrates.

The calcareous shells of marine invertebrates transformed by the processes of fossilization and embedded in rocks, as in the illustrations above, are evidence of the changes which have occurred on earth during the various geological eras. The activity of invertebrates and calcareous algae, building immense outcrops, has given rise to formations which, having risen up from the folds of the earth's crust, now form large mountain ranges, like the Dolomites. Opposite: the three peaks of Lavaredo.

THE FRESH-WATER INVERTEBRATES

In fresh-water environments the invertebrates are almost as numerous and varied as they are in the sea. However, survival in inland waters poses more problems than in the sea, and hence entire zoological groups are unrepresented.

In inland waters there is a scarcity of dissolved salts: thanks to a phenomenon known as *osmosis*, water thus tends to enter the cells and tissues and would burst them if the animals did not have recourse to timely mechanisms which prevent too much water from gaining access.

Many unicellular creatures like the paramecium accumulate this water in an appropriate cavity, the *vacuole*, from which they expel it at regular intervals through thin canaliculi.

Another problem which often confronts the fresh-water animals is the oxygen supply. The cold waters of swift-flowing torrents usually supply enough oxygen, but in low-lying rivers, lakes and to an even greater extent ponds and marshes, the small quantity of oxygen which is dissolved in their waters often simply disappears. In the summer months in particular the putrefactive phenomena which occur on the bottom extract all the available oxygen from the water. For the aquatic animals this is a serious problem. The larvae of the chironomids or midges, small flies with feathered antennae, solve it by introducing *haemoglobin* – the red pigment which is also present in human blood – into their circulating fluids: haemoglobin is capable of catching oxygen, even if this gas is only present in the atmosphere or environment in small quantities, and passing it on to the various tissues for respiration. Therefore the larvae of midges are bright red.

In small pockets of water the animal life may hang on an even more delicate thread: once the oxygen has disappeared, the summer heat may also cause the very water to evaporate. When this occurs, the inhabitants of ponds and marshes turn into forms which can stand up to dry conditions, like the gemmules of the fresh-water sponges or the cysts of the rotifers; these dormant forms resume their active life sometimes after several months have gone by, when the water returns.

These then are the main problems to be faced by the fresh-water invertebrates. But if we are to take a closer look at their life, we must consider the various environments separately: *river*, *lake* and *pond*. In the river, the water is subject to constant and often energetic exchange, as a result of which oxygen is on the whole fairly available. But the movement of the water creates a new problem: in order to avoid being dragged out to sea, the animals must live on the bottom or anchor themselves among the plant life growing along the bank: only the larger species, above all the fishes, can swim actively against the current for long stretches. On the bottom we generally find large numbers of debris-eating forms, such as many types of insect larvae (midges, mayflies, etc.), worms, crayfish and water-fleas, Large bivalve creatures live deep in muddy areas, constantly filtering the water and extracting the suspended organic particles: they feed on these, in the same way as oysters in the sea. Their larvae are very odd: they are very numerous and attach themselves to shoals of fishes where they live for a while as parasites. Then they detach themselves and burrow into the mud where they may grow to a considerable size and sometimes may produce small and usually irregularly shaped pearls.

Among the submerged vegetation we find many species of snails which spend their lives crawling lazily among the stalks and stems and voraciously nibbling the leaves of the aquatic flora. Some species have gills with which they

*In fresh-water environments the animal
life certainly is not as amazingly varied in
shape, size and habits as life in the seas
and oceans, insofar as several factors
make life more difficult in the former;
the problems may respectively involve the
presence of swift currents, and a scarcity
of available food and of dissolved salts in
the water. But all these factors have not
managed to discourage numerous living
creatures from successfully colonizing
environments which are even extremely
inhospitable, like underground waters and
the icy alpine lakes.*

breathe the dissolved oxygen in the water; others, though, have a lung, like the land snails, and have to head for the surface from time to time to breathe.

They usually attach their eggs to submerged plants; these eggs are transparent and laid in groups of a few dozen inside characteristic gelatinous masses. To start with, one can just make out minute white or yellow dots in the diaphanous involucrum: these are the embryos taking shape. After a few days, corresponding to each small dot one can see the outline of a delicate snail in miniature, already assymmetrical with a small shell which seems to be made of glass.

Every now and then the tiny creature revolves on its own axis, almost as if testing its capacity to embark on its active life. Then after a few more days, the small snails strike out through the gelatine which has become slightly opaque and start to crawl out into the vast world about them.

On the aquatic plants we sometimes find small green or brownish polyps like tiny plants with buds. They look more like vegetables than animals, especially in the case of the green forms. If touched or provoked, however, these organisms react by contracting, like animals. In the eighteenth century these creatures were called *zoophytes*, in other words plant-animals. One day the Swiss scholar Tremblay tried cutting one into pieces and to his amazement saw

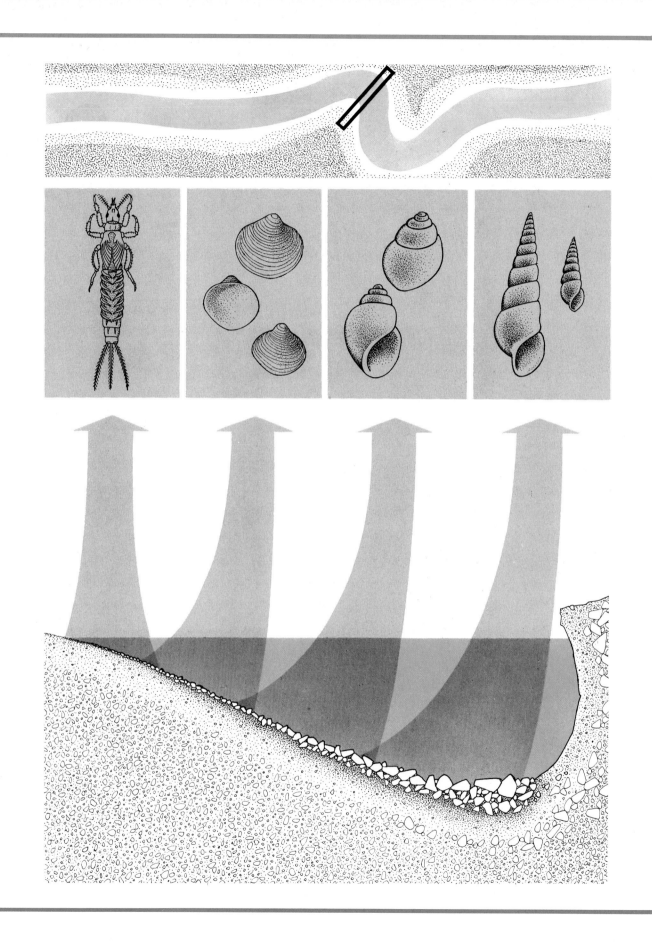

Opposite: a section of the northern Chicago River; from left to right we see the distribution of certain organisms, and more precisely a plectopteran insect and three species of molluscs. Top left: the larva of an aquatic insect (stone fly or plectopterid) which lives by clinging to the underside of stones and pebbles in watercourses. Below: a plectopter: this insect only lives a few hours as an adult and does not feed, whereas its larvae enjoy a long cycle of development in fresh-water habitats. Right: the fresh-water crayfish, the largest crustacean found in fresh-water environments in Europe.

that each small piece gave birth to a whole new animal: it seemed as if he was in the presence of the mythical Lernean Hydra whose heads, once cut, immediately grew again. The name 'hydra' still applies to this fresh-water polyp which, despite its extremely simple structure and harmless appearance, is nonetheless a voracious carnivore: the tentacles surrounding its mouth are armed with stinging cells, like those of the medusae, with which small prey are stunned. But there are far more active predators, such as the flatworms or planarians, leeches, dytiscids or diving beetles, and water bugs.

The planarians are small flatworms which like dirty water; they have two or more tiny eyes at the front end of the body and a mouth, strangely situated half-way down the ventral section: the pharynx issues from this mouth, being a sort of muscular tube which seizes and sucks in food, made up of living or dead animals. In some instances, if it finds itself in trouble, the planarian heads off and abandons its pharynx which re-forms within a few days.

In fact remarkable regenerative capacities are peculiar to many invertebrates living in fresh-water environments, but this does not include the leeches. These creatures move in a very characteristic way. On the bottom they use two suckers to achieve their caliper-like gait: with the rear sucker fixed, they

explore the area around them, looking for a spot where they can affix the front sucker; once this is done, they draw up the hind part of the body, attach it in a more forward position and then start all over again. But many species of leeches can also swim: in this case they extend and flatten themselves, and move forward by energetically wriggling up and down: the body thus remains on a single vertical plane. Some leeches suck blood from fishes, frogs and turtles; others prey upon molluscs, worms and insect larvae. They lay their comparatively large eggs inside membranous cocoons which they fasten to submerged objects or which they carry with them attached to the belly, until the eggs hatch. One African species is even provided with a marsupium or pouch within which it incubates the eggs.

From the hydrodynamic point of view, the dytiscid or diving beetle is one of the best designed animals: the elongated oval body which is smooth and flattened can move swiftly through the water, propelled by the central and hind legs which have a fringe of long hairs. In the male the front legs have an adhesive disc which prevents them from slipping during mating while they clutch the body of the female; a firm grip is also helped by the lengthwise grooves which run along the female's wing-covers or elytra. In the family

Marsh plants in Colombia. The marshes contain a wealth of different forms of life which far exceeds that of rivers. Here we find very numerous species of invertebrates: tiny animals which are almost invisible to the naked eye, consisting of a single cell (amoeba, paramecium) or several cells (rotifers, tiny crustaceans); molluscs with spiral shells (gastropods) or with two valves (lamellibranchs); insects which live on the bank, skate on the water or are born in the water and later as adults lead an airborne life above the surface.

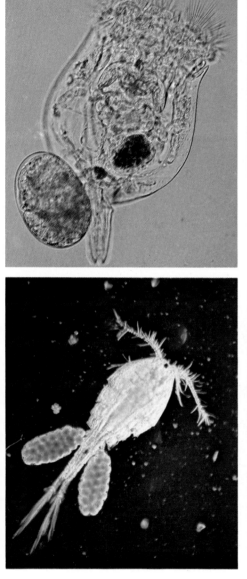

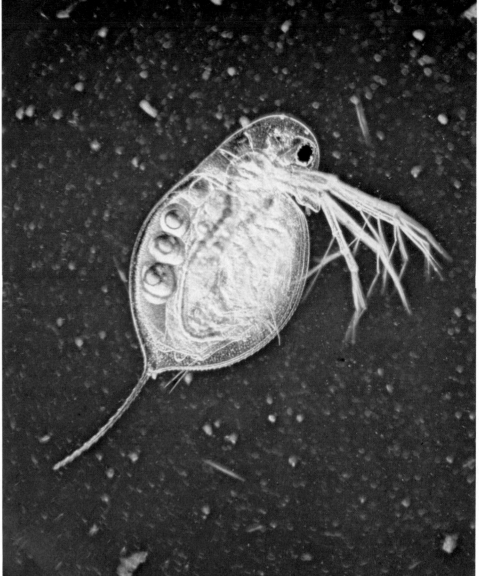

*Top left: a rotifer (*Branchionus*), a small creature with cilia surrounding the mouth, the movements of which enable the animal to catch food and move. Below: a female cyclops, a fresh-water crustacean with one eye; its name recalls that of the mythical Cyclops; attached to the abdomen are two sacs full of eggs. Right: another tiny crustacean, the daphnia; this strange little creature uses its antennae to move.*

Dytiscidae the smallest species – a few millimetres in length – feed on small worms, crustaceans and insects; the larger species, which may be as long as 2 inches (5 centimetres), will even seize small fish and tadpoles. Their role of nimble predator is shared by certain water bugs, one of which is the notonectid or back-swimmer, which swims with its belly uppermost. Other such water bugs (water striders, water gnats, etc.) skate on the surface of the water, supported by the tension of the surface membrane or film on which they rest with their long central and hind-legs. Other species, which are thickset and sluggish, live among the submerged plant-life. Here they find themselves competing with dragonfly larvae, an extremely voracious group of predators both in their aquatic larval life and in their adult life.

The biological cycle of a dragonfly is truly intriguing. The egg, which is laid within the tissues of aquatic plants or in gelatinous masses which are left on the surface of the water, produces a small and apparently innocuous larva, although it is in effect armed with a fearsome weapon for attacking other creatures: the mask. This mask corresponds to the lower lip of other insects and consists of a kind of spiny spoon with two moveable appendices which can close like pincers; a powerful set of muscles can suddenly launch the mask on

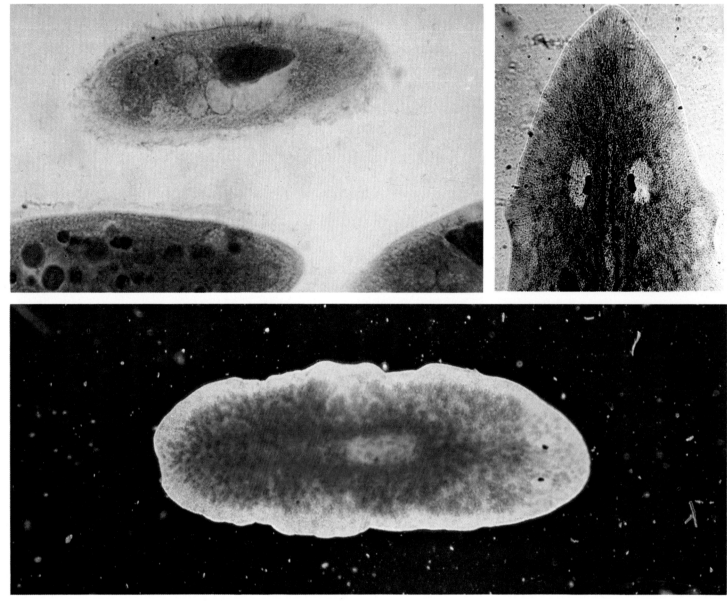

any prey within range and swiftly draw it back in: the tool used for catching prey is thus transformed into a sort of dinner plate on which the jaws and mandibles grind up the food. Watchful, ever-alert eyes, muscles with a cat-like spring mechanism, and a hook assuring a tight grip are the characteristic features which govern the day-to-day life of a dragonfly larva. Its aquatic life often lasts for two or three years, sometimes even longer, and undergoes some ten changes. In the latter stages the wings start to appear, and these grow longer with each successive change, until one fine day the animal is ready to make the final change from water to air. During this final transitional stage it starts to leave the water more and more often, for longer and longer intervals; eventually it finds a safe place, on a stalk above the water level or on the ground on the river bank. The legs anchor themselves to the substratum, the skin cracks at the thorax, and the adult insect emerges quivering from the old jacket which starts to fade and dry up. After lengthy and delicate acrobatics, it manages to free itself from the nymphal cast-off skin, but the wings still look like shrivelled petals and the body, still soft, tries to find its balance on the flexible legs. Already an hour has gone by since the start of the transformation; after another hour or two the dragonfly will be able to fly away on its iridescent

Top left: the paramecium, one of the most complex unicellular animals. The blue fringes which envelop the animal are the cilia, appropriately coloured, which enable it to move through the water. This small animal is quite common in stagnant water. Top right and below: a detail and an overview of planarians, flatworms which live mainly in clear water. Opposite: a hydra which reproduces by gemmation. This animal is one of the few fresh-water relatives of the medusae and marine polyps.

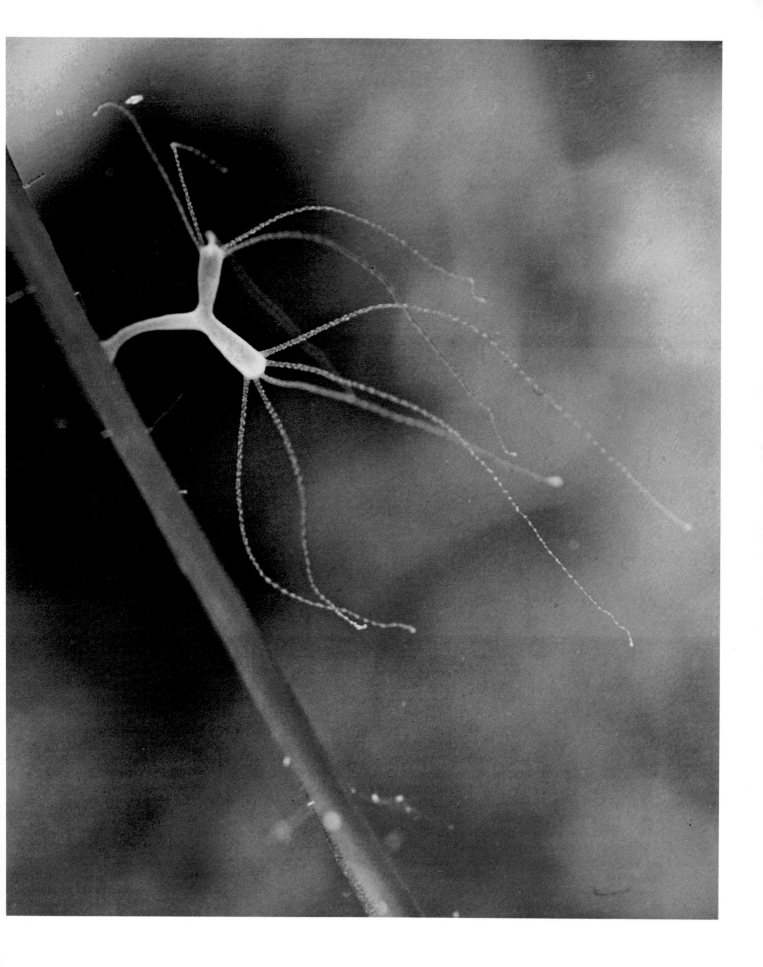

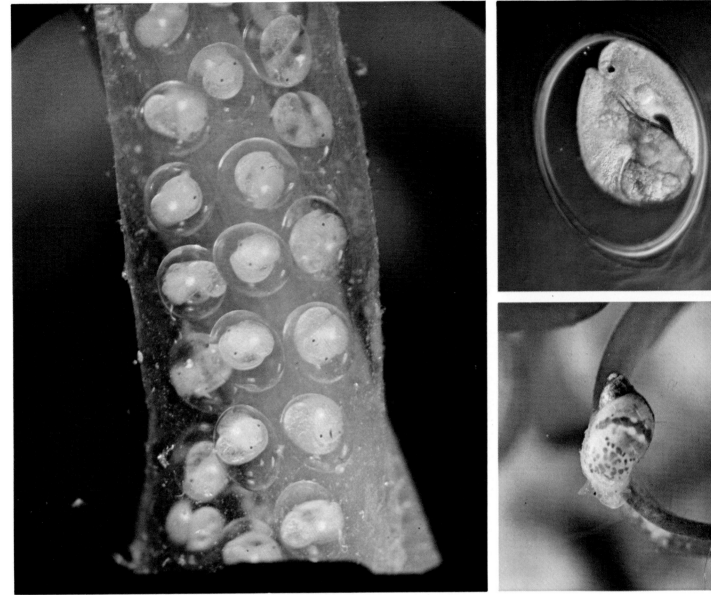

wings. After a few days it will mate, and then the female will lay her eggs, with the male often accompanying her devotedly to the water.

A similarly amphibian life is also led by many other aquatic insects, which include, for example, the mayflies, whose adult life lasts no more than a few days or even just a few hours – as long as is required for the nuptial flight.

Let us now leave the river and take a look at the typical pond. Using a very finely meshed net, it is possible to gather the plankton, which is made up of forms which are quite different from those in the marine plankton. In fact we find an absence of the innumerable bottom-dwelling invertebrate larvae, arrow-worms and medusae, and the eggs and larvae of fishes. On the other hand, there are large numbers of crustaceans like the cyclops with its single frontal eye – like the legendary Cyclops – and the water-flea (daphnid) with its bivalve cuirass or shell. But the most common animals in the fresh-water plankton are the rotifers, tiny filtering creatures whose entire life-span lasts just a very few days. With these animals – daphnids and rotifers – there is a curious reproductive process known as *parthenogenesis*: the males, in any event during the whole of the warm season, are absent, and the females lay eggs which develop without being fertilized and produce more females. But when the cold

*Phases in the development of a fresh-water mollusc (*Limnea*). Left: the eggs, contained in a gelatinous substance, which, being transparent, give a glimpse of the embryo developing within. Top right: a magnified egg with an embryo. Below: a young* Limnea. *These gastropods, which are often found in stagnant water, are sometimes carriers of parasites which are a danger to other animals and to man, such as the* Fasciola hepatica *which attacks the liver.*

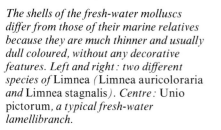

The shells of the fresh-water molluscs differ from those of their marine relatives because they are much thinner and usually dull coloured, without any decorative features. Left and right: two different species of Limnea *(* Limnea auricoloraria *and* Limnea stagnalis *). Centre:* Unio pictorum, *a typical fresh-water lamellibranch.*

season gets under way, the males often make their appearance and fertilize other eggs which, producing both male and female of the species, are destined not to hatch until the following spring. Beneath the surface of the water, in ponds and marshes, there are hosts of mosquitoes and midge larvae and nymphs. These animals also lead an amphibious life; the larva spends some weeks under water, but breathes the oxygen in the atmosphere; it then turns into a mobile pupa, in which the beginnings of wings are visible: next, during a period when the air is calm, comes the transformation to the adult state.

In the larval state, mosquitoes and midges feed on other small animals or, more often, on organic substances in a state of decomposition. The females lay their eggs, either individually or in groups according to the species, in any pocket of water, no matter how small, where, in the warm months, development takes no more than a few days. As a result these tiny insects sometimes multiply out of all proportion, forming thick clouds which seethe in the evening up and down rivers, canals and marshes.

These small pond-dwelling creatures are under constant threat from predators, usually fishes, but insects as well: diving beetles, dragonfly larvae, water bugs, etc. In this way ponds also have an alimentary chain, or better,

network, but the small size of the basin and the sharp daily and seasonal variations in the environment make the balance between species very unstable. It is this which causes periodical population explosions of water-fleas, or midge larvae, rotifers or protozoans; and the overall picture of the animal population which can be observed during a season may be quite different from the population in other years in the same season. The fauna of lakes is more constant because of the stability of lakes' physical and chemical characteristics.

The fauna of a shallow lake with abundant plant-life and large numbers of suspended organic substances is quite different from (and more varied than) the fauna of a deep lake where the water is clear and pure. In large lakes the plankton thrives in the same way as it thrives in the sea, even if the fauna is almost exclusively restricted to rotifers and those tiny crustaceans which are to be found in ponds. In some basins, particularly in the large African lakes, there are even medusae which remind one, though not very strongly, of the extremely numerous marine species. The most striking feature in the fresh-water plankton is the absence of the larvae of invertebrates which, in the adult state, live on the bottom. This is due both to the absence of many exclusively marine groups – echinoderms, tunicates, nemertines, etc. – and because the fresh-water

Everyone knows that mosquitoes are most common near stagnant water: this is owing to the fact that the larvae of these insects are born and develop in water. Top right: the first stage in the development of the common mosquito; the egg, laid in floating clusters just under the surface of the water. Left: a larva with the thin siphon which enables it to breathe oxygen from the atmosphere. Below right: the pupa, the stage of development where the animal, immobile, turns from a larvae into a winged adult.

These four creatures all lead a water-orientated life and are all predators, but they have conspicuously different ways of life. Top left: the water scorpion (or bug) walks slowly on the bottom and is a fearless predator, undaunted even by large prey like tadpoles. The water striders (or pondskaters) (top right) skate at great speed on the water with their extremely long legs. The water spiders (below left), on the bottom, carry with them a water reserve, held by the abdominal hairs. Below right: a cryptocerate hemipter: its smooth form enables it to swim very fast.

species tend to complete their larval development within the egg: the fresh-water crayfish and crabs, for example, no longer have the characteristic planktonic larval stages of their marine relatives.

Like the marine plankton, that of lakes moves both horizontally and vertically, in a way which is peculiar to each particular basin, and important in relation to the feeding of fishes which, as a rule, represent the last link in the alimentary chain of lakes.

On the bottom, and especially where the mud is richest in organic substances, there are numerous species of long, segmented worms like tiny earthworms. Some of these worms have a conspicuous red colour – due to the presence of haemoglobin – and move in an eye-catching, lively way; others, smaller and transparent, live at the base of aquatic plants or get themselves taken out for a stroll by small marshland snails. *Vegetative reproduction* is frequent in this group of invertebrates: in one or more parts of the body a crosswise constriction occurs, which gives rise to a short chain of individuals which can separate themselves and embark on their own lives; in some instances this reproduction by spontaneous fragmentation or gemmation alternates with normal sexual reproduction.

Still on the bottom, we also find a considerable number of crabs, although they do not stray far from the shore of the lake. Their variety in fresh water is greatly reduced as compared with the marine forms, and in many regions there are no crabs at all. Other crustaceans are much more frequent and widespread, for example the water-fleas, where the females carry their eggs behind them – the eggs being blue-green in colour – in a sort of ventral pouch. Molluscs once more abound in the underwater plant-life, with some species of bivalves and many species of gastropods, e.g. snails. Among these we find the amber snails, with their thin shells and very wide mouths, frequently to be seen on marsh grasses, which lead an amphibious life, half underwater and half on dry ground. On the contrary, the great pond snails with their long shells, are purely aquatic, and are also extremely voracious eaters of tender vegetable tissues, as well as being very prolific. In their vicinity one often finds planorbis or ram's horn snails with a flat, disc-shaped shell, and marsh snails with their globular shell; these are viviparous and from an early stage their shells have curious rows of long bristles which drop off successively.

The larvae of mayflies and dragonflies also abound in lakes, with quite a variety of species. The former usually have respiratory appendices in the form

Dragonflies and mayflies are considered to be among the oldest animals; in the Carboniferous period, some two hundred and fifty million years ago, dragonflies existed with a wing-span of some 28 inches (70 centimetres). These animals develop in water, and only leave it when, as adults, they have acquired functioning wings (left). Top right: a mayfly. Below: a female dragonfly (colopteryx) laying her eggs on a stalk.

The large order of coleopters includes animals which live in virtually every environment except the sea. Thus we find coleopters in fresh-water environments, such as the dytiscid or diving beetle (left), 1 to 1½ inches (3 to 4 centimetres) in length and a ferocious hunter, and the whirligig beetles (right), a family of small and curious creatures which move at great speed in groups on the surface of the water.

of oval or fringed laminae, arranged at the sides of the abdomen and frequently moving. These larvae feed on decomposing vegetable substances as a rule and undergo many changes during their aquatic life. In the transition to an airborne life, the mayflies show a feature which is unique among all the insect world. Usually, in fact, the insect does not undergo any further changes once it has acquired its wings, which are synonymous with the adult state. However, the mayfly is still not adult even when it has just acquired its wings; it has to undergo another change in order to acquire a lighter, diaphanous jacket which it will wear for a short period of time, but long enough all the same for mating and egg-laying to occur, both of these being necessary acts for the continuation of the species. Adult mayflies do not eat at all, and their buccal apparatus does not function. But dragonflies continue to be voracious carnivores.

Animal life, which is abundant in every basin and watercourse on earth, also occurs in inhospitable underground waters. Thus in the phreatic strata there are protozoans, crustaceans and tiny cylindrical worms known as nematodes. The so-called *hyporrheic interstitial fauna* living between the sand and gravel in river beds is similar, but richer still; these invertebrates have an elongated, almost worm-like body and are completely blind.

THE INVERTEBRATES AND PLANTS

Invertebrates which are associated with plants are to a large extent insects. Countless insects find sources of food in plants, and, equally, a large number of plants need insects for *pollination*, which is a necessary process if the species is to multiply. Starting from the Carboniferous period, therefore (three hundred million years ago), the history of land plants is interwoven with the evolutionary history of insects. The huge expansion of the modern-day flowering plants (the Angiospermae) is matched, from the beginning of the Tertiary era, by a huge expansion of the insect world. Today at least three hundred thousand different species of insects rely on plants for their food: some eat the roots, others the stem, leaves and fruit; and others feed on the pollen or sacchariferous fluids which issue from wounds in trunks.

Many insects spend their whole life on a plant. This is the case with the aphids, which from the moment of birth, insert their pointed stylet into the tender tissues of a bud or leaf, only withdrawing it on rare occasions when they must undergo a change or when they lazily shift positions. Every so often individuals with wings appear, destined to spread the species. These mate and usually lay ordinary eggs, whereas previously there were successive generations of wingless aphids which reproduced without mating: in fact they were all parthenogenic females, whose eggs, in other words, developed without fertilization; what is more, these females were viviparous.

Other insects which are associated with plants are more mobile and energetic. On the flowers of daisies and wild carrots it is common to come across colourful coleopters with long antennae, distinctive slender bodies, and strange and restless habits. These are the long-horned (or longicorn) beetles, whose adult life unfolds within a very short season on the flowers where they feed on nectar and pollen, and mate. Before dying the females return to the place where they developed and here they lay their eggs: in the cracks in the bark of some old tree trunk small yellow or whitish larvae will hatch, armed with strong mandibles, and continue to burrow their way through the wood, where they find food and shelter, for several years.

During their life cycle, many invertebrates change both environment and diet one or more times; others remain on the same plant throughout their life. Among the phytophagic, or plant-eating, animals, some adapt to a remarkable variety of vegetable species, whereas others live in association with just one or a few plants: the Colorado beetle, for example, is usually only found on the potato.

Every insect has its preference for such and such a plant, where food is concerned, but it also has its specific way of feeding. The grubs and caterpillars of butterflies and moths and of saw-flies (tenthredinids) usually devour the blade of a leaf, starting from a point on the outer edge; the larvae of many coleopters nibble away at the leaf from beneath, somehow respecting the upper surface. Other larvae take a more tricky path, but one which, at the end of the day, is safer and more elegant: they actually penetrate inside the leaf and dig themselves a tunnel or burrow without eating into either surface. These tunnels or burrows follow very irregular routes, mainly twisting and turning in zigzags, or in star- or flower-like shapes, depending on the species to which the larva belongs.

Many insects which are equipped with strong mandibles dig tunnels in wood. This applies to the larvae of certain large butterflies and moths, and many coleopteran larvae. We have already mentioned the long-horned beetle;

Phytoecia coerulescens, a coleopter of the family Cerambicidae, generally wood-eating creatures in the larval stage, as adults they usually live on flowers.

the buprestidae green or blue beetles have similar habits, too: their larvae often live off roots or stalks, and the adults, often resplendent in their metallic livery, are frequently seen on branches or flowers during their short life. Then there is a whole legion of tiny coleopters which even as adults make burrows in wood, often creating amazingly designed tunnels. These are the scolytids or bark beetles, which have deservedly earned the name of 'typographers' because of the fine chiselling work which in no time embroiders the underside of the bark. There will often be a main, wider tunnel, excavated by the mother; from this tunnel others branch off in a regular pattern, excavated by the young larvae. The eggs are in fact laid at regular intervals and the small larvae which hatch from them start excavating on their own account to begin with, without ever intersecting with the tunnels of their sisters. Once the larval phase is over, they turn into chrysalids at the end of the tunnel and here turn into small black or brown woodworms which make a fleeting appearance in the world outside, once they have entrusted their issue to a particular tree trunk.

These wood-excavating larvae are preyed upon by certain wasps of the family Ichneumonidae (ichneumon flies), which live off them during their young life. Sometimes one can see an elegant, long- and slender-bodied wasp

Top: the network of narrow interwoven tunnels created by small larvae of scolytid coleopters. These insects, which install themselves between the bark and the wood of trees, cause serious damage to plants. Below: the goat moth larva, one of the most harmful of wood-eating insects: in fact it attacks numerous species of trees, particularly apple trees, pear trees and poplars. The adult that develops from this larva is a grey, squat-bodied moth.

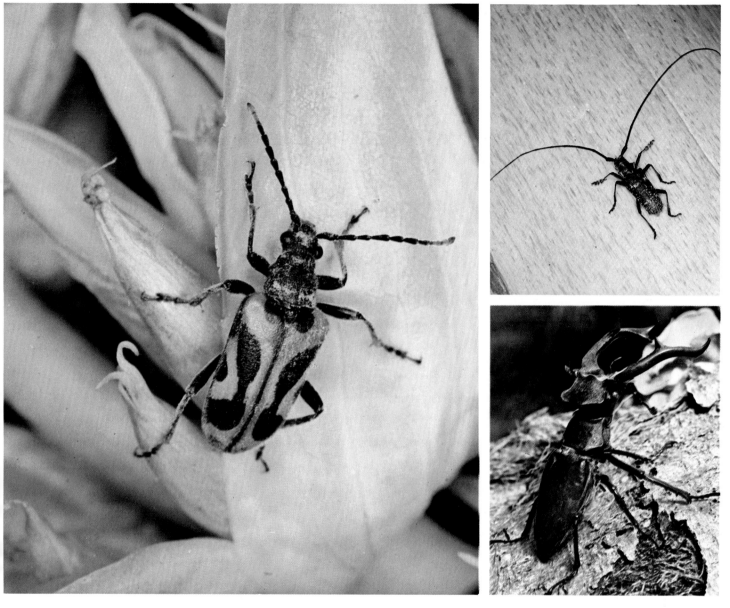

*Three species of wood-eating coleopters. Top right and left: two species of long-horned beetles (*Evodinus interrogationis *and* Monochamus*); these insects are characterized by the length of their antennae which can be two or even four times the length of the body; if disturbed or annoyed they produce a doleful groan. Below right: the stag beetle, the largest European coleopter which can reach 3 inches (8 centimetres) in length. The horns which grow on the male's head are in fact disproportionately developed mandibles which the creature cannot even use for feeding with; the larva of the stag beetle lives in the wood of old oak trees where its complete development takes five years.*

insert its thin, hair-like ovopositor through the bark: this is an extremely delicate weapon for attacking a thick protective shield! By closely reconnoitering among the branches, the wasp manages to find where the larvae are digging their tunnels: they are difficult victims to flush out, but they certainly cannot flee once discovered. The ovopositor is thus skilfully driven into the wood until it reaches the body of the larva: here the wasp lays an egg, from which a legless creature will hatch, which in turn will feed on the tunnelling larva and later turn into a wasp.

In many cases the fresh wood of healthy trees is hard to attack; quite a few insects prefer to feed on unhealthy or fallen plants, and even make use of the fungi growing on them. At a certain point in the demolition of a tree trunk, the vegetarian kingdom is replaced by creatures living in the ground: when this occurs the fauna changes radically and assumes the characteristics dealt with in the next chapter. We can nevertheless include among the vegetarians insects which feed on dead wood, such as the larvae of quite a number of long-horned beetles and the larvae of stag beetles.

Stag beetles are large coleopters, the males having a very developed head with enormous mandibles which are too large for catching food with but come

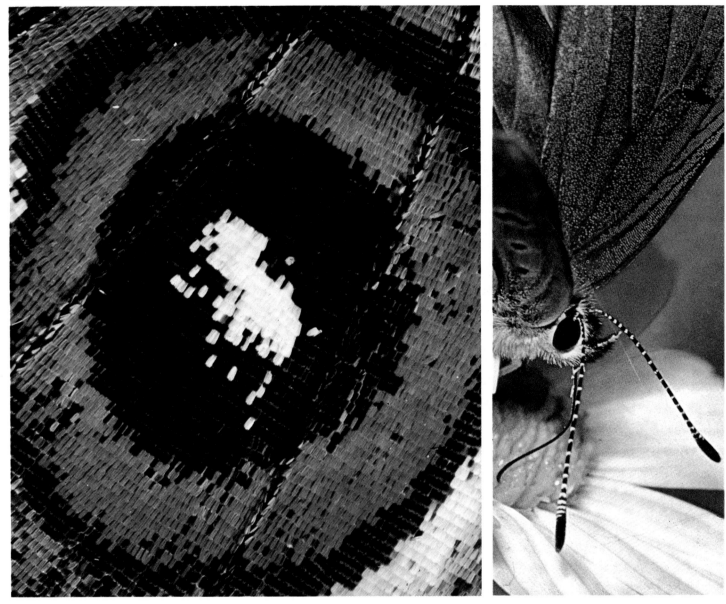

in handy in intra-specific combat. As well as the ordinary males, it is not uncommon to find smaller males which, with less well-developed heads and mandibles, look more like the females of the species.

The most colourful invertebrates associated with plants are nevertheless those which frequent flowers: bees, wasps, coleopters and, most of all, butterflies and moths. There are sometimes very close alliances between insects and flowers, because the former find the sugary liquids on which they feed at the bottom of the corolla and at the same time help the plant by carrying the pollen of one flower to the stigma of another flower, thus fertilizing it. Some flowers are in fact designed specifically on the basis of pollination via insects. The sage flower, for example, opens with narrow fauces: when a bee approaches the flower, it is obliged to support itself on two small levers at the base of the corolla; this pressure bends the curved stamens downwards, and they in turn leave a small amount of pollen on the furry back of the insect. When the bee enters another sage flower it rubs the stigma with its back, and thus leaves behind the precious specks of pollen. Flowers attract insects by means of their bright colours and scent; in order to attract flies – which can also act as pollination agents – the inflorescences of many Araceae use a smell which

Among the creatures which frequent flowers, the butterflies and moths are the best known and most colourful; their large wings are covered with small scales in a wide variety of colours, and it is these scales (left, magnified) which form the fine dust which is removed from the wing at the slightest contact. Butterflies extract the small amount of food required by them from flowers; they suck out the nectar by means of the thin sucking tube which forms their buccal apparatus (right: a butterfly feeding).

*Butterflies have many-coloured wings, a slender body and antennae which end in a clava. Top left: a satyrid (*Pararge aegeria*), a European species which frequents shady places. Centre: a small lycaenid (*Lycaena phlaeas*), particularly common in all parts of Europe. Below: the well-known monarch butterfly (*Danaus plexippus*), the American migratory species. Right: an apaturid (*Apatura ilia*); the violet coloration seen in the illustration would disappear if the butterfly were seen from another angle.*

recalls that of rotting meat.

Butterflies and moths are also courteous flower-frequenters. In order to suck the nectar they use a long proboscis, the sucking tube, which may be even longer than the overall length of the animal; when not being used it is rolled up beneath the belly. Certain moths like the large hawk- or sphinx-moth can suck the nectar from a stationary position in the air close to the flower with their long evaginated proboscis by continuously vibrating their strong wings.

The wings of butterflies are covered with a host of small coloured scales, arranged so as to make extremely elaborate patterns: round or elongated markings, eyes, zigzag designs and even letters of the alphabet and figures can be seen on the wings of the various species. The metallic greens, blues and violets are caused by the diffraction and interference processes which the light undergoes when it contacts the complex lamellate structure of certain scales; this process produces changing colours which miraculously alternate with the variation of the direction from which the light ray arrives. The large South American *Morpho* is especially colourful with bluish-violet shades and tones of colour, while smoky hues grace the lycaena and the uraniids.

These *physical* colours are matched by the more frequent *chemical* colours

caused by the presence of special substances – called pigments – with a specific colour in the scales which cover the wing. These chemical colours do not change with the direction from which the light arrives, and their range covers white, yellow, brown and black. Butterflies and moths usually mate in flight or tend, in any event, to start their amorous advances in the air; once mating has taken place, they do not usually live longer than is necessary for laying the eggs.

These are as a rule left on plants which will provide food for the caterpillars which will hatch from them. Some butterfly eggs are more or less spherical, but quite a few are also in the shape of small barrels or cigars; the small shell is often decorated with reliefs, crests and protuberances; quite often the eggs are covered with a layer of down from the abdomen of the mother.

Some species spend the winter inside the egg, but most quickly embark on their larval life and hibernate in the form of grubs or pupae; only a few species, in temperate parts, hibernate in the adult state.

Many caterpillars resemble dried, brown, gnarled twigs: they remain in a rigid position at an acute angle with a real branch with incredible dexterity. Others resemble the inflorescences of the plants on which they feed.

A butterfly grub usually has sixteen legs: six fore-legs, of the normal type,

*Top left: a typical moth (*Briophila muralis*) with its front wings turned back so as to cover the rear wings and the body. Below: the chrysalis of a moth; inside this container the changes take place which turn the grub or caterpillar into a winged adult. Right: a group of tropical butterflies; these insects gather on the ground in very large numbers when they find pockets of moisture.*

An anomola or chafer (a scarabeioid coleopter) in flight. The front wings, transformed into rigid elytra, remain fixed; the rear wings beat so fast that the photograph has almost failed to catch the movement. This animal, like the cicada and the cockchafer, lives as a larva in the ground and as an adult frequents plants, around which it is quite common to see it flying sluggishly at sundown in the summer months.

corresponding to those of every other insect, and ten very short hind-legs, shaped like papillae and ending in a crown of small serrated teeth. In the family Geometridae, however, the hind-legs are considerably reduced, as a result of which the caterpillar finds itself equipped with just six small front legs and a strong pair of legs at the hindmost end of the body. Thus equipped, it moves in a caliper-like manner, like a leech; with its hind-legs fixed in position it stretches forward and attaches itself with its front limbs; then it arches itself, drawing up the rear part of its body.

Caterpillars and grubs are very voracious and grow quickly. The cuticle covering them is nevertheless not very flexible, as a result of which the old skin is every now and then replaced by a new and fuller one.

This periodic renewal of the cuticle is the *moulting* which recurs several times during the animal's development. At the end of its larval life the caterpillar seeks out a suitable place for its metamorphosis and in some cases, for protective purposes, constructs a silk cocoon. Inside the cocoon, or in the sheltered nook it has found, the caterpillar undergoes a slow but radical transformation. In the static *pupal* stage many of the larval organs are destroyed and structures peculiar to the adult form are constructed. The wings

develop from four bases situated at the sides of the second and third segments of the thorax; the abdominal legs disappear, while the front (thoracic) legs are re-fashioned. In place of the old buccal apparatus with its masticatory function which was used to grind up leaves, a tube is formed for sucking in sugary fluids.

Like a great many caterpillars, a lot of coleopters also feed on leaves, both as larvae and adults. Two particular families are common in this environment, the leaf beetles (chrysomelids) and the corn weevils or snout beetles (curculionids). The leaf beetles often look like ladybirds, with metallic colours (although they may also be yellow, red or black) and short, strong mandibles, designed for nibbling at leaves. They are shy little creatures, and will drop to the ground feigning death when an approaching unknown body knocks against the branch on which they are perched. The well-known Colorado beetle (or potato bug) belongs to this family.

The habits of the snout beetles are quite similar; these have a characteristic head which extends into a pointed rostrum – sometimes of considerable length – at the end of which there is a masticatory buccal arrangement. Some species of leaf and snout beetles have burrowing larvae which excavate small tunnels inside leaves or, more often, in the thin root-stems.

The two animals illustrated above are totally different: the soft, shell-protected body of the snail (left) has nothing in common with the nimble, slender shape of the moulting grasshopper on the right. For both creatures, however, the source of food consists principally of leaves which they devour in such amounts that in some instances they pose a serious problem.

226

Top left: a dynastidan coleopter of tropical parts; this animal is related to the cockchafer, whose habits it shares. Below: a red slug; like snails (right) slugs are equipped with a specific abrasive buccal organ called a radula which is used to attack plants which they feed on.

Typical members of the curculionids are the leaf rollers; here the females bite cleverly into the blade of a leaf, but avoid the main nervation, and roll it up like a cigar within which the larva will later develop.

Almost all the orthopters, locusts, grasshoppers and crickets are vegetarian. Often mimetic in their green, brown, or sand-coloured livery, these animals indicate their presence, even at some distance, by producing sounds. The so-called cricket's song is in reality a sort of music for strings, produced by rubbing together two rough surfaces. The two most used mechanisms consist in the rubbing of the hind-legs on the wings of the first pair (as grasshoppers do) and in creating friction between the front wings (as crickets do). Every species has its particular melody, the notes of which are transmitted by heredity, like the structures of the legs or the colour of the wings. It is interesting to note that the grasshopper's ear is situated on the rear tibiae or shin bones.

One behavioural pattern for which no clear explanation has been found, but which has consequences quite obvious even for the layman, is the alternation of a solitary phase with a gregarious phase in the life of some species of grasshopper. Every so often, at irregular intervals, generations appear which

do not have the usual appearance and which, instead of leading the usual solitary life, gather in gigantic migratory swarms which move *en masse*, devouring every leaf and every blade of grass in their path.

Another group of vegetarian insect is represented by the cicada and harvest-flies (Cicadidae). After a long underground larval life, the cicada spends its adult life in trees, from which it sucks the sap by means of a short strong rostrum, from which it also emits its characteristic chirring. The abdomen of a cicada is a sort of large sound box which amplifies the noise produced by the deformation of a membrane which is rhythmically moved by two powerful muscles. The harvest-flies, which are smaller in size, but often more brightly coloured, do not have devices for producing sounds; they render themselves conspicuous instead by the long dances they do when they are disturbed, and by the curious habits of many of their larvae. These are usually covered with a protective foam to keep predators away, in particular ants; these are the creatures which produce that 'spittle' which is commonly seen in late spring on grass in fields. With the rostrum sunk into the stalk of the plant, they avidly suck the lymph or sap and make it froth so that they remain covered by it.

A somewhat distant relative of the cicada and harvest-fly is the cimex, which

Resting on leaves (below centre) or motionless in their webs made of sticky threads, spiders are a constant threat to insects which frequent grasses and bushes. Top left: a garden spider. Right: a crab spider or thomisid, shaped like a crab. Below left: a striped spider. Right: Singa heri *with its curious swollen dome.*

The poisonous sting of the bee does not intimidate this crab spider which awaits its prey hidden in the petals of a flower. For the bee there is no way out. The spider has paralyzed it with its venom-loaded claws (chelicerae), which are also used to suck blood from the victim.

also has a rostrum with which to suck. Nearly every type of cimex is vegetarian, that is, feeds on lymph drawn in through the rostrum. The eggs of these animals are like small barrels, usually with a small round lid which flies off when the egg hatches; the newborn cimex already resembles the adult, except for the size and development of the wings, and frequents the same environments. Carnivorous types of cimex are also found amidst the vegetation, with a shorter, strong rostrum, which hunt other insects, or suck the blood from dead insects.

But not all the vegetarian invertebrates are insects; there are also many mites, nematodes, slugs and snails, all sharing the available vegetable food.

Mites, like tiny spiders with a sharp rostrum, come in many gall-producing species, as well as species which feed by pricking tender leaves and buds.

The presence and activity of the molluscs – snails and slugs – is far more conspicuous. These animals are not well protected against dry conditions, and are mainly to be seen on wet days or in rainy spells. The strong radula, formed by numerous small horny teeth, enables them to rasp the vegetable tissues on which they feed. Slugs and snails are *hermaphrodites*; however they have to fertilize one another and it is not hard to find a mating pair. Their gelatinous

eggs are laid in the ground, from which hatch miniature slugs or snails.

Slugs are proverbially slow creatures, but their slowness is more than matched by that of the cochineal insects (or scale insects). We all know these insects, even if we are often unaware that they are animals. It is quite common to see small scales on indoor plants, which seem to be smooth in appearance or covered in wax-flaxes, and occur on the least healthy-looking branches; and sometimes, on oranges or lemons, one finds small black bacillus-like commas attached to the peel; the intruder in both cases is the female scale insect, which lives a static existence with its rostrum sunk into the tissues of the plant; its body often dries up, and becomes a shelter for the eggs.

The eggs produce tiny moving creatures which scatter along the branches until they find a place where they can suck the sap: and here they will remain for the rest of their lives. The adult male, which does not always appear regularly, is often winged and flies for a short distance from the spot whence it has emerged from the chrysalis until it finds a female, which it then fertilizes.

Plants can put up little defence against such a numerous array of enemies. The prickles which keep a ruminant away from a thistle are useless against an insect which gnaws away the stalk, and the very chemical substances which are

Camouflaged like grass, thanks to their shape and colour, praying mantises, predatory insects which are distantly related to crickets and grasshoppers, wait for some unfortunate creature to come within range of their strong front legs, which are held ready to pounce on any prey. The ravenous and cruel hunger of the mantis is symbolized by the fact that, after mating, the females often eat the males.

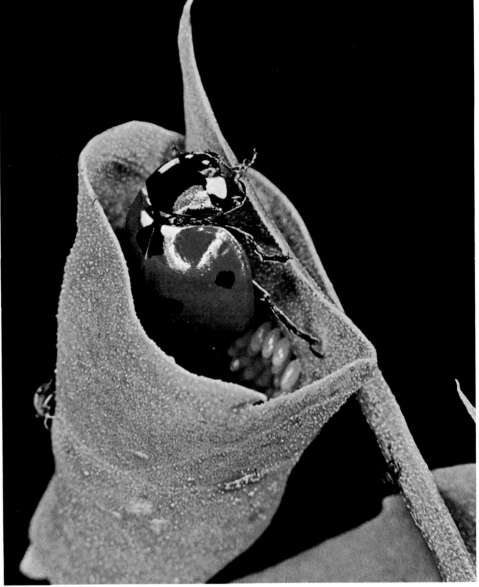

Ladybirds (right) and chrysopids or stink flies (below left) are among the prettiest animals which frequent plants: no one would imagine that they are in effect ferocious predators. Their victims are aphids (top left), in other words small plant lice. The larvae of the stink fly and ladybird have no trouble feeding on these small and vulnerable animals.

poisonous for certain animals do not stop others from eating the plant which produces them. However, the multiplication of these vegetarians is checked by the presence of large numbers of predators and parasites which live on them, among which we can mention frogs, lizards, many types of birds and some small mammals: most of them end up in the jaws of other invertebrates: small parasitic wasps, predatory flies, spiders, praying mantises, etc.

The ambush is the hunting technique of the praying mantis too; this is a grasshopper-like insect with its front limbs transformed into strong claws ready to seize hold of prey. The mantis will remain motionless in the grass for some time, its fearsome legs tucked up like a jack-knife, ready to pounce on the first insect that comes into range. A ruthless hunter, the female mantis also shows her bloodthirsty nature during mating: in fact she will not infrequently devour the male once mating has taken place.

Ladybirds lay their long, yellow or orange eggs in small clusters without any special protection on leaves and buds. These produce small, elongated larvae with three pairs of quite large legs, a head which is armed with fairly strong mandibles, with bright patterns consisting of yellow, red, violet or black markings, depending on the species in question, which are almost as colourful

as the shells of adult specimens. The larval stage lasts for a few weeks and is followed by a short dormant period, during which the animal – in the pupal state – stays suspended from a leaf, attached to it by its last larval skin. When the adult emerges from the chrysalis, it continues to live in the environment in which it developed, although it can make flying sorties away from it. Not all ladybirds feed on aphids: some are vegetarian, eating leaves and pollen and – more usually – the microscopic fungi which often cover the leaves of many plants with a whitish, talc-like film.

Caterpillars, likewise, have many enemies, and no defence is offered by their mimetic colouring, the spines with which they are often covered, or the toxic substances which they sometimes have at their disposal. Many wasps feed on them during their larval life, and many predatory coleopters eat their fill of them and raiding parties of ants carry them off *en masse* to their nests.

The population growth of slugs and snails is also checked by coleopters – fireflies, silphids and carabids – while the numbers of grasshoppers are controlled primarily by wasps which destroy their eggs.

Within the vegetation of an environment, therefore, we find complex alimentary relationships established which in turn create a balance between

To merge into the natural environment is, for many animals, the most effective way of escaping from predators. The insect shown on the left is so like a plant that even the legs are enlarged so as to resemble leaves. Top right: a hemipter which resembles the thorn of a rose. Below: a butterfly grub which is perfectly camouflaged with its host plant which has withered flowers.

Top left: Gastropacha quercifoglia, *a European moth which is named after its resemblance to an oak-leaf. Below: a grasshopper camouflaged like a rock; if you saw it flying you would see its blue or red lower wings. Right: two caterpillars of the family Geometridae: motionless on plants, they bear an amazing resemblance to dry twigs.*

vegetarians, parasites and predators. It should be added that the relationships between invertebrates and plants sometimes take on very specific aspects. For example, there are certain insects which use the blades of leaves to build their nests: the leaf-cutting bee, a small solitary bee, cuts from rose-leaves perfectly circular small discs and long segments, with which, respectively, it makes the plug and walls of the small cells in which its young are reared.

Other solitary bees use already existing hollows in wood, which they enlarge and fit out to their requirements: this is done by the carpenter bees, brilliantly coloured bees with violet and blue mantles, which are among the largest insects in Europe. The Osmylidae, smaller and decorated with copper-like reflections under the thick reddish down, fit out small cells for their young in bramble branches or elder branches, from which they take the medulla or pith. All bees depend on flowers for their food supply, even though some species simply burrow tunnels in sand to build their nests – like the anthopores – or build cells in walls – like the mason-bee; some even use the empty shells of dead snails.

There are solitary bees and social bees in every conceivable transitional form, ranging from the most individualistic existence to the highly practical organization which hallmarks the domestic bee. In the case of this latter, a new

233

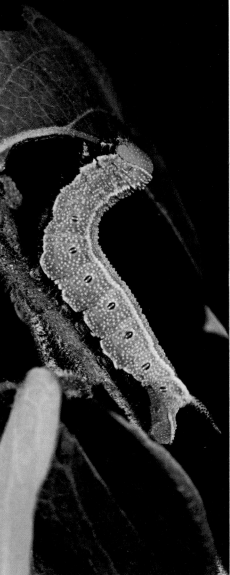

*There are essentially two ways which animals use to escape from their enemies: invisibility, by hiding or camouflaging themselves, or frightening the possible aggressor. This latter is the technique of the caterpillars shown here: the species on the left (*Papilio trolius*) has two colourful patches which to all intents and purposes resemble the eye of a large animal; the species on the right is brightly coloured with a horn at the top of the abdomen.*

nest is started during the warm months by a swarm of worker bees which emigrates from the native hive, following the trail blazed by a queen bee; there is always just a limited number of queens in each nest, and the same thus applies to the males (the so-called drones) which die shortly after mating.

The worker bees, which make up the main section of the population of a hive, are females in which the ovaries are not yet completely developed and which, unable to reproduce, devote themselves to keeping the matters of the hive in order. During its adult life, which lasts for about a month, each worker bee changes trades several times: in the initial period it is allotted the most menial tasks, such as ventilating the nest and cleaning it; and the time it spends being a nectar-collector is very brief. Then the larvae have to be attended to and fed, and the cells have to be sealed off when they turn to pupae; new cells have to be built, and old ones repaired. Because a nest will always have worker bees of different ages, there is always a team to take care of each task to be done.

When a worker bee has found flowers containing a lot of nectar, it communicates its find to its colleagues; a better-nourished team of bees might move in and exploit the source of food that has been found. When the flower discovered is near the hive, the worker bee, on its return, simply indicates the

Top left: a scolytid, a hymenopteran insect which as a larva lives a parasitic life, but as an adult frequents flowers to find sugary fluids which make up its diet. Below: a solitary bee; this species gathers pollen and nectar not for itself, but for the larvae: likewise the bustling activity of worker bees (right) on flowers helps to accumulate food for the newborn.

species which can be identified by the smell which still clings to the bee's back. What is more, the bee will have left its own smell on the flower, and the other worker bees will pick this smell out as soon as they fly off with it in search of other flowers of the same type. They in turn will leave behind their 'visiting-card' and when they return to the hive they will rustle up other worker bees for the job. When the flower is farther away smell and sight are not efficient enough guides: by doing a characteristic dance, the worker bee which has made the discovery will indicate the direction and distance of the flowers that are waiting to be visited. The dance is always in the form of a figure-of-eight and is carried out by wriggling the tail over a vertical comb. When the flowers in question are in the direction of the sun, the two halves of the eight are arranged symmetrically in relation to the vertical; if the opposite is the case, the crosswise arm of the figure is inclined, to left or right, a certain number of degrees which corresponds to the angle between the hive-flower direction and the hive-sun direction. When the flowers are comparatively near, the bee energetically wriggles its tail and traces the line which divides the two circles of the figure-of-eight; when they are some way off, it wriggles more slowly. Hence the position of the target is located perfectly.

THE INVERTEBRATES IN THE GROUND

The *ground* is the part of the earth's surface where the action of living organisms is evident; here they dig tunnels, live and decompose, combining with water and chemical agents to transform sandy and gravelly areas, wear away the rock and create fertile land.

Life in the ground is extremely active, hidden though it may be. Innumerable bacteria undertake the highly important task of fixing inorganic substances in forms which can be used by other organisms: the nitrobacteria, for example, enrich the soil with nitrogen to be absorbed by plants in the form of *nitrates*.

Furthermore, the presence of roots is vital, because they effect a mechanical action, even at great depth, and supply many invertebrates with food.

Organic debris – shed leaves, fallen tree trunks, dead animals – is constantly accumulating on the surface of the ground. This debris is attacked first of all by bacteria and fungus-hyphae, but it also constitutes a rich source of food for small creatures living in and on the ground. The same fungus mycelium which is found among dead leaves and on old tree trunks provides insects, mites, and so on with food. In order to get a picture of the invertebrates living in the ground, let us look at a broadleaf wood, and pick up a handful of dead leaves, possibly from a small hollow where conditions remain moist for longer.

Our haul will certainly include a few earthworms with their segmented bodies. Some rings will be larger than others, with a distinctive coloration: this is the region of the *clitellum*, with plenty of glands, the secretion of which forms the cocoon – like a small pea – inside which the eggs are laid. The earthworm is a hermaphroditic animal, nevertheless, it must mate in order to achieve cross-fertilization. In this mating each of the two worms behaves as male and female.

When it rains, earthworms head towards the upper strata of the ground and can sometimes be seen out in the open: when it is dry, on the other hand, they burrow deep down, dragging small shreds of leaves behind them. When burrowing its tunnel in the soil, the earthworm swallows everything in its path: its intestine absorbs the useful organic substances, and the rest is eliminated. When the tunnel is burrowed at a shallow level, the earth digested is sometimes off-loaded on the surface: this gives rise to those typical small mounds.

Because of their incessant activity the earthworms are thus important for the continual mixture of the earth which they produce. No less important, although less evident, is the work carried on by mites and spring-tails (Collembola), which are found in vast number in all types of soil, but mainly in the humus in forests. Spring-tails are tiny, wingless insects which jump by means of a ventral appendix, the furca. This forms a lever on the ground which enables this species to move almost as well as the more lively flea. Spring-tails come in all colours – yellow, pink, orange, green, violet and brown; those that live deeper down, however, are usually white. They feed mainly on fungus-hyphae or vegetable debris, like many species of mites which have squat, cuirassed bodies.

The mites are related to the spiders and scorpions, but have a pointed rostrum designed principally for sucking, and are usually very small creatures, a millimetre in length or even less. They are very common in the ground in two main groups: the Oribatidae, like tiny coleopters with a hard and often shiny body, which feed on fungus-mycelia and decomposing plants, and the mites, light and lively creatures which prey on other tiny animals.

Woodlice, centipedes and millipedes are much larger. The first are small crustaceans which have made their way to environments on the land; they feed

The floor of the forest, where large amounts of decomposing debris (dead leaves, dry branches, etc.) accumulate, provides potential food for thousands of small animals: the patient activity of these forms of life profoundly alters the organic matter, and enriches the earth with substances which can be re-used by plants.

236

on vegetable debris and are usually capable of curling into a ball when troubled. Under deeply buried stones certain types are found measuring half a centimetre or less, white or pale pink in colour, and eyeless: their habits and appearance are akin to those of cave-dwelling creatures.

There are many different types of millipede. In most cases they are well protected and slow-moving, and vegetarian like the wood-lice (with which they are at times confused), while the centipedes, which are lighter and more slender, are fearsome predators. The millipedes have a large number of legs, two pairs for each of the segments into which the animal's trunk is divided: when they move, a wave-like motion runs along the tightly-arrayed rows of little legs, similar to the movement of grass in a field blown by a gust of wind. Some of these millipedes have long, cylindrical bodies, others are flat and articulated like a railway train; and other types still are convex and can curl up into a ball. When they are disturbed, they emit often coloured irritant and toxic substances, which stick to the molester, and warn predators to leave these innocuous creatures of the ground in peace and quiet.

Just as the millipedes are usually slow and lazy, so many types of centipede are energetic and swift-moving. A pair of long antennae, a mouth armed with

Among the animals which inhabit the soil, the earthworms are of prime importance. By treating the various decomposing organic matter with their digestive systems, they help to render the earth fertile, and by continually digging narrow tunnels, allow it to remain soft and well-ventilated.

Small or tiny in size, the spring-tails (an Orchesella is shown in the illustration) live in the ground, often in very large numbers. Their manner of jumping when frightened is rather curious: in fact for this purpose they use not so much their legs as an appendix called a furca situated at the end of the abdomen and folded back beneath the body.

two powerful poisonous small pincers, a long and very flexible body, supported by some twenty pairs of legs: this is the typical portrait of a scolopendra; subtract a few legs, and we have a lithobius. The geophilomorph centipedes are different again, with their very long bodies with sometimes more than a hundred pairs of legs, and common among dead leaves in woods; these creatures often throw off a faint glow in the dark.

The bite of the large scolopendras can be very painful for man, as well as for other animals. The smaller species, which cannot harm us, are nevertheless feared by many small invertebrates in the ground.

Spiders which live on the ground and in it are usually less colourful than those which weave webs in grass or on bushes. They are often small errant creatures, reddish, brown or black in colour, which catch insects which come within range in the turf or among the dead leaves. But some species are large and hairy, like the bird-eating spiders of tropical countries and the Mediterranean tarantulas. Many of these species dig a tunnel in the ground and lie in ambush for passing insects. Others dig a den and line it with a silk tube, the tip of which projects into the world outside and acts as a trap.

Other animals which hunt with the ambush technique are the larvae of tiger

beetles, common coleopters in sandy and exposed areas of all temperate and hot countries. These larvae burrow dens in the ground which differ from one species to the next, and close off the aperture of the den with their heads which are strong and equipped with efficient mandibles. Tiger beetles are extremely nimble and active insects, which can switch from a run to being airborne with incredible dexterity; they have enormous toothed mandibles covered by a cuirass decorated with bright metallic colours and white or yellowish stripes.

The larvae of ant lions also ambush their prey. The adults – akin to dragonflies – are, on the contrary, peace-loving winged insects which often flit around lights, like moths, in the evening. These squat, hairy larvae dig a sort of tunnel in the sand and settle at the bottom of it, in the centre of the small crater; the body is entirely hidden in the sand, and just a small part of the head with its two long, sharp mandibles emerges. Every so often an ant or other passing insect happens to approach the edge of the crater and starts to slither into it: once this happens it is doomed. In fact it is quickly smothered by a shower of grains of sand which the ant lion larva throws at it, causing it to slip farther and farther into the tunnel until it comes within reach of the predator's mandibles. The larva then bites the victim, injecting it with digestive enzymes which soften

The scarab beetle belongs to the numerous group of stercoral insects, animals which manage to make use of matter considered of no use by the intestine of vertebrates. They frequent the warm regions around the Mediterranean, and are famous above all for their habit of rolling the ball of excrement selected for eating to the den in which they live.

Among the various flying animals, the brachycerous dipterans (flies and bluebottles) are among the best equipped: their wings enable them to move at great speed and apparently tirelessly. And yet, their life is inextricably earth-bound, in as much as, in the larval stage, they are among the most active agents of destruction of rotting organic matter (faeces and rotting carcasses). Top left: dipterans which, as adults, frequent flowers. Below: a fleshfly, a common bluebottle which frequents corpses. Right: the head of a fly in which one can see the very large, minutely faceted eyes.

the tissues; the hapless creature is then sucked dry. The ant lion stops too much indigestible dross from entering its intestine; in fact throughout its larval period it never emits faeces and only frees its digestive tube of all the undigested residue after a year or two when it prepares to emerge from its chrysalis.

The carabids or ground beetles constitute another handsome family of predators. These are elegant coleopters, often boasting shining metallic colours, or otherwise gaudily attired with many-coloured patterns; they may also have a rather lugubrious appearance with very distinctive black carapaces. Like the long-horned coleopters, these are great favourites of insect collectors. Almost all the carabids are voracious predators, both as larvae and adults. Many species, and above all the larger ones, spend the day cowering beneath stones and only emerge from their hideouts in the moist, cool evening air. In some cases they are hearty eaters, happy to feast on any prey which comes their way; in others, they have fussy tastes; this applies to the carabids which hunt grubs, or large long-headed ground beetles which feed exclusively on snails.

Dealing with a large snail is in fact no easy task. Although the snail may be very slow-moving and thus unable to escape, it does have two efficient weapons: a shell within which it can withdraw, and a healthy dose of sticky

froth which it can spray over the aggressor. In spite of this, these molluscs have many enemies, each of which uses a different technique to win the fray. The ground beetles, for example, use their strong mandibles to cut away the last coil of the shell, until they can reach the animal hiding inside. Firefly larvae, on the other hand, attack the snail with series of poisonous bites.

Among the invertebrates living in the ground, the most awesome hunters are nevertheless the scorpions. They earn this position, of course, because of their size and their poison, for, in other respects, if one excludes their nuptial rituals, these animals have shy habits and their behaviour is of little interest. The scorpions are very old animals, dating back, in the fossil state, to the Silurian period, some four hundred million years ago. These prehistoric scorpions were in every respect like their present-day counterparts: the same short poisonous chelicerae, the same two long appendices ending in a pincer, the same four pairs of legs, and the same long tail with its sting at the end.

There is, however, a curious fact about them: the oldest prehistoric species were aquatic animals, whereas the present-day species are all decidedly land-animals and are indeed very common in deserts. Clearly, the fossils known to us represent a transitional form between a primitive family of marine animals

The life of the cicada (top left) and the cockchafer (below) have remarkable similarities: in fact both these animals start life as larvae in the ground (right: a cockchafer larva); then as adults they live among plants, the cicada sucks the sap with its thin rostrum, and the cockchafer nibbles the buds and shoots with its strong mandibles.

The mole cricket, an insect well known to farmers because of the damage it causes to plants, is something of a perfect little excavator; the front legs are in the form of very powerful shovels equipped with shears, the front part of the body is shielded in such a way that it is not damaged by making its way through its tunnels; the abdomen, on the other hand, which does not run any such risk, is soft and unprotected.

and the current land-based scorpions; the organizational plan of these animals is still defined in the aquatic environment, but it was obviously prearranged for the shift to the land.

In the mating season the males become more active than usual and go in search of a partner. Pairs are formed under cover of darkness. The male seizes the female's pincers with his own; each arches its abdomen upwards and the stings touch. Then they start a strange, slow dance which lasts for several hours. Then the tails are lowered and the male draws his mate towards him, moving backwards as he does so. Next actual mating occurs, and next day the female can be seen devouring the remains of the male, a tragic conclusion to the wedding night which recalls the violent practices of many spiders and of the praying mantis. The baby scorpions develop within the mother's body, which carries them for a long period; at birth they cling to her back and remain there for a while before striking off on their own. The scorpion has eyes, but sees little with them, even in daylight. One species which lives in a cave in the Pyrenees is totally blind, but still manages to survive.

The book scorpions or pseudo-scorpions are like tiny scorpions, with two colourful pincers, but without the poisonous sting in the tail; they are common

beneath the bark of trees and among dead leaves, where they catch spring-tails and other small creatures.

Let us now turn our attention to a somewhat special world, the *rhizosphere*. The name is strange and slightly high-flown, but its meaning is quite straightforward: it is the world of roots. There is in fact a fairly varied fauna living in this world, made up once again of vegetarians and carnivores. Sometimes the roots belong to cultivated plants, and in such cases man delves into the rhizosphere to seek out his enemies. These include the larvae of cockchafers, squat little creatures with such a large hump that they seem to be in the shape of a C. They have three pairs of well-designed legs, but use them little once they have found some succulent root, where they stop while they voraciously gnaw at it. Depending on species and climate, they take one or two years to develop. For this reason the number of adults emerging from the chrysalis are subject to periodical variations: if their development requires twenty-four months, for example, emergence from the chrysalis will probably be most abundant every second year.

The cicada is typified by an even slower development; one North American species is known to spend seventeen years underground, sucking on roots,

*Predatory invertebrates which live in the ground. Top left: an ant lion larva; this lives at the bottom of a tunnel dug in the sand where it awaits for some unfortunate insect to come within range of its long mandibles. Top right: a golden carab beetle. One of the most colourful and elegant coleopters. Below: a rove beetle (*Paederus baudii*). This species lives exclusively in the Alps and Apennines near watercourses.*

A fine specimen of Carabus solieri, *with its characteristic green livery with copper reflections. The carab or ground beetles are coleopters which are rather sought after by collectors of insects because of their elegance and, often, for their rarity. As a result some species are so ruthlessly hunted that there are fears that they may become extinct.*

before reaching the light of day for just one short summer. To enable them to burrow into the ground, the cicadas are equipped, in early life, with very strong front legs shaped like toothed blades; during their last stage of moulting, which occurs on the surface of the ground, they abandon their nymphal skins and therewith their digging tools, and acquire their wings and a sound-producing organ shaped somewhat like a drum.

Another large insect is likewise equipped with legs for digging; because of its habits and appearance it has earned the name of mole cricket, and is no favourite among peasants throughout Europe.

If these creatures burrow in search of food which the soil might offer, other invertebrates dig to bury the food they have found on the surface, and then, once underground, eat it at their leisure or – more often – use it as a food supply for their young. These tiny creatures are certainly not hard to please: they have found that even the faeces and rotting carcasses of larger animals can be eaten and in this way, thanks to their activity, such dangerous decomposing matter is quickly eliminated.

Another rich source of food for animals living in the ground is the fungus, whether it be of the ligneous type which grows on old stumps, or the fleshy type,

many of the latter being edible species for man. The fleshy fungus (or mushroom) has a short life, and apart from certain midges, there is thus not enough time for the completion of the biological cycle of a species within their pulp or flesh to take place; the fauna which, as it were, drops by, consists of slugs, rove beetles (Staphylinidae), mites, and spring-tails. The fauna of the ligneous fungi is more varied and interesting; these fungi live for months and even years, thus providing a place where thousands of insects, which can even number several consecutive generations in a single fungus, can proliferate.

The invertebrates living in the ground so far discussed are thus subdivided into different areas of influence, with different sources of food to be tapped. There are, however, two groups of insects, the presence of which is so colossal and the way of life so varied that they are of great interest even to the layman. These are the ants and termites. Ants and termites have a rather unusual feature in common: they are in fact *social insects*, organized in communities where the various tasks are efficiently allocated. Herein, quite obviously, lies the basic reason for their successful colonization of the earth's surface.

A new ants' nest is founded by a queen ant which, leaving her native nest, mates with a male in a short but stormy nuptial flight. The smaller male does

Present on the earth from as far back as the Palaeozoic era, scorpions are predatory animals living in hot, dry regions. Using its chelae (below right), a kind of strong pincer attached to the buccal apparatus, it seizes hold of its prey and then stuns it with the poisonous sting in the tip of its tail. Left: a Buthus. *Top right: a* Euscorpius.

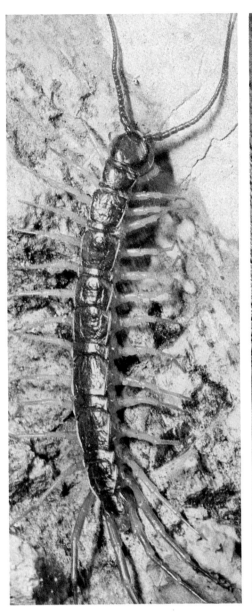

Left: a Lithobius, *commonly known as a centipede; this small predator is quite frequently found among foliage and beneath stones, especially in wooden areas, where it can move with great agility even in narrow cracks, because of its flat body. Right: a tarantula seen head-on, and considerably magnified; its six eyes make it a somewhat unnerving sight. Note the poisonous claws.*

not live beyond the mating season, while the female, returning to the ground, loses her wings or tears them off and then goes underground. Before long the first eggs are laid and after a few weeks she is surrounded by an initial nucleus of worker-daughters: from this moment on she can count on effective backing as far as the organization of the nest is concerned: she will then confine herself more and more to her reproductive function, while her offspring enlarge the nest, procure food and look after the larvae. The *worker* ants are females in which the ovaries do not achieve full development. In rare cases they may reach sexual maturity, and replace the vanished queen; this happens more frequently in termite populations. The new queen does not always found an ants' nest on her own; the new colony is very often a simple derivation from an old nest and a squad of workers may take part in its foundation.

In many species of ants, some of the common workers may be larger in size, with a large head armed with powerful mandibles: this is a special caste known as the *soldier* ants, whose job is to defend the ants' nest and – not uncommonly – to organize fierce raiding-parties in the neighbouring area. In the case of the Amazon ants, in fact, there are no real workers but just fierce soldier ants which enslave the workers of another species, and get them to run their nest. This is a

kind of social parasitism, a familiar phenomenon in the family of ants.

What is more, not all ants live in underground nests. The warrior ants of tropical contries lead an errant life, constantly moving from place to place and stopping to bivouac in the open, in an impressive seething mass of limbs, heads and mandibles. An army of warrior ants may number as many as a million individuals: and it is to be expected that wherever such an army passes no trace of any living thing will remain.

Other species of ants, on the other hand, practise the cultivation of fungi. When leaving on her nuptial flight, each queen loads herself with a little of the mycelium from the fungus culture of her nest: this is a dowry which will be valuable for her daughters. The fungi are then grown by the ants in appropriate underground 'glasshouses', on top of a nutritive composed of shredded leaves. The saliva with which these tiny bits of leaves are moistened probably serves to prevent the proliferation of other species of fungi, which might develop to the detriment of the cultivated variety. This is regularly subject to pruning by the ants which use the takings to feed both themselves and the larvae. Ants are also acquainted with the raising of livestock. In fact some species have found that the sugary honeydew produced by aphids is good and nutritious, for which

Termites are typical social insects; in fact they live in communities in which they are divided into workers (top left, the small specimens), which are specialized in the construction of the nest and the procurement of food; soldiers (the large specimen at top left), whose task is the defence of the nest; kings and queens (below left, the queen can be identified by the enormous abdomen filled with eggs). Centre: a typical termites' nest in the shape of a column. Right (top and below): two cross-sections of termites' nests.

Among ants, as among termites, there is an allocation of tasks within the community; once again we find the workers (top left) and the queen (top right, surrounded by workers). Sometimes the work of the common ants is somewhat specialized: this is the case with the South American leaf-cutting ant which nibbles off shreds of leaves (below left and centre) and then carries them to the nest where, once ground up, they can be used as a kind of humus for the culture of fungi (below, right).

reason they see fit to govern and use rationally several herds of these idle lymph-suckers. The aphids are thus protected from predators, taken to better pastures, and sheltered in ventilated stables.

The underground nests of ants are organized on a rational basis, with separate cells for provisions, eggs, larvae in different stages, pupae, and the queen. The workers bustle continually from one part of the nest to the next, digging new passages, and making sure everything works efficiently.

But the life of the ants' nest is never altogether calm and safe. Enemies abound, from insectivorous vertebrates (mainly birds) to predatory insects and spiders. What is more, there is a whole world of small arthropods living within the nest, sometimes in hostile relations with the owners, sometimes with friendly understanding, but more often than not with apparent indifference. These small creatures – wood-lice, millipedes, mites, coleopters, etc. – do not find themselves in the ants' nest by chance. The nest provides them with cheap food, either by taking the actual larvae of the ants, or consuming the food provisons laid in by the worker ants. Many of these guests pass themselves off as ants. Some have assumed an appearance that resembles that of the legitimate occupants of the nest, and can thus be taken for them. But this is a

rather external resemblance, which in no way guarantees total safety. It is more important for the lodgers to insinuate their way into the complex system of chemical communications used by the ants. The emission of a particular substance means the signalling of a track to be marked or the warning of some imminent danger, and many other types of message. A small animal wishing to live undisturbed in the thick of a swarm of ants must know how to interpret such messages without looking conspicuously like an intruder. Sometimes, furthermore, it is enough for such a creature to have a skin oozing sugary liquids and the ants will leave the intruder in peace.

Many aspects of the complex life of ants are also to be seen in the life of termites. Here, too, there is a caste system within the community. Here, too, the reproductive function is in the hands of just a few individuals: here, too, we find chemical methods of communication and variously bedecked communities of guests. The termites' nest, however, is not built by the queen alone, but by the royal couple together, flanked closely by replacement royal individuals ready to take their place should the need arise. The termites' food is usually made up of wood-pulp, an indigestible type of food because of the presence of cellulose which the termites on their own could not use: the job of breaking down this

Among the ants we find instinctive characteristics which are among the most complex and highly developed in the whole animal kingdom. There are cases where one species enslaves another (left); then we find individuals with their abdomens swollen like balloons with sugar liquids which are used to feed the community (top right). These features always apply to the worker ants: the males are winged individuals designed exclusively for reproductive purposes, and have a very short life (below).

Extremely plucky when up against intruders (top left), good at exploiting other insects (below, extracting the sugary liquids from aphids), patient hoarders of food reserves (right), ants are one of the most interesting subjects for students of animal habits.

substance falls, however, to the micro-organisms which live inside the intestines of the termites in a very characteristic symbiotic relationship.

Among the various guests found in the termites' nest are many types of rove beetle, often in strange and unusual forms. The most extraordinary is possibly the trilobite rove beetle, with its flat, rough body, both wingless and elytron-less, which, on a reduced scale, reminds one of the trilobites, marine creatures living in the Palaeozoic era. While ants have colonized virtually every inhabitable corner of the land, termites are rare outside the tropical and subtropical belt.

Social organization has been separately achieved by these two groups of insects in the course of their evolution, and certainly has taken some time: the Baltic amber deposits, almost fifty million years old, have in fact preserved various species of ants which are in every respect similar to present-day ants. The shift to a social life was a major step ahead in the history of these animals, and has enabled them to achieve ways of life which are more complex and varied than those of the other insects.

THE PARASITES

Animals have a job surviving on their own. The life of the meekest organism involves continuous encounters with other individuals of its own or of different species.

Relations with individuals of the same species are based either on *competition* (for territory, food and a female mate) or on *co-operation* (formation of the pair-bond, parental duties, the development of an organized society).

Relations with members of other species are more varied and range from indifference to the establishment of the most intricate forms of *symbiosis*, in other words, cohabitation. In the relationship that exists between the sea-anemone and the hermit crab, there are advantages for both parties: this is the phenomenon of mutual symbiosis. There is a different relationship between the mole and the hundreds of tiny insects which inhabit its underground tunnels; in the mole's den these insects find the most favourable conditions in which to live, and the mole is not even aware of the presence of its lodgers. There are similar relationships between ants and the many small coleopters which frequent their nests. In such cases the word *commensalism* may be used. There is not much difference in the behaviour of those small creatures which attach themselves to other larger animals, with greater mobility, to hitch long-distance rides from them: many mites do this, and we can often see them in the form of tiny red dots attached to butterflies, coleopters, etc., and so do acorn barnacles which fasten themselves to whales and turtles.

A commensal animal which exceeds the limits of discretion becomes a *parasite*. In this instance, the relationship between the two animals becomes considerably unbalanced: one of them benefits, the other suffers.

There are really lots of parasites among the invertebrates. Sometimes they may be temporary parasites, which confine themselves to sucking a little blood from the victim, and then make off – like mosquitoes, for example – at others they may be permanent parasites, which install themselves in the alimentary canal, in the blood, and in various tissues of all kinds. Many parasites change hosts during their development, thus achieving a complete cycle, often with larval stages which differ considerably from the adult.

An example of a parasite with two hosts is offered by the taenia or tapeworm. Man becomes infested with this worm if he ingests raw meat containing encysted larvae of the parasite: these are small whitish vescicles which, when acted upon by the digestive juices, produce a protuberance – the *scolex* – which attaches itself to the intestinal wall and becomes the foremost end of the taenia. This portion proliferates, producing, by a continuous process, a series of *proglottids*, that is, short segments full of eggs which progressively detach themselves and are emitted in the faeces. When rooting and grubbing in dirty places (which does not apply to modern industrialized pig-farming), a sow will usually become infected by swallowing the germs of the worm. The larva of this worm, armed with six distinctive claws, spends a certain active period and then becomes encysted in the muscles; its hope, at this point, is to end up still alive in the stomach of a human being. The cycle of the taenia thus involves two phases: the first, as a larva, is spent in the body of a sow, the second, as an adult, in that of a human being. The sow is the *intermediary host*, and man the *final host*.

Another case in which man acts as the final host for a worm is that of the liver fluke. This animal has a flattened, leaf-like body, some centimetres in length, which settles in the bile ducts of certain herbivorous mammals; it is frequent in sheep, but has also been found in man. It has a very complex cycle of

One of the simplest ways in which one animal can exploit another is to use it as a means of transportation. This habit, common among certain animals like the small red mite attached to the belly of a harvestman (opposite), is called by zoologists phoresy.

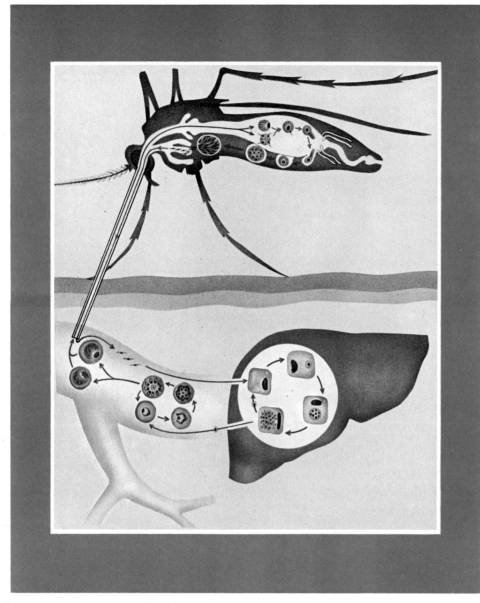

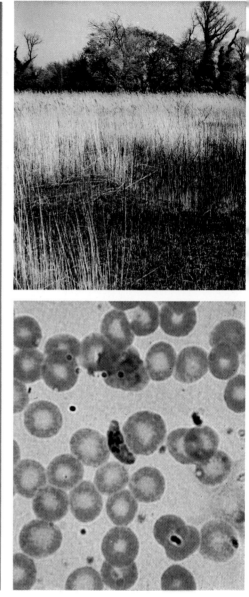

development. In fact its larvae develop only if the eggs end up in water: here the small ciliate larva – the *miracidium* – heads off in search of a small marshland snail which might serve as an intermediary host; having found the snail it turns into a second type of larva – a cilialess *redia* – which multiplies by producing, eventually, a third and final larval form, the *cercaria*. This is equipped with a tail by means of which it can move in water, once it has abandoned the intermediary host, and fix itself to an aquatic plant. The herbivore which grazes amongst such marsh-plants may thus find that as well as swallowing food, it will also swallow a worm, to which it will then become the final host.

In the complex cycle of development of a parasitic worm, man may sometimes also act as the intermediary host. In the case of the echinococcus, for example, man (once again alternating with sheep) is the intermediary host, while the adult worm – if it manages to complete the cycle – develops in a carnivore such as a dog. The adult echinococcus is a worm some millimetres in length, like a tiny tapeworm. It is not uncommon to find hundreds or even thousands in the intestines of dogs, particularly in the country. The eggs, of course, are emitted with the faeces and it is sadly not that difficult to become infected by them, given the close relationship between man and his proverbial

Malaria is a widespread disease especially in marshy and swampy districts (top right). The malarial plasmodium, namely the protozoan that produces this disease, spends part of its own life inside a particular species of mosquito, the anopheles. The cycle is shown in the diagram at left: the insect's bite enables the parasite to reach the human liver and from there to invade the red corpuscles. After various divisions the plasmodium is ready to return to another mosquito while this next one bites the diseased person. Below right: a plasmodium by red corpuscles.

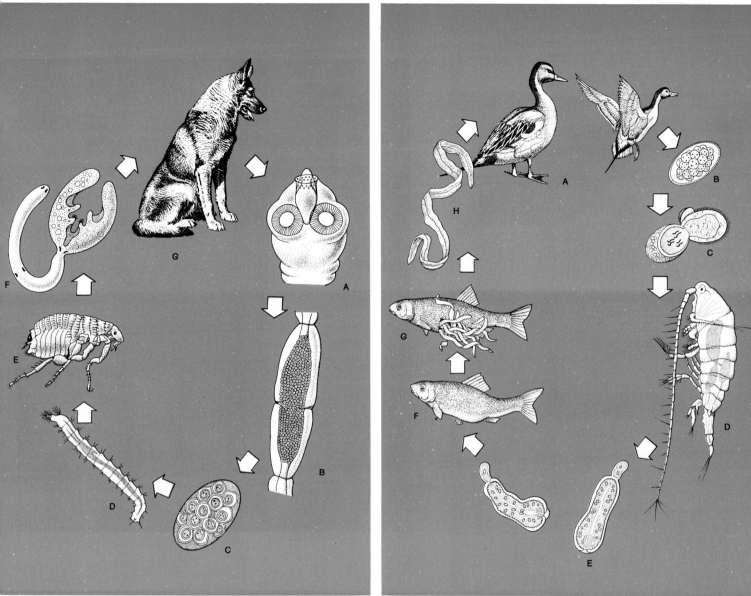

In nature there are many cases of parasites which carry out various phases of their life in different hosts. Left: the cycle of the cestoda or tapeworm, Dipylidium caninum: its usual hosts are alternatively the dog (G) and the flea (D: larva; E: adult); letters A, B, C, and F show various stages in the worm's development. Right: the cycle of another tapeworm, Ligula intestinalis, which passes from a water-bird (A) to a small crustacean (D) and to a fish (F–G) and then returns to the bird.

friend. In the intermediary host the echinococcus likewise produces a cyst, like the common tapeworms, but in this case it is an enormous cyst, containing as much as a litre of liquid, as well as the making of thousands of future adult taenias. These cysts may also form in very important organs, like the liver and the brain.

Parasites living on man also include invertebrates with a very simple cycle of development; they do not go through specific larval stages and have only one host, within which they achieve their complete biological cycle. Among these we find the pinworm or seatworm, although, occurring as it does even in advanced countries, the accompanying complications are luckily only mild.

All these human parasites show clearly the main problems that any parasite is up against in its day-to-day life. It may seem odd or out of place to put oneself in the shoes of a tapeworm, but from the zoological viewpoint this is normal procedure. Once the host has been reached, the most serious problem for the parasite is to avoid being digested and thus eliminated; it is vital to find a nook or cranny in which the host may be exploited and from which the eggs can be easily released. For the intestinal worms this latter problem is easily overcome: the eggs are released by the faeces; the eggs of the liver fluke, abandoned in the

bile duct, follow a similar, if slightly longer, route. The tapeworm, which remains anchored to the wall of the alimentary canal, does not even have feeding problems; it can use the food ingested by the host which has already been thoroughly attacked by the host's digestive enzymes. It is a fact that the tapeworm is totally mouthless and has no alimentary canal either, and confines itself to absorbing the substances which it requires through the body wall. In other respects, however, living in the intestine of another creature may be dangerous: for this reason the tapeworm is protected by a well-designed cuticle from attacks from the host's digestive enzymes.

The other major problem is represented by the difficulty of infesting a new host. Once the egg of the parasite has reached the outside environment, the likelihood of infesting another victim is quite slim. One of the first devices at the parasite's disposal is the increase of fertility. Every proglottid of a tapeworm contains thousands of eggs and proglottids are produced and emitted constantly. Another device is the intercalation, between the final hosts of two successive generations of worms, of an intermediary host which, on the basis of its habits, might act as a go-between: the advantage scored by the parasite by this solution is considerable, especially when it is not very mobile itself. Let us

Another example of parasitism working through a cycle on two different hosts. Fasciola hepatica (top right) is a flatworm that lives in the liver of a sheep; the intermediary host is a gastropod mollusc living in fresh water (below right) which is reached through the sheep's faeces. The return to the sheep takes place while the animal is drinking.

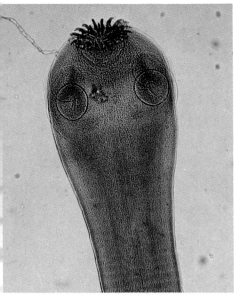

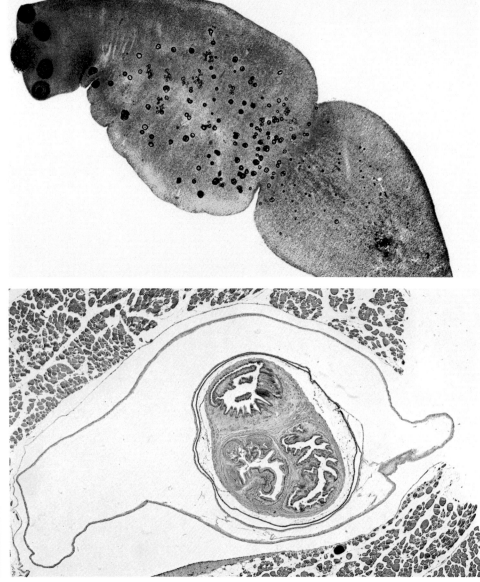

The taenia is a strange parasitic worm without any digestive apparatus; it in fact absorbs food through its cuticle. Left (top and below): two scolices, which may be considered to be the animal's head, equipped with suckers and hooks. Below right: a section of a bladderworm seen through the microscope showing a larval phase in which the worm, which is still in the early stages of development, is upturned on itself, like the finger of a glove. Top right: the complex appearance of Taenia serrata.

look at the reproductive cycle of the malarial plasmodium.

Man receives this parasite from the bite of certain mosquitoes, the so-called anopheles. In order to bring its eggs to maturity, the anopheles, like its peers, must suck a little blood from a mammal; it thus pricks it with its slender stiletto, injecting a little saliva containing a substance which renders the blood more liquid and prevents it from coagulating before the insect has finished its meal. It is precisely with the saliva that the mosquito can inoculate the plasmodium which develops first in the liver and then in the red blood corpuscles. Another mosquito may become infected by sucking the blood of an individual with malaria. In the intestine of the insect the plasmodium achieves its sexual reproduction, and then passes into the salivary glands: here it is ready to be inoculated into another person. A direct transmission from mosquito to mosquito or from person to person would be virtually impossible; but the alternation of the two phases of development between different hosts, though with inevitable losses, guarantees the survival of the parasite.

The damage that a parasite can cause to its host can be very varied and include both the direct extraction of food and injuries to various organs and tissues, and the indirect consequences – both mechanical and chemical – of the

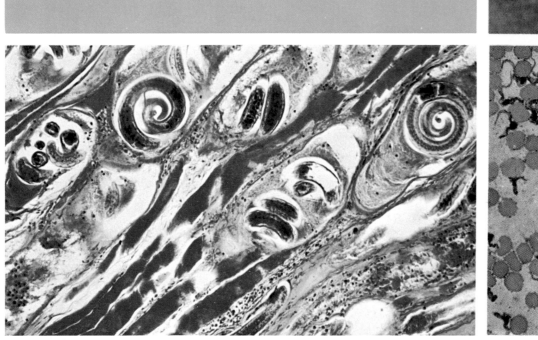

close proximity of an intruder. A tapeworm, for example, does not essentially cause any injury to the vertebrate which it has found to be its final host, but it takes food from it, and renders it more difficult for the food to pass through the intestine, as well as leaving in its body some of the debris of its metabolism. The liver fluke, on the other hand, damages the tissue of the liver and impedes the discharge of bile, with obvious consequences to the health of the host.

In some cases the parasite works so hard that it will completely devour its victim. When this occurs, if the truth be told, we are at the very borderline between actual parasitism and a specialized form of predation. For example, we know of a large number of wasps and small wasps which spend their larval life inside eggs, larvae or adults of other insects, which they devour until all that remains is the empty corpse. But the out-and-out destruction of the host is nonetheless a bad strategy for the parasite to adopt, because it is the same as destroying the food reserves of generations to come. In effect, a parasite and its host normally tend to achieve balanced relationship, in which the former exploits the latter as much as is compatible with the survival of the host. For this reason one cannot expect an insect which is harmful to agriculture to be completely wiped out when its specific parasite is introduced into the areas

One of the best-known tropical diseases is sleeping sickness, transmitted to man by the tse-tse fly (top right). The agent responsible for the disease is a parasitic protozoan, the trypanosome (top left) which also lives in human blood. Below right: a microscopic photograph in which one can see numerous trypanosomes by red corpuscles. Below left: section of a muscle, seen through the microscope with numerous parasitic worms (trichinae) which are encysted and can be identified by their curled shape.

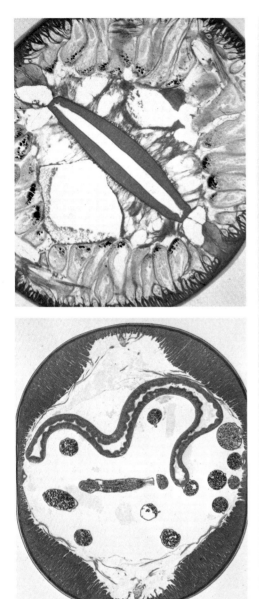

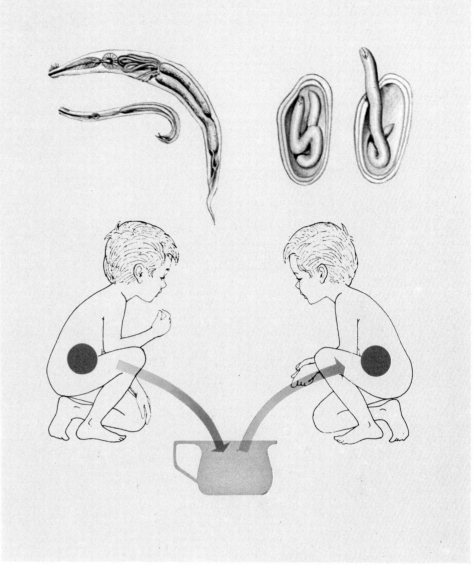

Left (top and below): two microscopic photographs of sections of a parasitic worm (round worm): these enable students to identify the anatomical structures of the animal (musculature, digestive system etc.). Right: the biological cycle of the seatworm: the male reaches a length of $\frac{1}{8}$ to $\frac{3}{16}$ inches (3 to 5 millimetres), while the female reaches a length of $\frac{5}{16}$ to $\frac{9}{16}$ inches (9 to 15 millimetres). Infection may occur when basic hygience is not observed.

where it is rampant. This technique, known as *biological warfare*, can only partially restore a balance altered by the extensive cultivation of a single vegetable species and by the explosive penetration by a vegetarian insect of a new area in which it does not come across its natural enemies. Massive infestations by parasites can in practice only occur where there is a large concentration of individuals (possible hosts) within a restricted area, as is the case in stock-farming.

Considerable havoc is caused by many parasites to their victims by the transmission of other parasites which may be smaller but are often more dangerous. This occurs in very many cases via external parasites which feed on the blood of their victims. By pricking the skin, the anopheles mosquito inoculates, as we have seen, the malarial plasmodium. In the same way the Egyptian plague flea transmits the plague, the tse-tse fly transmits the trypansome which causes sleeping sickness (which is quite widespread in the African countries), and ticks transmit other pathogenic germs, among them that of Rocky Mountains fever, which is extremely harmful to cattle.

The bodies of many parasites seem to be extremely simplified, and reduced to whatever is required for the functions of sucking and reproduction.

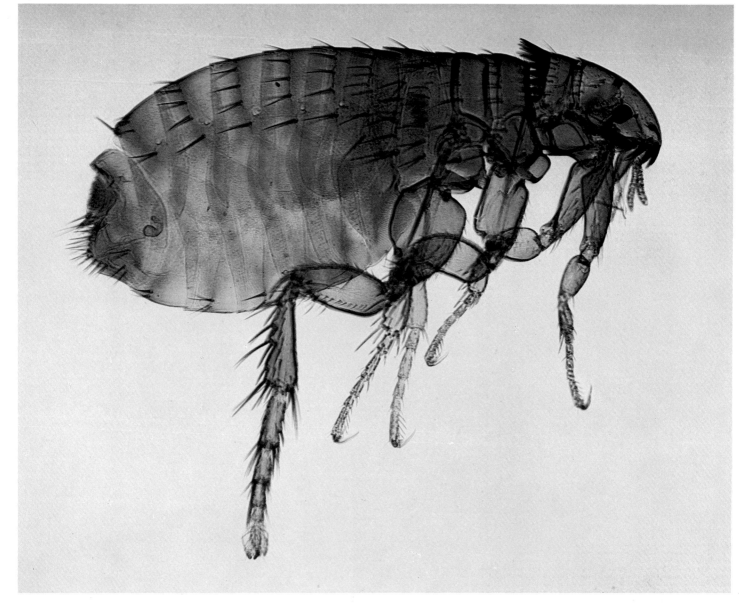

Tapeworms, for example, are totally without an alimentary canal, while planarians – their predatory relatives – have this feature.

Entoconcha, for example, is a small gastropod mollusc which lives inside other molluscs: in its structure it is quite hard to pick out a foot, visceral sac and mantle as they appear in other molluscs; it is totally shell-less and the overall appearance is more that of a worm than a mollusc. The real nature of this creature is revealed by the structure of its larval stages which are identical to those of normal sea-snails; the progression to a parasitic life has in fact resulted in the simplification of the structure of the adult, reducing it to those organs which are strictly necessary for the propagation of the species.

Certain parasitic isopods show similar modifications; these are marine crustaceans which live on fishes and other crustaceans and have somewhat unconventional forms which in no way resemble their relatives living an independent life. The sensory organs are also considerably reduced, except for the case of external parasites which identify the presence of the victim by sight or, perhaps more often, by smell. Ticks, for example, are very sensitive to the smell of butyric acid, which they can detect on the skin of passing mammals.

A great many parasites can tolerate prolonged periods of not eating. This is

The flea is one of the commonest external parasites of man, from whom it sucks blood. It can easily be got rid of by careful hygiene, but not so many decades ago it was fairly widespread, even in the advanced countries of the world. Its larva lives among debris and rubbish and thus develops without any trouble wherever there is any dirt. This insect can transmit very dangerous diseases, like, for example, the plague.

The cimex is another very common human parasite which feeds, like the flea, on blood. It is active at night, when it emerges from cracks where it remains hidden during the day, and seeks out warm-blooded animals (mammals and birds). Like many parasites it can tolerate long periods without food. If squashed it emits a distinctive, strong and unpleasant smell.

an important factor for many external parasites, ticks and leeches, which cannot forecast how long it will be before they enjoy their next meal of blood. To combat the possibility of a long period without food, leeches make sure that they suck as much blood as possible from their victims; the liquid imbibed is stored in special sacs at the side of the intestine and later digested at leisure.

Some parasites are particularly un-choosey, and make do with victims belonging to different species; others are highly specialized and depend entirely on one or just a few host species. In reality this specialization is very often due to the particular correspondence existing between the habits of the host and those of the parasite, for which reason any encounter between the parasite and other kinds of victim is quite unlikely in the natural order of things; in the laboratory, however, parasites show no reluctance in living off other species.

Rarely attractive creatures, and often odd-looking, the parasites are a basic component of natural balance in every environment. It is the task of legions of protozoans, tapeworms and round worms to check the numerical growth of many animal populations. Their presence is certainly onerous for their hosts, but it is actually no more than further proof of the adaptability of animals to the most disparate conditions of life, by using all the food resources available.

THE INVERTEBRATES AND MAN

The eyes of the layman are not attuned to picking out the vast majority of the invertebrates, and yet a considerable number of species play a rôle in man's day-to-day life, either as useful animals or, more commonly, as harmful animals. In some instances the invertebrates have left a lasting mark on the economic and cultural history of various countries. In the Mediterranean regions and in the Middle East, migrations of locusts have always been a scourge, and are mentioned more than once in the Bible.

The extraction of the purple substance from marine molluscs was a mainstay of the economy of the Phoenicians in Tyre and Sidon, while pearls – which were once haphazardly gathered from natural oyster-beds – have become the object of a thriving industry, especially in Japan.

Every year, all over the earth, enormous sums are spent in the fight against the various tiny animals which threaten crops.

But the relationship between man and invertebrates does not always take the form of battles or exploitation. Man's initial encounter with such forms of life quite often takes the form of curious observation or collection. A collection of brightly coloured butterflies or exotic tropical seashells often provides a good glimpse of the immense and highly intriguing world of the invertebrates.

In fact invertebrates are even to be found in our homes. Certain dangerous human parasites, like the louse, flea and cimex, are gradually disappearing in our cities, but for all this not even modern hygiene can hold the proliferation of cockroaches and moths at bay, let alone put a stop to it altogether.

Dwelling-places undoubtedly offer fairly comfortable conditions for several invertebrates, which can complete their biological cycle more swiftly in heated rooms than they can outside. It is thus not hard to understand the preference that many cockroaches show for coffee-machines in bars, where they sometimes abound to a worrying degree. A distinctive feature is the way they lay their eggs, which are abandoned by the female in clusters of about twenty, enclosed within a strong protective involucrum – the ootheca – shaped like a small pouch. But not all the animals which frequent our homes are bothersome to man, victuals and furnishings; on the contrary, some invertebrates are our allies in the fight against mosquitoes and other harmful insects, like the quick-moving scutigeromorpha and the diaphanous ploiraria.

The scutigeromorpha is a centipede with very slender antennae and fifteen pairs of legs, of which the hindmost are very nimble. A creature of nocturnal habits, it can sometimes be seen running quickly across walls in pursuit of small insects. Its natural environment is, typically, a heap of stones, but it clearly finds the environment of the home quite suitable for its needs. The ploiasia on the other hand is a small, wingless cimex.

Less common in houses is the presence of insects whose larvae burrow tunnels in wood: these are very often introduced to the home in pieces of furniture and unseasoned wooden fixtures, but there are also species which prefer wood that is already worked, like the insidious termites and more than one type of moth.

Flies and mosquitoes are man's constant companions, and we use every conceivable physical and chemical means to do away with them. Other invertebrates, on the other hand, have found a rôle in economic activities as producers of useful substances, for which they have been exploited since time immemorial. Two insects, the bee and the silkworm, have been reared for some time on a large scale, even though they cannot properly be called domestic

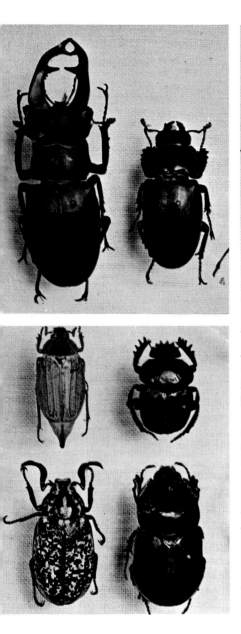

Numerous invertebrates are gathered and collected by many enthusiasts for their beauty and their strange forms and colours. Left (top and below): some coleopters prepared for collection. The classification of these animals, which is often very tricky because of the vast number of living species, has been keeping scholars busy for several centuries. Right: a page from a treatise on insects dating back to 1589.

animals. We have discussed the complex social life of bees, the efficient allocation of tasks among workers, and the curious dance-language which enables them to relay messages. The habits of the silkworm are not nearly so interesting, and the same goes for the habits of the moths in most cases. The silkworm spends its entire life feeding on mulberry leaves and the adult's life-span is a matter of a few days. Although they have wings, the females cannot fly, but the males can cover short distances in the air. The precious silken cocoon is simply the shelter that the mature worm prepares for itself in order to be able to spend the days of its difficult transformation into a silk-moth in greater peace, quiet and safety. The silk is produced by enormous glands which open into the mouth: the silkworm thus has to carry out an extraordinary number of oscillatory movements with its head, in order to organize suitably the very numerous loops in a single thread. Insects which produce silk are many, and attempts have been made to husband more than one species: but the only one to prove economically viable is the ancient silkworm of the Chinese emperors.

Other insects from which useful substances are taken are the carmine and lacquer cochineal (or scale) insects, the females of which are covered with these respective precious substances which are periodically gathered in the form of

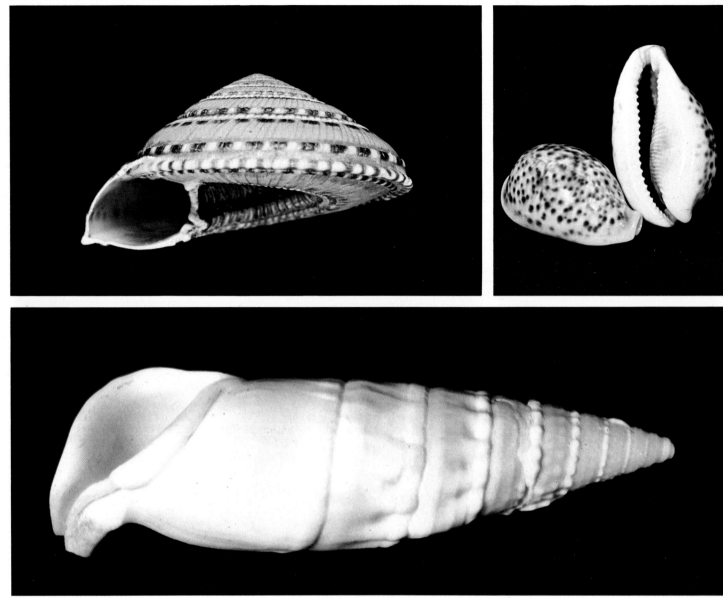

small scales.

Considerable economic importance attaches to the farming of oysters and mussels for food and for the production of cultured pearls.

The farming of mussels is widely practised along the coasts of Europe, both in Mediterranean and Atlantic waters. For this purpose particularly calm creeks and inlets are chosen, where the degree of salinity – reduced by the influx of fresh water from rivers – achieves levels suitable for the reproductive activity of this mollusc. During the cold season, every female produces a large number of eggs – sometimes as many as a million – from which hatch tiny larvae which enjoy a brief period of active life before fastening themselves to submerged surfaces. Man intercepts these swarms of larvae, and provides them with the chance to attach themselves to specially designed artificial substrata which are then arranged in the conditions most favourable to the development of the mollusc.

Oyster-farming is similar, though practised with different techniques in the various countries. The artificial production of pearls, started in Japan on the initiative of a professor of zoology, is achieved by the introduction, between the mantle and the shell of a pearl oyster, of an extraneous body around which

Extremely elegant and bizarre shapes and the often bright colours make shells one of the prized objects in wild-life collections; some species, because of their beauty and rarity, also have a high commercial value. Top left: Architectonica perspectiva. Top right: Cypraea panterina. *Below:* Cerythium virgatum.

Our own dwellings shelter a large number of small animals, which are sometimes harmful to man. Top left: the cockroach is common in damp dark places, remains in hiding by day and roams round the home by night. Below: a Ploiaria domestica *; despite its completely different shape, it is related to the* cimex *; it lives by preying on small insects. Right: a small moth; these creatures cause considerable damage to clothes and materials.*

the oyster deposits regular layers of mother-of-pearl. A quick X-ray makes it possible to distinguish between natural pearls and cultured pearls, although the beauty of the latter is in no way inferior to that of the former.

Although the molluscs include several species of considerable economic importance, no group of invertebrates is of more interest to man than the insects. Extremely numerous and varied, and present on every inch of *terra firma*, the few species of useful insects are vastly outnumbered by the legions of species which are harmful to crops.

Farmers have in fact continually waged war with every means at their disposal against voracious and insatiable swarms of locusts, against grubs and caterpillars which devastate trees and fruit, against aphids, and against scale insects with their treacherous rostrum. The fight has taken many forms. The simplest measures, which still apply to small orchards and gardens, consist in the direct capturing and killing of the harmful insects, or in the straightforward burning of the breeding-ground. Measures of this type have the advantage of not causing unpleasant indirect damage, but in practice they are almost never applicable. As a result we have seen the development of chemical warfare against plant parasites, using more and more refined methods, but methods

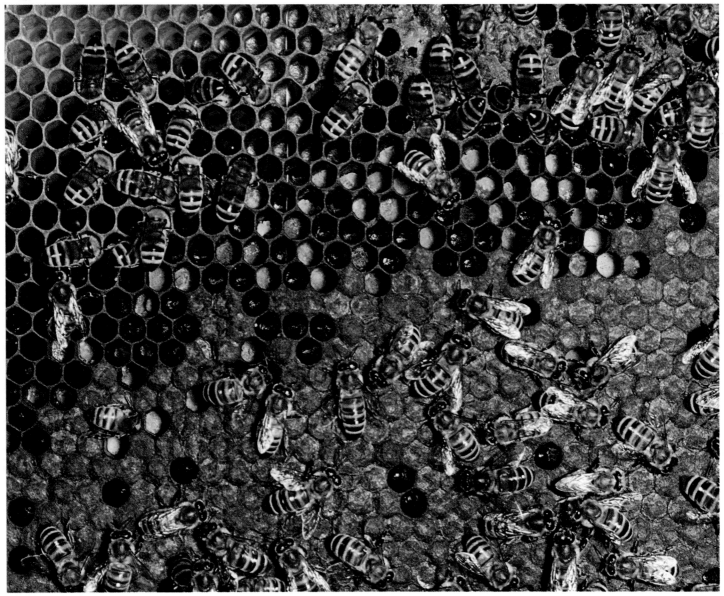

which also entail greater and greater risk. The antiparasitic substances used on a wide scale in the fight against insects which damage crops have progressively polluted the water-bearing strata which provide drinking-water for hundreds of millions of people who, as a result, regularly absorb their daily ration of insecticide, not to mention the coating from sprays remaining on fruit and vegetables which have been saved from the voracious jaws of insects.

The chemical war against the foes of our crops has thus shown itself to be a double-edged weapon. For this reason experts have been seeking an alternative method of combat, the application of which would not entail dangerous indirect consequences. Certain students of the problem have thus posed themselves a simple question. How does it happen that, in natural conditions, there are never (except in the case of the migrating locusts) those massive numerical population explosions of any one species which occur so frequently in fields or greenhouses? There are probably two reasons for this phenomenon.

In the first place, just a single vegetable species, or two species at most, are grown in any given field, whereas in natural environments dozens of different species find themselves shoulder to shoulder in quite small areas. A specific parasite of the cultivated species thus finds at its disposal an amazing quantity

Man has raised bees since time immemorial: the use of honey to sweeten food and beverages alike was widespread among the ancient Greeks and Romans and lasted until sugar-cane became a common commodity. However, the usefulness of bees is not confined to the production of honey; in fact they carry out the vital function of fertilizing the flowers that they visit by transporting the pollen and thus enabling the fruit to develop.

The silkworm moth is a clumsy white creature which came originally from China (centre). The very fine thread produced by its grub (top right) to construct the cocoon in which it wraps itself, has been used from earliest times to make very fine and shiny silk materials. Top left: laying the eggs. Below: a cocoon inside which the chrysalis is visible. Below right: a cluster of cocoons.

of food, which seems expressly designed to sate its hunger, whereas in natural circumstances the path from every plant that is used for food to the next such plant may be long and uncertain.

This difference between field and pasture, between orchard and wood exists beyond any doubt; but its *raison d'être* lies in the economics of agricultural crops: the consequences of monoculture are being fought against, but they cannot be cancelled out by simply not planting a single vegetable species on an intensive level. There is, however, another slant to this question. In our agricultural areas, many harmful insects multiply out of all proportion because there is a lack or absence of those parasites which, in natural conditions, usually tend to check such growth. In natural circumstances, a balance is established between vegetarians, predators and parasites; in cultivated areas and greenhouses, this balance is broken. It is thus possible to conceive of building a new balance by introducing into areas infested by harmful species those parasites which are specific to these species. Many insects which damage our crops come from distant countries, and have crossed oceans without being followed by their natural enemies. In their native country it must nevertheless be possible to find the small carnivores which controlled their population

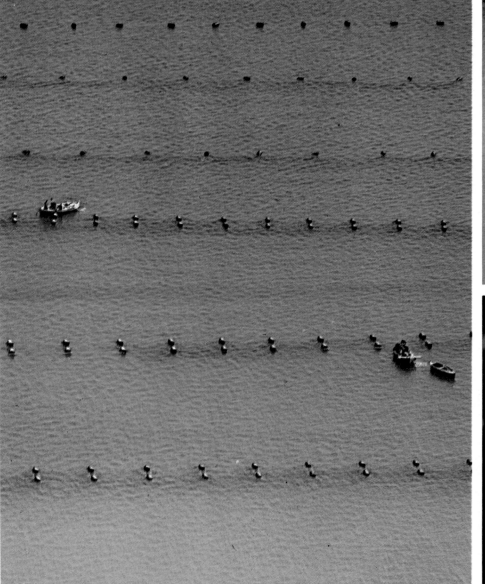

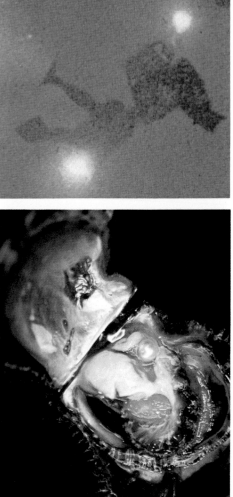

growth and these we could introduce into our own countryside, thus launching a method of *biological warfare* as an alternative to the chemical fight.

Naturally enough, a predator or parasite will never totally destroy the species on which it relies for survival and evolution: extinction of the victim would be signing their own death-warrant. We can nevertheless hope that the balance between the two species will work itself out at a numerical level which is compatible with the requirements of agriculture. Moreover, we can be sure that biological warfare will not give rise to negative side-effects, as happens with the use of insecticides.

During this century the biological struggle against harmful insects has chalked up some conspicuous successes, along with more than a few serious failures. It is in fact no simple matter to reconstruct a biological balance, even if the species in question are few in number.

Some fearsome citrus fruit scales – the olive fly and the Mediterranean fruit fly – have nevertheless been successfully combated by predatory ladybirds and parasitic wasps expressly introduced to this end. Another form of biological warfare is that which uses red ants to check the spread of grubs of species which are harmful to forest vegetation. Another strange form of alliance between

The farming of oysters may be practised for two quite different reasons: for purely gastronomic purposes, because of the highly prized flesh of the animal (left: oyster-farming); and to obtain pearls. The latter type of farming is common above all in Japan. Top right: pearl-oyster fishermen. Below: an oyster with its pearl.

The species of invertebrates considered harmful to man are far more numerous than those considered useful. Top left: migratory locusts. Below: a female laying her eggs. Recorded as early as the Bible, these animals are a veritable scourge in hot regions. Right: a European red mite, a small mite living on plants and which is one of the most damaging vegetable parasites and also one of the hardest to eliminate.

man and an insect is that struck up in Australia with a moth, thanks to which it has been possible to contain the invasion of a weed within acceptable limits.

But the species of invertebrates that man carries with him as a result of his movements and traffic are much more numerous than those intentionally introduced as allies in the biological struggle against harmful species.

The spread in Europe of the Colorado beetle and the grape phylloxera or vine-pest, both of which arrived accidentally from America and spread like wildfire from small initial breeding-grounds, bears a striking resemblance to the victorious advance of Napoleon's army or Caesar's legions. The Colorado beetle, with its distinctive yellow and black colouring, came originally from the Rocky Mountains, Kansas and Texas where, in natural conditions, it feeds on hard, thorny plants peculiar to the arid plateaux of those regions. It was discovered in 1823 in Missouri, sweeping through the potato fields of much of North America, from Florida to Canada. In 1874 the Colorado beetle reached the Atlantic coast of the United States, ready to cross the sea. Three years later it in fact turned up in Germany, and reached England in 1901. In 1922 it was observed for the first time in France, and in 1944 in Italy. At the present time it is to be found in almost every corner of Europe.

Many species of invertebrates thus have man to thank for their widespread presence on the earth's surface and, paradoxically, the fact remains that man has been largely responsible for spreading those species against which he now finds himself obliged to wage war. Other species, conversely, have been practically wiped out, even if the animals involved have been totally harmless or even, in some cases, useful. The extinction of a butterfly or coleopter certainly attracts less attention than the extinction of a large mammal, but in ecological terms both should be seen in an equally negative light.

It is not difficult to see that the species most directly threatened with extinction are those insects which are most keenly sought-after by collectors for their beauty or for their rarity. The government of Madagascar has recently had to issue a decree forbidding the indiscriminate collection of uraniids, those marvellous butterflies with iridescent colours which fetch very high prices on the specialized markets. Among the coleopters, many splendid ground beetles with bronze- or copper-coloured liveries which are found in limited areas in the Pyrenees and Alps have been shamelessly decimated, ending up – like sad jewels – in the display-cases of unscrupulous collectors. The fate of many cave-dwelling insects is even more in the balance; some of these live in small groups

The Colorado beetle is a coleopter originating from the United States and has recently spread throughout Europe. It causes much damage to the leaves of potato plants on which it feeds as both larva and adult. Top left: cluster of eggs. Right: two larvae. Below left: three chrysalids in various stages of maturity. Right: the adult.

The homopters are insects with extremely varied and odd forms. This group includes many vegetable parasites like the phylloxera (top left), the aphids (below), which are very common on rose bushes, and scale insects; the two dark markings we see highly magnified on a piece of orange peel (right) are nothing else than females of this species of parasite, the body of which turns into a simple egg-sac.

in just one or a few caves, where collectors use lures which are capable of depopulating a whole cave in no time at all. All that remain of some of these rare and intriguing species are specimens gathered over many years, and it can often be justly suspected that there is not a single living individual left.

What is more, in this day and age there are more serious problems than the extinction of the odd species of butterfly or coleopter. In many – too many – cases an entire environment finds itself fighting for survival. Whether it is an aquatic or land environment, the invertebrates always play a basic rôle in it. Man cannot, with impunity, modify the balance that governs such an environment without a rapid and often not easily reversible alteration of the once favourable conditions. The hosts of mites, earthworms, millipedes and insects which work away demolishing the dead leaves and turning them into fertile humus are among man's most reliable allies on earth. The same goes for spiders, which are constantly on the alert to catch tiny prey, among which there are many insects harmful to man or crops. It has been calculated that in one year, in England alone, spiders catch a quantity of insects comparable to the weight of the entire human population of that country.

The number of the species of invertebrates on which our existence depends is

thus much higher than it would appear at first glance and is in no way confined to the short list of edible species, or other species which are commercially exploited in different ways. We are inextricably linked to the dynamics of an overall natural environment in which every species has its place, and its rôle in the overall balance.

It should therefore seem unjust, after this brief journey among the invertebrates, to consider this large section of the animal kingdom as a world of inferior or unimportant creatures. The same process of biological evolution which has enabled the vertebrates – with man among them – to adapt themselves in spectacular ways has also equipped the bee with its complex instincts and amazing social organization, and the octopus with its appreciable intelligence. And there are a million more different species, each one capable of surviving perfectly well in its particular environment, and each one showing a high rate of success, although its strategy for survival may be quite different from that of all the other species. This fact alone should prompt us to show equal respect for all the animal species and to acknowledge, among the varied invertebrates, all the many ways of life, interlocked by the logic of an environment in which we too live.

The serious damage caused to crops by plant parasites has given rise to the production of more and more powerful antiparasitic substances and more and more effective means of mechanical application: this wide use of poisons entails implicitly the danger of pollution and damage to man's health, although the actual gravity of the situation cannot yet be evaluated.

THE LIFE OF VERTEBRATES

THE VERTEBRATES –
General Overview

A fitting start to this chapter might perhaps be 'once upon a time'. So, once upon a time – and we are talking about four hundred and fifty million years ago – there was a strange animal called an ostracoderm. Clumsy in appearance and hampered in its movements by its heavy cuirass of bony scales covering its body, it lived on the ocean-bed, sucking in any food which the water brought within range of its jawless mouth. But the ostracoderm (its name literally means 'shell-skin') played a fairly important part in the history of animal life on earth. In fact for the first time a new structure appeared within its body, a sort of *frame* around which the various organs of the body were duly arranged in a precise order. The ostracoderm was the original founder-member of the vertebrates.

The name itself indicates the basic feature of this group: the presence of a *spinal column* which is a bony structure formed by various segments, to which other bone structures are harmoniously joined, forming in turn the so-called *skeleton*. The skeleton is solid and strong, but at the same time elastic and flexible, and constitutes the rigid scaffolding which, like a tree trunk, supports the body, provides a fixture for the muscles, protects a number of delicate organs, and also acts as a warehouse for substances (calcium salts) which are vital for the life of such organisms.

In all the vertebrates (and to put our thoughts immediately in order, it should be mentioned that this name refers to seven classes of animals, namely: cyclostomata, cartilaginous fishes, bony fishes, amphibians, reptiles, birds and mammals), the skeleton can be broken down into three parts; the head, the trunk and the limbs. Each of these parts is in turn made up of bones which, on the basis of their shape, number, type of reciprocal position and connection, will form the framework of the head rather than that of the trunk, or vice versa.

The skeletal framework may vary in terms of its composition in the different forms of the vertebrates: in the oldest forms (cyclostomata and cartilaginous fishes) the skeleton is in fact formed not by bones but by cartilages, which are thinner and more flexible and which, in the other classes in this group, are present almost exclusively only during the period of embryonic development. Of course on less close examination the skeleton might seem to repeat the same basic elements in all the vertebrates (*cranium*; *spinal column* which may or may not culminate in a tail of varying length; a *rib-cage* which protects the lungs and heart; *limbs* which are called *fins*, *wings* or simply *legs*, depending on the animal in question), but these supporting structures are in turn adapted to the natural environment of the class of animals to which they belong. And this is not the result of some quirk of fate. As in all her works, here, too, nature has had to take into account the existence of certain inexorable and inevitable physical laws. It is for this reason that whereas an average elephant (weighing 4 tons [3.5 tonnes] and some 9 feet [3 metres] high) must have enormous legs with a diameter of 19 to 23 inches (50 to 60 centimetres) in order to support itself and move with a certain agility, the gigantic blue whale which can weigh up to 130 tons (120 tonnes) and reach a length of almost 111 feet (34 metres), thanks to the aquatic environment in which it lives, has a comparatively reduced skeleton, with legs which are so rudimentary that they are not even externally visible.

In water it is in fact simpler to support the weight of the body than it is on land: water and protoplasm – which constitutes the cellular matter, and thus the organism as a whole – have a similar density, with the result that, in practice, the whale does not weigh anything in its environment. It also follows from this that aquatic creatures need know no limits in terms of their

Fossil skeleton of Edaphosaurus, *a reptile living in the Palaeozoic era in the Texas region. The presence of an internal bony skeleton, giving the body its rigidity and supplying fixtures for the muscles, is a feature peculiar to the vertebrates.*

dimensions: a whale and an anchovy move with the same agility. Things are quite different on the land, where an excessive increase in size may bring about the disappearance of a species. Suffice it to mention those colossal dinosaurs which ruled the earth unopposed for a hundred million years and then in a flash became extinct. One of the main causes of this was the drying-up of the swamps which in those distant times covered much of the low-lying flatlands: the dinosaurs on dry land were no longer capable of supporting themselves and moving their bodies which had without warning become so bulky and cumbersome.

The skeletal modifications of the snakes and birds are also explained by questions of adaptation to the environment: the former on the whole have no legs and have very mobile vertebrae (as many as four hundred in the case of the pythons); the latter have reduced the number of skeletal ingredients and have long, internally hollow bones, in order to be lighter.

In the organization of the body of the vertebrates, and closely allied to the skeletal apparatus which acts as its passive supporting structure, we then find the *muscular system*, which develops from the inside outwards. The muscles which form it are the active organs of movement because they can be contracted to between two-thirds and a half of their actual length, if stimulated

The movement of animals is closely allied with their diet. The cats, hunters which actively pursue their prey, have a body structure and musculature suitable for running. Top: a sprinting cheetah; this, the swiftest cat of all, can easily travel in excess of 60 m.p.h. (100 k.p.h.). Below: having brought down a zebra, a group of lions shares out the booty.

The ability to fly, which is the privilege of the birds among the vertebrates, and of a few species of mammals, is guaranteed by special body structures. In the birds these are most evident in the presence of wings, the lightness of the skeleton and the huge development of the musculature of the thorax.

to do so by a nervous impulse. This contraction produces a shortening and a stiffening in the muscles which, being attached to the bones at either end, enable them to shift and thus produce movement. The muscles can only be contracted and released; they cannot lengthen. For this reason they are nearly always in pairs: one determines movement in one direction, the other in the opposite direction.

As a general rule one of the muscles in the pair is more developed than the other: in man the biceps, which bends the arm, is stronger than the triceps which stretches it; in the frog the extensor-muscles in the hind-legs which enable the animal to jump, are stronger than the flexor-muscles which relax the limb. When the organism is resting, both the *antagonistic* muscles remain in a state of slight tension which keeps the body in position. Apart from this musculature, which because of its structural and physiological characteristics is called the *voluntary striated musculature* (the muscular fibres which form it show, under the microscope, a regular alternation of light- and dark-coloured bands and, in addition, the movement determined by this musculature depends on the direct will of the individual), the vertebrates have, along the walls of all their hollow organs – stomach, intestine, blood vessels – a set of muscles with

277

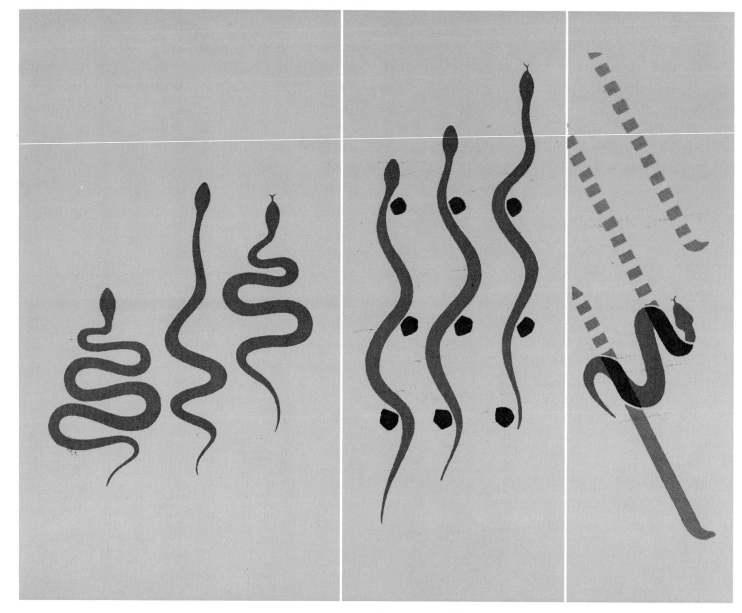

non-striated fibres (known as *smooth muscles*) which function without being controlled by the will of the individual, thus at any given moment guaranteeing the movements of the organs designed to carry out the main vital tasks.

In exceptional cases the task of some muscles becomes removed from the basic job of producing movement; this applies to the heart which, while being a striated, red muscle like all the voluntary muscles, behaves as an *involuntary muscle* (with the extremely important task of pushing blood into the vessels which then distribute it to all the cells in the body); and it also applies to the so-called *electric organs* of certain fishes (the electric ray, common ray and electric eel) which are provided for purposes of defence.

Outside the musculature, representing the limit between the organism and the external environment, and at the same time in immediate contact with it, the body of the vertebrate has a continuous covering known as the *cutis*, which varies considerably in structure and differentiation depending on the species in question. The cutis is formed by two principal layers: the outer one is called the *epidermis*, which is in turn stratified, and the inner one the *dermis*, which contains a large number of vessels and nerves. In the land vertebrates, the surface layers of the epidermis consist of dead cells (*stratum corneum* or *horny*

The way snakes move is specifically related to the absence of limbs. Left: the so-called accordion movement which enables the animal to anchor itself with its ventral scales. Centre: the snaking movement which uses lateral points of support. Right: the typical movement of desert snakes, a sideways motion which uses the spiral principle.

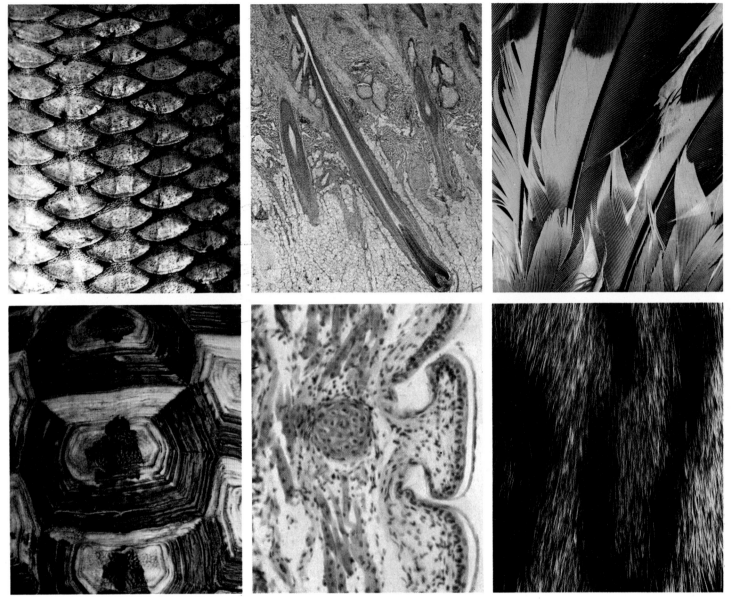

A look at various tegumentary apparati of vertebrates. Top left: fish scales. Below: detail of a tortoise shell. Top centre: microphotographs of cutis sections, with evident hair folicles. Below a microphotograph of the scales of Tarentula mauretanica. *Top right: bird feathers. Below: detail of the hide of an ocelot.*

layer). These form a thick outer coating which can stand up to damage and bacterial invasion, and reduces the loss of water by evaporation. The horny cells are continuously eliminated and replaced by new ones which form in the area below. In the reptiles the process of hornification is accentuated with the production of very frequent formations known as the scales. The wings, plumage, beak and talons of the birds and the hide, hooves and horns of the mammals have similar origins. In the aquatic vertebrates the cutis is covered with tiny scales formed originally – depending on the group – with the help of just the dermic cells or both orders of cells (dermic and epidermic). The only vertebrates whose skin has no cutaneous formation are the amphibians: in fact in these animals the cutis has a predominantly respiratory function.

Except for the reptiles and birds, the skin has *glands* adjacent to it which are capable of producing various substances depending on their rôle. In the fishes and amphibians they produce a *mucus* which, in the case of the former, facilitates movement through water; in the amphibians it keeps the skin constantly moist and viscous so as to make respiratory exchange possible. In the mammals there are *sweat-glands* which have a thermo-regulatory function and also handle the excretion of waste substances; *sebaceous* glands, usually

connected with the fur which they keep soft and impermeable; and *mammary glands* which produce milk, and are vital for the nourishment of the young.

Feeding, breathing, the elimination of useless and harmful substances, movement, the reaction to environmental variations, healing wounds and, lastly, giving birth to similar forms of life: these are the features which distinguish a living being from a non-living being. These functions are carried out by specialized organs with precision and perfect timing.

From the remote moment when the organism is formed by a tiny cell, the *egg-cell* fertilized, up to the time when as a perfectly made adult specimen it is on the point of bringing its own life-cycle full-circle, the search for food is the primary activity of every animal on our planet. Fights, danger, long and tiring migratory journeys are, and always have been, the price that animals, large and small alike, must pay to earn their daily bread. As we have mentioned, the most ancient ostracoderms living in the sea had a simple, jawless mouth without mandibles which was permanently held downwards so as to be able to suck in the food contained in the mud on the sea-bed (the lampreys, which are still to be found today in rivers and lakes, are reminders of those remote forbears with their round boneless mouths). The oldest vertebrates equipped with a proper

The feathers of birds carry out many jobs. First and foremost they have a protective function and prevent the dispersion of heat; they are vital for flying, in as much as they are the supporting structure; lastly, they are part of the reproductive function because plumage during the mating season often serves to attract a mate.

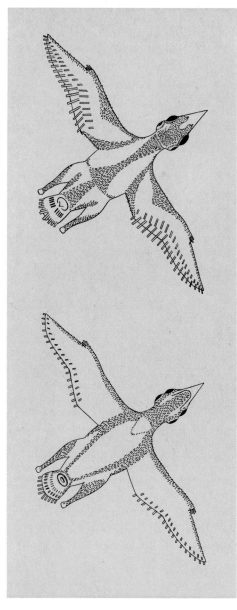

Contrary to appearances, the feathers of birds do not grow all over the cutaneous surface, but only in certain areas known as pterylae or feather tracts, as can be seen in the figures on the left which show the dorsal surface (top) and the ventral surface (below) of a bird. Right: various types of feathers: (A) filoplume; (B) plumule or down-feather; (C–D) coverts or tectrices; (E) a typical feather; (F) detail of a feather with the series of barbs and barbules with hooks; (G) double feather of Dromaeus.

mouth were the placoderms, with their bodies protected by articulated bony plates. In these fishes the mouth, formed by an inter-hinged jaw and mandible, was in a ventral position, like the mouths of the dogfishes. The mouth of the placoderms then gave rise, after a series of adaptations, to all the various mouths with teeth in present-day vertebrates, as well as to the beaks and bills of the birds. In the mouth there is a tongue which is supported in the lampreys by a special skeletal formation, and transformed in the fishes into a simple protuberance, and which in other animals is a muscular-cutaneous organ.

Beyond the mouth we find the *pharynx*, *oesophagus* (or gullet) and *stomach* which, depending on the type of diet, may come in a variety of forms (typical examples are the stomachs of the birds and ruminants, which are respectively subdivided into two and four compartments). In the last part of the digestive apparatus, that is, in the intestine which is always of a remarkable length (longer in herbivores than in carnivores), it is possible to identify various areas which have different functions. In fact, while the task of the stomach is essentially to secrete substances necessary for the chemical 'dismantling' of the long molecular chains which make up the food itself, the intestine also has the job of absorbing the nutritious substances which have been simplified by the

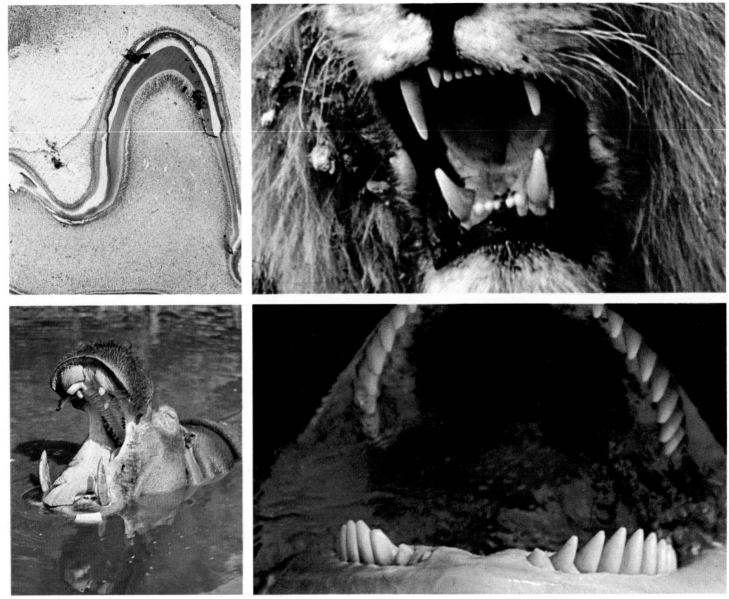

digestive processes, and it is for this reason that its inner surface is greatly increased by the presence of folds having microscopic blood vessels.

Generally speaking the digestive apparatus has its own outlet through which it eliminates the waste products of the digestive process; in the reptiles and birds, on the other hand, this outlet opens in a cavity common to the urogenital apparatus (the *cloaca*). Nevertheless the food introduced to the organism serves no purpose whatsoever unless it manages to release the *energy* contained in the molecules of the food. This energy-releasing process, which is called respiration, is achieved by the intake of oxygen and the production of carbon dioxide. Oxygen enters the body of an animal from the air or water in the immediate vicinity: in the oldest vertebrates, as in the present-day fishes and in the amphibians during the first stages of their life, respiration takes place through the *gills*, which are groups of lamellae or tufts of filaments situated at the sides of the head, which are directly washed by the water taken in through the mouth. This is why fishes must always move quite quickly: in fact it is only by so doing that an uninterrupted influx of water through the gills is guaranteed. This is why it is not infrequent to find cartilaginous fishes, – dogfishes, rays – whose gills open directly outwards, dying of asphyxiation if

The mouth of the vertebrates is almost always equipped with teeth (exceptions being certain birds which do not have a set of teeth), with various features relating to the diet of the animal. Top left: a section of an embryonic tooth bud seen under the microscope. Below: the teeth of a hippopotamus with its enormous mandibular eye-teeth or canines which grow continually. Top right: typical teeth of a carnivore (lion) with canines designed for seizing prey. Below: the mouth of a killer whale, a carnivorous cetacean with extremely strong, pointed teeth.

The tongue of the vertebrates is part of the digestive system which can assume fairly different shapes and functions in the various species. In the lamprey (top left) it has horny plates which enable this parasitic animal to rasp the flesh of the fish attacked; in the anteater (below) it is thin and viscous. Snakes' tongues (top right) are forked and have a tactile function; the blue tongue of the saurian (below) is particularly brightly coloured.

caught in a net with no chance of escaping, even if they remain beneath the water (the same does not apply to the bony fishes in which the *operculum* or *gill cover* protecting the gills carries out the task, by its specific movement, of pumping water from the mouth to the respiratory organs, and thus of supplying oxygen).

Three hundred and fifty million years ago, when the waters of the earth were seething with life and plants were beginning to develop on dry land, a clumsy animal, whose body resembled that of a fish, left the sea and ventured on to the land. There were many dangers and unknown factors threatening its survival: a less uniform climate than in the sea, a bleak and impassable surface, a possible food-shortage, but most of all the risk of not being able to breathe. In fact, oxygen can reach the cells only in the form of a solution: for a land animal, which is constantly threatened with seeing the water contained in its tissues and in the fluids circulating through its body evaporating, the possibility of breathing thus represents a limiting factor in its life. The amphibians, which derived from that early explorer of the land, preferred from then on to remain in contact with the aquatic environment from which they hailed: in the initial stages of their life they in fact breathed through gills which were only replaced

by *lungs* during a second stage. These are sac-like organs, somewhat akin to sponges which, by being situated within the body, manage to remain constantly covered by a very fine layer of water. It was only with the process of evolution whereby the skin thickened and protective measures were added (scales, feathers, fur, and so on), reducing the dangers of evaporation, that water was permanently abandoned, and a totally pulmonary respiratory system developed, even in the newborn.

But with this development two questions immediately come into play: how does the oxygen absorbed by the gills of a whale-shark or the lungs of a python reach the cells in the tail which are so remote? And how do the alimentary substances, suitably transformed into simple nutritive substances, manage to be used by all the parts of the body?

The missing link in our chain is therefore the *vehicle* which conveys to the cells the necessary fuel (digested substances) and oxygen, freeing it at the same from the harmful waste products. This vehicle is the *blood* which, by flowing uninterruptedly in special vessels, reaches every cell in the body.

When life made its appearance on the earth, there was no problem as far as the conveyance of matter, whether useful or harmful, was concerned. The first

Birds' beaks are remarkably adapted to the diet of the various species. Top left: a macaw's beak which can break up tough seeds and is also used to support the animal. Below: the beak of the spoonbill is designed to hold darting prey. Top right: a cormorant's beak, designed for catching fishes, with a crop which can be dilated. Below: the beak of another sea-bird which feeds on fish, the frigate-bird.

The vulture's beak (left) is perfectly designed for the diet of this creature and enables it to rip and tear flesh; the function of the jabiru's beak, on the other hand, is somewhat mysterious (top right) and it seems to be an encumbrance for this bird. Below right: the head of the duck-billed platypus, that strange Australian mammal.

living beings were unicellular organisms which lived in an aquatic environment capable of supplying them with whatever they needed: water, food and oxygen.

Difficulties arose with the emergence of the first pluricellular organisms: but here, too, up to the point where the number of cells was not too large, the exchange of substances occurred on the basis of simple physical phenomena such as diffusion. It was only when the increased size of the animals entailed long-distance transportation that, among other things, there had to be a rapid development (diffusion is such a slow process that a drop of blood would take at least six months to travel a distance equivalent to the stature of a man), and the circulatory organs started to appear. In all the vertebrates the circulatory apparatus is of the *closed* type, in other words, the blood flows within a system of *vessels* (*arteries* and *veins*) from which it never exits. Red in colour because of the presence, in the corpuscles, of a red substance (*haemoglobin*), it conveys the oxygen, with which it has re-supplied the cells on the branchial or pulmonary level, handing it on to the tissues, and absorbing the carbon dioxide in it. As well as the *red corpuscles*, the blood also contains *white corpuscles*, which handle the task of combating possible infections, and *blood platelets* which are specialized in the task of coagulation. The blood combines its respiratory function with

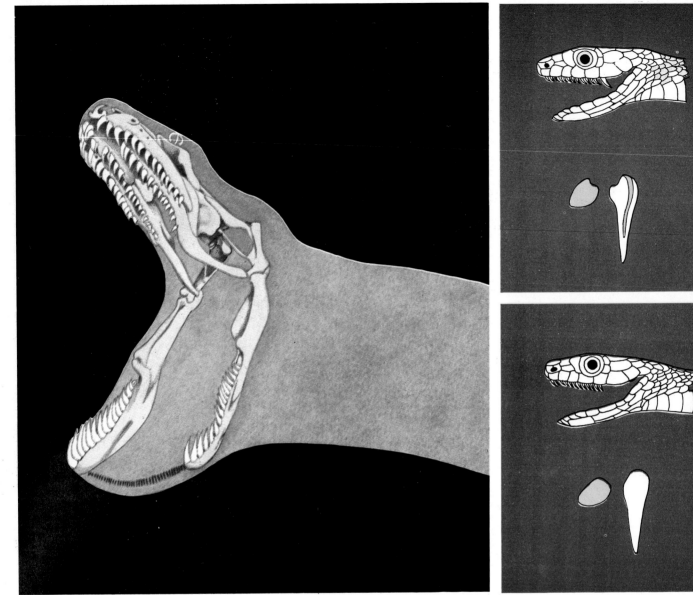

that of conveying substances absorbed at the intestinal level to all the cells in the organism. While the structure of the blood vessels is similar in the various classes of the vertebrates, the *heart*, that is, the organ which propels the blood, shows modifications which correspond with the evolution which has taken place from the gills of the fishes to the lungs of the land vertebrates.

At this point it simply remains for the organism to eliminate the waste and the unusable products which are inevitably formed by its activity, in such a way that a constantly balanced situation is maintained within it. This task falls to the *urinary apparatus* which, by using structures with varying levels of complexity as it proceeds up the evolutionary ladder (from the fishes to the mammals), handles the job of filtering the blood, extracting everything from it which is no longer serving any purpose, and disposes of such waste products outside the organism.

But to remain alive, it is not enough just to eat, breathe, and have a perfectly functioning heart and a precise and efficient system of purification (or depuration); at any given moment it is necessary to be informed about any changes in the outside world and in the actual internal situation of the organism. In the most simple animals, this informative system (the *sensory*

Snakes which swallow their prey whole without chewing them first need an exceptionally large buccal aperture; this is provided by a special articulation of the mandible with the rest of the skull (left). These creatures may have poisonous teeth or they may be toothless. Top right: the head of an opisthoglyph colubrid with grooved poisonous teeth. Below: the head of an aglyph colubrid.

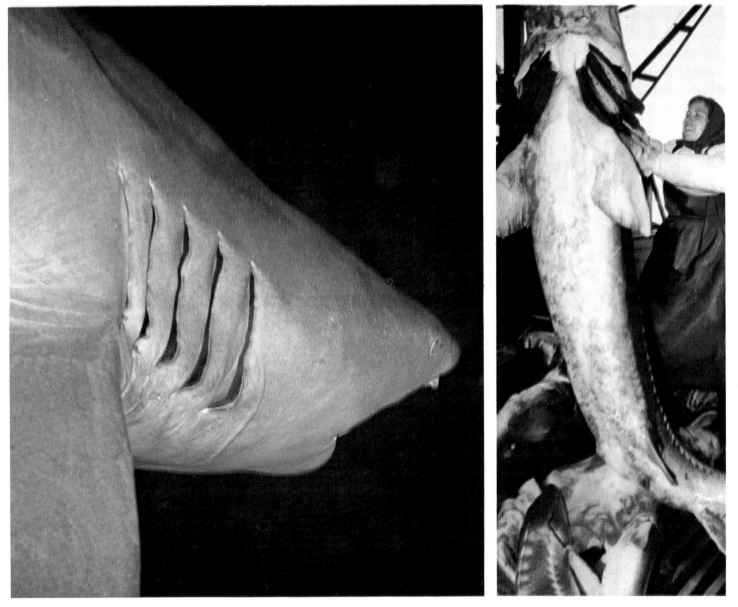

Fishes breathe by means of gills which enable the animal to use the dissolved oxygen in the water. Left: a shark (Carcharhinus) with five visible gill slits. Right: a Caspian Sea sturgeon whose gills can be seen.

organs) can only receive very general stimuli, such as light, darkness, warmth and cold. The vertebrates on the other hand (although highly specialized structures of this type do exist in some classes of invertebrates) can receive more detailed information such as distance, dimension, the colour of things, and so on. The fishes in the water, on top of their basic five senses, have nothing short of a real 'sixth sense' which, in the form of a series of canaliculi situated immediately beneath the epidermis, tells them about the most imperceptible movements of the water around them. The eye of these creatures is not only very large in relation to the size of the head, but can also see in different directions at the same time; the ear, however, has an imperfect structure even if, despite this, fishes can hear every slightest sound, as fishermen and anglers know only too well. Their sense of smell is highly developed as well: this fact is confirmed by the way sharks can detect the smell of blood at great distances and blue-fish are lured by shreds of fresh flesh belonging to other fishes. The same cannot be said about taste, which is entrusted to organs situated outside the mouth: fish therefore taste their food even before swallowing it! The sense of touch is in the hands of the barbels which hang from the chin like whiskers. Amphibians and reptiles have fairly poorly developed senses: touch, taste and

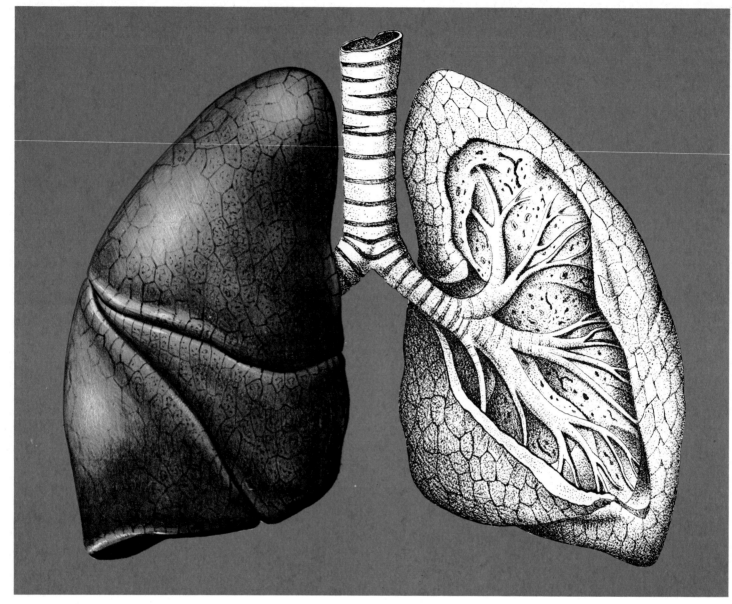

smell are virtually non-existent; hearing likewise leaves a lot to be desired, although it is somewhat more effective. Only their sight works reasonably well, with large eyes with contractile pupils which enable some species – the gecko for example – to detect objects in the dark. A feature of the rattlesnakes and a particular species of viper (the pit-viper) is their capacity to detect the presence of a warm-blooded prey, even in pitch-black conditions, by means of some tiny thermosensitive pits situated on the head between the eye and the nostril. As far as perfect vision is concerned, first place goes to the birds; a swallow can catch an insect while flying at full speed; and a grebe can pursue the prey it has picked out under water; but these are just two examples of a thoroughly general situation. And this is not only because the birds have remarkable eyes, but also because they are able to see objects at very close quarters, because they have an exceptionally wide field of vision and because they can detect movements at a distance. As far as their other senses are concerned, the only one which is reasonably developed is the sense of smell.

In the mammals the sensory organs are better developed than in any other class of vertebrates. In this animal group the sense of touch is variously positioned and developed: the elephant touches things with those tiny finger-

The lungs are the organs which enable the land-vertebrates to breathe. Inside the lungs, in tiny cavities called alveoli, oxygen is introduced into the blood: this being the case, however, the oxygen must first be dissolved in the liquid in the alveoli, a procedure which harks back to the land-vertebrates' derivation from aquatic forms of life which breathed the oxygen contained in the water.

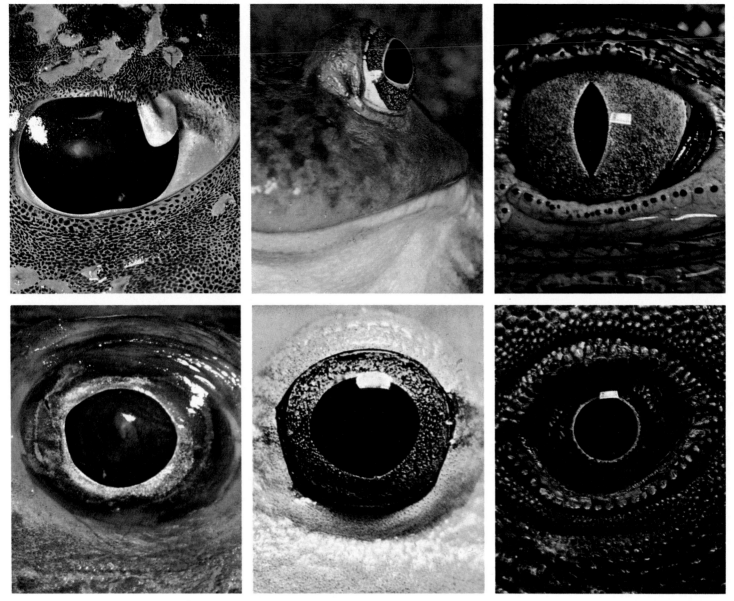

The eyes of the vertebrates have a special and complex structure, which is matched in the animal kingdom only by the molluscs. Top left: the eye of a cuttlefish. Below: the eye of a perch. Centre (top and below): the eyes of anuran amphibians (frogs). Top right: a crocodile's eye. Below: an iguana's eye (reptile).

like appendages situated at the end of its trunk; the cat is guided by the highly sensitive vibrissae which cover its upper lip; in flight, the bat receives tactile data from the patagium, that very fine membrane that stretches between the toes of its front limbs. As far as the sense of smell is concerned, many mammals (carnivores, rodents and insectivores) are equipped with extremely developed systems, and others are almost totally without this sense. The weaker and more timid the animal, the more acute its sense of hearing: as a general rule, the ear is externally protected by a pavilion which, in certain species (dogs and cats) is extremely mobile and can be turned to pick out the direction from which sounds come. Many mammals (mice, rats, moles, bats) can hear ultrasounds (in excess of 20,000 vibrations per second). Bats use the echo of their ultrasonic cries, which they emit from the larynx in flight, to detect obstacles and prey. The otter and the beaver have pavilions which can be closed; and the aquatic mammals do not have them at all. The eyes are generally more developed in mammals which lead nocturnal lives; some (e.g. the mole) which live underground have very small eyes. Almost all of them can see colours; only those with nocturnal habits have either very reduced or no colour sensitivity.

All the sensory organs and the organs of movement are controlled by a sort

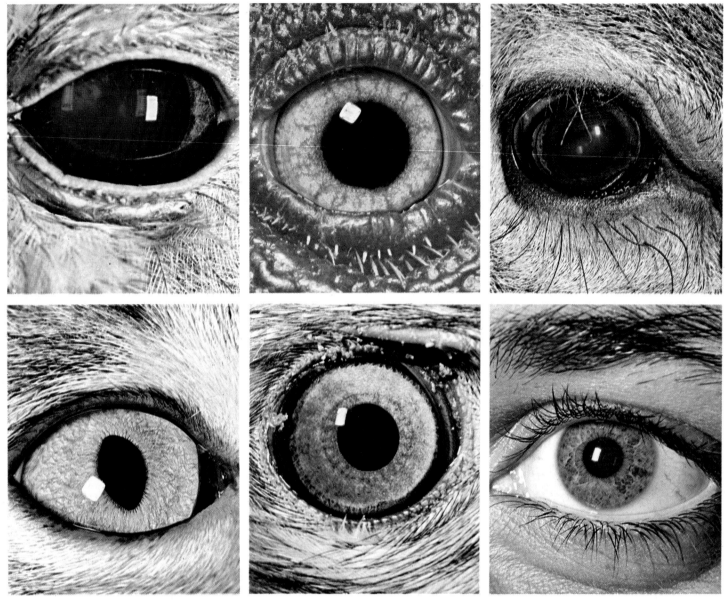

The eyes of higher vertebrates. Top left: a tawny owl. Below: a cat. Top centre: a hen. Below: a buzzard. Top right: a deer. Below: a human eye.

of telephone exchange (the nervous system) which in the vertebrates consists of three elements: the *encephalon*, formed by the brain and the cerebellum which are protected by the skull; the *spinal cord* or *medulla spinalis*, a long element contained in the spinal column; and the *nerves* which fan out throughout the whole body. Of course, the development of the individual cerebral areas varies depending on the degree of evolution attained by the class in question and on the type of life lived by the animal. This means, for example, that in fishes the cerebral formations responsible for the sense of sight and the sense of smell are dominant; in birds the sense of sight reigns supreme; and mammals in general have a larger and more highly developed brain. Also mammals and birds are able to keep their body-temperature constant and can thus remain alive in the most varied environmental conditions.

The last task which nature delegates to all living things is the propagation of the species in time and space. And for each living being, in the environment where it lives, there is a specific technique for finding a suitable mate, building a nest, and guaranteeing food, protection and instruction in what life holds in store, first for the eggs and subsequently for the young. Once again the aquatic environment requires more simple procedures. Fishes, and amphibians which

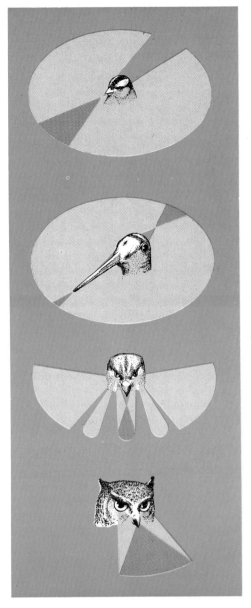

Sight is the principal sense of the birds, especially the birds of prey (centre and right), which use their eyes to detect prey. Left: from top to bottom: a sparrow, with mainly front vision; a woodcock, which can also see backwards; a falcon and an owl, both birds of prey with highly developed sight.

actually return to the water precisely to mate and rear, simply scatter eggs and spermatazoa in the water: chance decides if the two will meet to give birth to a new life. Hence the need to lay very large numbers of eggs to defy all the untoward circumstances awaiting them, and the need to equip them in such a way that they are movable rather than anchored, or to camouflage them against possible predators. Among both fishes and amphibians, however, we find more responsible parents who, as well as building proper nests, carefully sit on the fertilized eggs until they hatch: this is the case with the sea-horse, the stickleback and the pipe-fish, and, among the amphibians, the so-called South American toad or pipa which carries the tiny eggs and the tadpoles which hatch from them in special small cells on its back.

Where reproduction is concerned, the reptiles introduce three new features: *internal fertilization* (and thus a considerable saving in the number of eggs produced); the *shell* protecting the egg, and in some specimens only, *viviparity*, which hallmarks the typical reproductive system of the mammals. As a result of these devices, the reptiles have eliminated the problem of adequate temperature, which sets a fairly sharp limit where the free laying of eggs in water is concerned (the temperature limits for the survival of the eggs and fishes

291

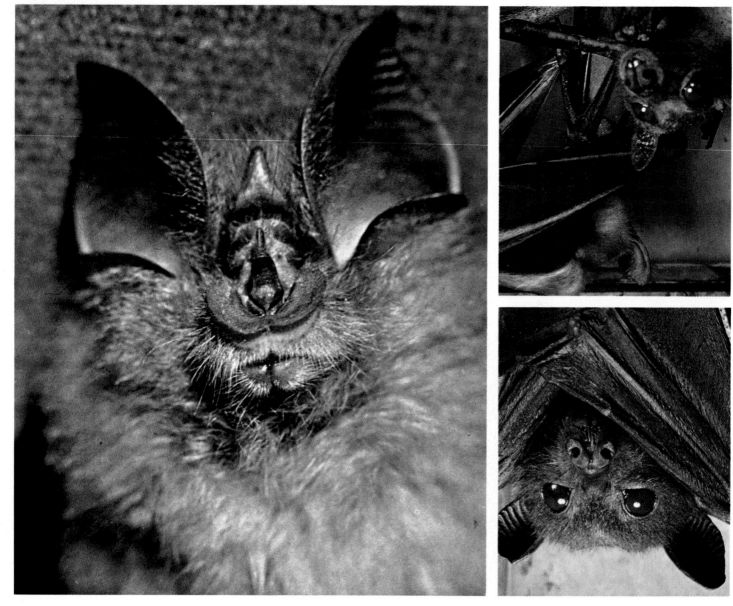

and amphibians are in fact quite tight). As a direct consequence of internal fertilization and hence of the encounter of two individuals of the opposite sex, there is all the more need, from this class upwards, for courtship rituals, with the possibility of fierce combat for the conquest of the female.

The next step is taken by the birds which, in a way, are the first creatures to set up the family unit. Once the future wife has been won (and here, too, there are countless techniques used, from the dances of the herring gull to the love-songs of the nightingale, to the brightly coloured plumage of the female red-necked phalarope), the nest is prepared; here the eggs are laid and sat on, and the young reared and defended until they can fly off on their own in search of food.

Birds, however, are not the only creatures to lay eggs, as we have already seen. In the mammals too – and this is the most highly developed class of vertebrates – there are egg-laying animals: these are the monotremes, nowadays represented by just two animals, the spiny anteater and the duck-billed platypus. These are followed by the marsupials, which are equipped with an incubator pouch or marsupium in which the mammary glands are situated and where the young are kept until they are fully developed. The final rung on

Bats, which are twilight and nocturnal creatures, use their hearing to detect obstacles and prey (they feed on insects). In fact they emit ultrasounds through their nostrils with a frequency of about 30,000 vibrations per second; when these encounter an obstacle they bounce off it and are then received by the ears; the system used is like radar. Left: the lesser horseshoe bat. Right (top and below): the common bat.

*Various types of birds' eggs: (A) pitta;
(B) wren; (C) black skimmer; (D) fork-
tailed king-bird; (E) migratory thrush;
(F) mistle thrush; (G) small crested
kingfisher; (H) peregrine falcon; (I) king
penguin; (J) white-tipped plantcutter;
(K) great horned owl; (L) foolish
guillemot; (M) roseate tern; (N) cuckoo;
(O) American oriole; (P) MacGregor's
bird of Paradise; (Q) emu.*

the evolutionary ladder of the mammals is represented by the so-called
placentals. Here the young develop and grow within the mother's body, and are
nourished via the placenta, a spongy organ with considerable vascularization
which is directly connected to the circulatory system of the mother. Generally
speaking, in this class the parents look after their young, for a period which is
extremely long in man's case – children are not recognized as independent until
they are twelve years old – and which varies between maximum and minimum
values in the different systematic orders, depending on the habits involved.

Mice, rats, antelopes, giraffes, and so on, which, like all herbivores, are
constantly being preyed upon by carnivores, must waste no time in learning,
from a very early stage, how to take to their heels. The same does not apply to
the carnivores which, as hunters, require a longer period of instruction and
training from their parents or from the adult members of the herd.

The vertebrates in particular represent one of the two apexes of this
organization of life (the other apex is represented by a class of invertebrates,
the insects), which is basically set forth via three types of behaviour which
are characteristic of only these animals: the establishment of *power-based
hierarchies*; the establishment of *personal bonds*; and the establishment of

forms of *solidarity*. And this state of affairs is bolstered by many examples. It is common knowledge that in a chicken-coop some chickens peck others, especially when there is food around, and that the pecked chickens withdraw from the fight without a struggle. Likewise among groups of fishes, reptiles and mammals these differences in importance – in other words hierarchies – are created. From adolescence onwards a stronger and powerful leader is acknowledged to whom the others submit themselves in a precise hierarchical order. The highest-ranking animal will eat first and alone, and have a larger area around him, and be able to choose his wife before the other members of the group. Immediately afterwards the seconds will enter the contest, while the least important animal in the group will receive only the leftovers and the least courteous attentions (blows from horns, pecks, etc.)

If the birds are to be credited with having invented the family formed by two parents and their own young, the first instances of personal bonds were to be found as early as the oldest vertebrates, namely the fishes. Among the jewel-fish, for example, pair-bonds are formed which last all their life and which will not even be broken by the death of one of the pair, inasmuch as the widowed party will not look for a new bond. Among the mammals it is not rare to find

Nest-building is one of the essential activities which animals carry out to look after their young, and thus to ensure the survival of the species. Top left: a blackbird's nest. Below: an eider duck's nest, lined with feathers taken from the female's breast. Centre: cormorants' nests. Top right: a waver-bird's nest. Below: the grebe's floating nest which is anchored to the bank.

In some cases the remains of animals which lived millions of years ago have been preserved in rocks in such a perfect way that it is still possible to identify their anatomical structure. This is the case with the Mene rombea *shown in the illustration, a fossil from the famous Bolca deposit in northern Italy.*

friendships established between animals which are traditionally regarded as enemies: dogs and cats, cats and mice have often become inseparable companions simply because they have shared their very first days of life together. The last typical feature of the behaviour of the vertebrates is solidarity. Whereas in the insect dynasties it is in fact not infrequent to come across privileges as far as the distribution of food is concerned (those who have, giving to those who ask for it), in the case of the vertebrates we find cases of real mutual aid. Dolphins and sperm whales regularly help wounded companions by assisting them to the surface to enable them to breathe; the walruses invariably help those among them who find themselves in difficult circumstances; elephants look after members of the herd who are injured by supporting them on both sides and helping them to walk. The life of these animals is influenced by the *environment* or *habitat* which in turn will be influenced and modified by animals. The environment is first and foremost dominated by the *climate* with its factors of temperature, humidity and pressure; by conditions of light and aeration; and by the presence or shortage of given nutritive substances. The way in which different animals may adapt themselves to one and the same area or the transition from one habitat to another which is often

295

achieved by many animals in the course of their lives, is evidence of how every living being has a precise task to fulfil in the position allocated to it by nature.

From the geographical point of view the vertebrates are distributed in a way which corresponds with the subdivision of *terra firma* into six regions.

There is a *Palaearctic region* which includes Europe, northern Africa and northern Asia. Typical birds here include duck and pigeons, and typical mammals are donkeys, bears, sheep, and deer. The *Nearctic region* of North America is very similar to the Palaeartic region. The American bison resembles the central European bison, which was virtually wiped out during the Second World War. We also find bears, but the amphibians are considerably different. Central-southern Africa, India, southern China, Indochina and the adjacent islands form the *Palaeotropical region*, undisputed realm of the large Catarrhina (orang-utans, gorillas and chimpanzees), elephants, rhinoceroses and the big cats (lions, leopards, etc). Hippopotami and giraffes, ostriches and guinea-fowl are African in origin. Peacocks, tigers and, among the crocodiles, the gavial are Asian in origin. Central and South America form the *Neotropical region*, typified by the Platyrrhina, monkeys with long prehensile tails, the marsupial opossum, and armadillos, three-toed sloths and anteaters. There are

Top: the fossil skeleton of Eryops megacephalus, *an amphibian discovered in the Parmian strata in Texas which lived two hundred million years ago. Below: the fossil remains of* Leptolepis knorrii, *from Solnhofen in Germany. This animal is the connecting link between the group of Holosteian fishes, now almost extinct, and the Teleostei, which cover most present-day fishes.*

*An overview of the main classes of vertebrates currently inhabiting our planet. Left: two illustrations showing fishes (top, the butterfly fish; below, a sea-horse). Centre: two amphibians (*Rana dalmatina *and* Rana temporaria*). Right: a reptile (coral snake).*

no insectivores at all, and birds include humming-birds, toucans and parrots. The *Notogaeic region* includes Australia, New Guinea, Tasmania and New Zealand. As far as the mammals are concerned, the monotremes are very widespread indeed here (spiny anteaters, duck-billed platypus) as are the marsupials (kangaroos, Tasmanian wolves); where the birds are concerned the Notogaeic region numbers among its species the emu, the lyre-bird and the cassowary. New Zealand has no snakes, nor does it have any original land-mammals; on the other hand it has birds which cannot fly, such as the kiwi. Lastly, the island of Madagascar and its neighbouring islands form the *Malagasy region*, characterized by the presence of large numbers of lemurs and, conversely, by the absence of large carnivores, elephants and rhinos.

Let us summarize the basic features of the vertebrates divided into classes based on the chronological order in which they appeared on earth:

Cyclostomata: 45 species. Jawless with a circular mouth like a sucker. Cartilaginous skeleton. Either predators or parasites. The predatory species include the river and sea lampreys.

Cartilaginous fishes: 600 species. Mouth with teeth and jaws. Cartilaginous skeleton. Transversal, unprotected gills. They include carnivorous and

The illustrations show the most evolved classes of vertebrates: the birds (left, a robin), which are very different from the reptiles but nevertheless retain many similar anatomical structures; and the mammals, to which man belongs, which are at the top of the zoological ladder.

swimming species (dogfish, sharks) and more sedentary bottom-dwelling species (rays, electric rays). Internal fertilization.

Bony fishes: 20,000 species. Bony skeleton. Gills protected by a gill cover. Swimming bladder which regulates the gases. Usually external fertilization.

Amphibians: 2,800 species. The larvae, which have gills, live in water. The adults have lungs and live on the land. They have four limbs and an unprotected skin. They include frogs and toads (tail-less), salamanders and newts (which have tails).

Reptiles: 5,900 species. Skin covered with horny scales. Internal fertilization. Include tortoises, crocodiles, snakes, lizards and the tuatara.

Birds: 8,600 species. Covered with feathers with front limbs transformed into wings. The mouth has a horny beak. They have a constant body temperature. They sit on the eggs they lay and rear their young.

Mammals: 6,000 species. These are animals covered with hair or fur, having mammary glands and a highly developed nervous system; they have a constant body temperature and show a great variety of body structures, depending on their habitat. This class also includes *Homo sapiens*, who evolved from the order of primates or simians about a million years ago.

THE MARINE VERTEBRATES

The sea is where life started (the oldest fossil remains belong to marine organisms), and the sea is also the largest environment known to us, with its shifting surface covering about 71 per cent of the earth's surface.

Thousands of billions of years ago, as soon as the earth's crust had cooled sufficiently, the first rains began to fall. Day and night for many long months, years and centuries, it rained and rained, filling every hollow and basin. This rain increased salinity and gave the sea its bitter taste. In fact, from those remote times onwards, the rain-water started eroding the rocks which showed above the surface, and extracted from them minerals which it then carried down towards the sea.

But a more impressive event was still to come, which was the moment when the first molecule capable of reproducing itself was born from those waters – in other words, the first *living being*. After this momentous development, things happened step by step. First of all, the birth of microscopic *plants*, with large amounts of green *chlorophyll*, capable of utilizing solar energy in the sea-water and carbon dioxide in the air to produce the organic substances required by them; next, the appearance of organisms (the first *herbivorous animals*) which discovered that they could feed themselves by eating the plants; and lastly, the development of other animals capable of surviving by eating their own kind (the first *carnivorous animals*). Thus the first *food chain* came into being. Even when life started to colonize dry land, the sea lost none of its original importance: fishes, amphibians, reptiles, and even warm-blooded birds and mammals still carry in their veins a saline fluid, the ingredients of which (sodium, potassium and calcium) are based on almost equal ratios to those of sea-water; similarly, the skeletons of all the vertebrates, which are rigid because of the presence of calcium carbonate, apparently hark back to the waters of the oceans, which contain large amounts of this compound, which filled the Palaeozoic basins. And the same goes not only for the protoplasm of which every single cell is made, but even for the development of every human being which, in the womb, repeats the stages through which our species has evolved, from the earliest aquatic forms up to our first actual ancestor.

In the course of time some land-animals returned to the ocean: these were the mastodontic dinosaurs with their webbed feet and long snake-like necks, of which we are still reminded even today by the giant sea-turtles which are to be found many miles out at sea. Much later, perhaps fifty million years ago, even some mammals abandoned their life on dry land and plunged back into the aquatic habitat: here we are talking about the sea-lions, seals, dolphins and whales which nowadays inhabit the sea. Lastly, even man made a move towards the sea. As he was unable to return there to live in a physical sense, like the seals and walruses before him, he constructed boats and ships which can ply the surface; and man has also ventured into the very abysses in the oceans, helped by effective instruments capable of making up for his own shortcomings: oxygen tanks to breathe with; mechanical eyes and ears to interpret colours and sounds; hydrodynamic fins to enable faster movement through the water.

'Big fish eats little fish' is said to be the law of the sea: a rather cruel but necessary law designed to maintain a constant balance among marine creatures which all have one specific feature in common: they are all tremendously prolific. As a result, the organization of marine life is like a pyramid at the base of which there is a great multitude of plants, followed by a lesser number of invertebrates and vertebrates' larvae which feed on the plants and which, in

turn, are easy prey for the even smaller numbers of carnivorous vertebrates.

The ways of life of the marine animals make it possible to describe a *benthonic environment* which applies to those animals which live to a greater or lesser extent allied to the bottom (examples being the angel-fish, sole, rays, moray eels, electric rays, not forgetting, of course, a whole host of invertebrates), and a *pelagic environment* which applies to those animals living in the open sea, either swimming freely in it or getting themselves carried hither and thither by the swell and the currents because they have little locomotive strength. The benthonic species are often nicknamed *fake* species: in fact, instead of swimming upright like most fishes, they are, as it were, upturned and swim on their side, with the result that in order to be able to see their eyes have had to change position and are situated on the side of the head on the upper surface. Lying flat on the bottom, their fins shift the ooze or sand so as to blur the outline of their own body and merge it with the environment around them; indeed they often change colour as they gradually move from one type of sea-bed to another during the daylight hours.

Based on the various conditions of salinity, temperature, light and pressure, the marine environment may be divided vertically and horizontally into

The tendency towards gregariousness, i.e. towards living in fairly numerous groups, is a feature common to many species of fishes. This photograph shows a shoal of Plotosus arab, a poisonous catfish living around the Seychelles.

The selachians, which include the dogfish and rays, are the most conspicuous group of fish with primitive characteristics. They have a cartilaginous and non-bony skeleton like almost all fishes, and live in the sea. Top and below left: the small spotted dogfish (Scylliorhinus canicula *). Below centre: a raiid, a selachian with pectoral fins which are so wide that they even stretch around the mouth. Right: a dogfish.*

various zones, namely: the *surface waters*, the *coastal waters*, the *open waters* and the *deep waters*.

To all appearances lifeless, the surface waters on the contrary seethe with an incalculable abundance of creatures (and plants) which often simply drift about. The factor responsible for this abundance is the generous amount of sunlight which it receives. Along with a vast host of microscopic invertebrates, shoals of tiny silvery fishes (mainly herrings, mackerel and sardines) spend their lives permanently fleeing from more voracious predators (tunny-fish, wolf-fish and even sharks) which locate the shoals thanks to the impatient cries of the seagulls wheeling above them.

In the sea the alternation of the seasons is especially evident. In spring the eggs of many bottom-dwelling or deep-sea fishes head upwards towards the warmer, sunnier surface waters; shoals of salmon leave the deep waters of the Pacific Ocean and swim upstream through the busy waters of rivers where they lay their eggs; cod swarm around the shores of Iceland; many birds driven here and there across the wide oceans by the cold winter months reassemble their ranks on cliffs and crags and islands, and set about building nests for their future families. Whales likewise head for the surface, drawn there by the

swarms of eggs laid there by tiny shrimps, accompanied by huge flocks of phalaropes, migratory birds which vie with the whales for the same food.

Summer is the seals' heyday: the seals in the eastern Pacific converge on two tiny islands to deliver and rear their young. Not until winter approaches will they set off in search of new and more plentiful sources of food. The second smaller environment, the coastal waters, likewise seethes with life, for two very simple reasons: plentiful plant-life, again benefiting from the sunlight, and plentiful food from above. Near the coast life is affected by the tides and the nature of the coastline itself: in fact rocky coastlines offer better guarantees of anchorage and water (based on the ebb and flow of the tide) than sandy or muddy shores. The animals which inhabit this zone must naturally be able to tolerate the possibility of dry conditions. They can thus be categorized as *permanent inhabitants* (these are all invertebrates) which can also survive perfectly well out of the water; *visitors from the sea* (like the mudskipper, a fish which catches insects and crustaceans in the tropical salt-marshes but which, if the need arises, can beat an extremely hasty retreat because of its fine performance as a swimmer); and other *visitors*. These latter are not fishes, but reptiles, birds and mammals. The first group includes the giant turtles which

Top: two raiid selachians: a ray (left) and an electric ray or torpedo (right), which can emit powerful electric shocks. Below left: a smooth hound (Mustelus canis), a dogfish highly prized for its tasty flesh. Right: the fearsome tiger shark. Opposite: top left: the Garibaldi fish; centre: the clupeiform Agonus cataphractus; below: the raiid Rhinobatus productus; top right: the anguilliform Gymnothorax mordax; centre: the painted comber (or bass); below: Dover sole.

make their way up the hot sandy beaches in the tropics simply to hide their eggs until they hatch, and crocodiles; the birds include a large variety of species (seagulls, plovers, terns and petrels) which halt awhile on the cliffs above the shore to lay their eggs; and the mammals include many members of the order of Pinnipedia (seals, sea-lions and walruses). These latter generally prefer the coastlines in both hemispheres which are washed by cold waters; of course there are exceptions; the monk seal and a particular species of sea-lion, for example, live respectively in the Mediterranean and the Galapagos Islands, on the Equator. The adaptations which enable the Pinnipedia to live satisfactorily in both environments concern the limbs, which are webbed, and the tail which acts as a rudder in the water and as a support on land.

The real realm of the fishes is the so-called open sea, where light plays the leading part; it is in fact in the strip between the surface and the 300-feet (100-metres) mark (where it consequently encourages the growth of plant-life and hence the development of herbivorous fishes), and becomes progressively diminished below the 300-feet (100-metres) mark where carnivores abound.

Structured and equipped in an exemplary fashion for the environment in which they originated and where they live, the fishes present highly varied

Fish which spend part of their lives in sea water, part in fresh water are not uncommon. The best-known example is the eel (right), which spends its adult life in fresh water but when the time comes to spawn migrates to lay its eggs in the Sargasso Sea. The newborn eels then return to fresh water where they undergo a double metamorphosis (left: the 'blind' stage).

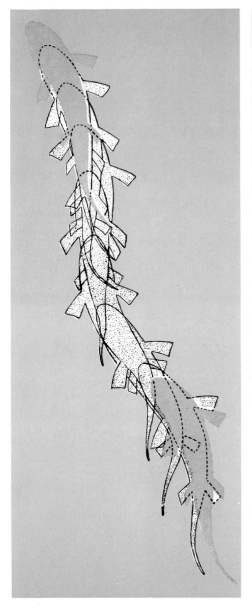

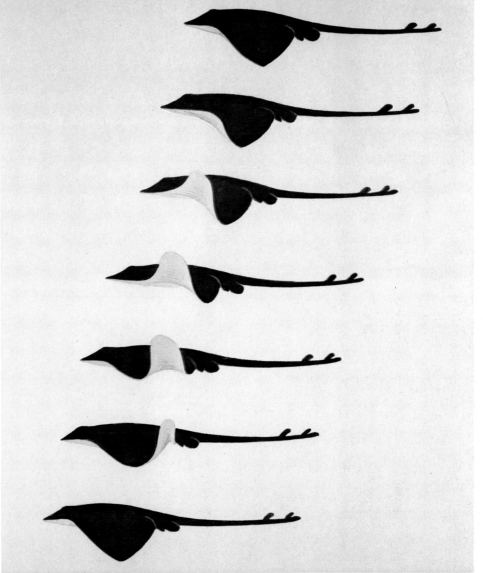

Fish usually swim by means of a swinging motion of the tail fin accompanied by a similar movement of the body (left). However, there are many exceptions to this, like, for example, the ray (right), which progresses by moving its wide pectoral fins.

forms, structures and colours. Depending on the nature of the skeleton, they are generally classified as *cartilaginous fishes* and *bony fishes*. Among the former we find the largest members of the category: the whale-shark and the basking shark which grow terrifyingly large (up to 42 feet [14 metres] in length) but are considered harmless, feeding solely on plankton, that is, on that multitude of tiny animal and vegetable creatures which drift freely about in the water in which they are suspended. The bony fishes never grow to this sort of size: the largest in fact are the tunny-fish and the sword-fish which reach a maximum length of 13½ feet (4.5 metres) and may weigh up to 1,430 lb (650 kilos). Even today, some of them have very primitive features: along with the well-known coelacanth (a specimen of which was caught in South African waters in 1938), which has been defined by experts as a living fossil, because they maintain that it has lived unchanged for the last three hundred million years, there are certain species of sturgeon living in particular in the northern hemisphere (Russia, Romania) which periodically migrate from sea-water to fresh-water rivers where they lay their eggs.

Mimetism, camouflage, merging of the body-contours with the immediate environment, temporary changes of colour, defence mechanisms consisting of

*The fishes which inhabit the abyssal regions of the sea have certain peculiar characteristics, like huge eyes and mouths, long appendages and luminous organs; they are always predatory. Left: the silver hatchet fish (*Argylopelecus hemigymnus*). Top right: Chauliodus sloanei. Below: a* Lasiognathus saccostoma.

armour-plating, stings and spined fins sometimes connected to poisonous glands, powerful teeth: these are just some of the devices supplied by nature in a world where 'might is right'. We should add to this list a unique defence mechanism, which is the capacity to emit electric shocks used by the electric eels, rays and catfish. But the feature which hallmarks aquatic creatures of this type living in open waters is their swiftness. Among the speediest of all are certain carnivores belonging to the scombroid family, which are equipped with perfectly hydrodynamic slender bodies and powerfully muscled tails; and we should not omit to mention the tunny-fish, bonito and king-fish. But the outright speed-merchant (averaging 100 k.p.h. [60 m.p.h.]) is the American scombroid of the family Histiophoridae, whose dorsal fin has the strange habit of folding into a sort of groove when the fish is travelling at high speed, and standing upright like a sail when the fish is resting.

As well as swimming, some open-water fishes can also fly. Such forms are the exocoetids or flying fishes, and triglids or flying gurnards. The former take to the air by means of a thrust from the tail which vibrates extremely rapidly just as the fish leaps out of the water; once airborne they are like gliders, and gently vibrate their stretched fins to alter direction; the latter on the other hand use

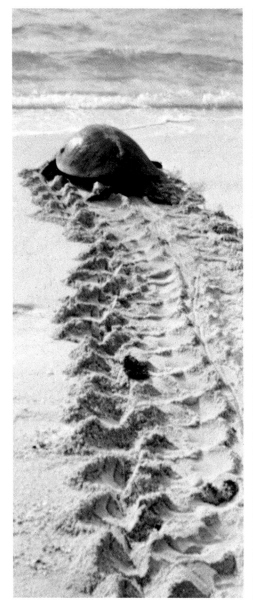

There are not many species of marine reptiles, and those there are often spend part of their lives on terra firma. *The sea-turtle lays its eggs on land (left). The sea iguanas of the Galapagos Islands (right), when not hunting for algae in the sea, join up in large groups on the rocks.*

their fins as proper wings once they have been launched in the same way by a thrust of the tail. If we look at the other classes of vertebrates, the only present-day reptiles which inhabit the oceans are the sea-snakes and sea-turtles (and we have seen how the latter return to dry land to lay their eggs). The sea-snakes, better known as hydrophids (Hydrophinae), mainly live in the tropical regions of the Indian and Pacific Oceans. As is to be expected, the environment has determined their somatic features: the tail is compressed at the sides to facilitate swimming and their colour is similar to that of the algae among which they hide. Often more than 6 feet (2 metres) in length, they are very dangerous for man, for whom their bite may be fatal.

There are not many birds which live in the vast skies above the oceans; but as far as the mammals are concerned we should mention the group of Cetacea, which includes whales, dolphins and porpoises. Colossal in size but nimble in their movements, the Cetacea breathe air but can remain in a state of apnoea (i.e. holding their breath) for considerable periods of time: forty-nine minutes in the case of a blue whale, seventy-five minutes for a sperm whale, and a hundred and twenty minutes in the case of a bottle-nosed dolphin. When submerged, the organism of these creatures is 'slow-revving': the heartbeat is

slowed down, and the oxygenization of the tissues is reduced in such a way that the air-supply stored in their large lungs is not used up too quickly. In addition they have a very thick fatty layer which uniformly covers the outer parts of their body, with the dual function of giving them a hydrodynamic shape and keeping their body-temperature constant.

The whales have different types of teeth: some (sperm whales) have enormous teeth, while in other species the teeth are replaced by whalebone or *baleen*, a sort of fringe with several layers of flexible, horny, long laminae, among which the plankton on which they feed is trapped.

Traditionally considered to be friends of man, dolphins live in shoals or schools which frequently follow shipping routes. Their hind limbs are so rudimentary that they are incorporated in the musculature; they have a crescent-shaped nostril on the top of the head which closes on contact with water; dolphins can also steer their way among rocks with great skill thanks to a highly efficient organ which can pick up the echoes emitted from nearby objects. They are also swift swimmers and this would seem to be owing to the structure of their skin which is covered with small, thin tubes filled with a spongy substance which enables a dolphin to adapt the profile of its body to that

The dog-like muzzle and vibrissae around the mouth (below centre) remind us that seals are mammals, but their finned limbs, the absence of a pavilion protecting the ear and their hydrodynamic shape render these creatures perfectly in harmony with life in the water. Below right: the polar bear which, second only to man, is the seal's worst enemy.

Unlike the seals, with which they are often confused, the eared seal or sea-lion (Otaria) has a small pavilion in the ear. Top left: the Alaskan fur seal (Callorhinus), a species whose numbers have been reduced by indiscriminate hunting. Below: a sea-lion. These animals gather in April, the reproductive period, in given areas of Alaska and the Pribilof Islands, where they deliver their young (right).

of the water through which it is swimming and thus reduce friction to a minimum.

Up to 1934 man's knowledge of the sea was thus confined to the three zones whose most intriguing aspects we have just discussed. There was still a vast region to be explored, which was some miles deep in places and which accounted for almost half of the area covered by the oceans: the abyssal region, beyond the range of the human eye. It was in 1934 that two explorers (William Beebe and Otis Barton) descended to a depth of 2,950 feet (932 metres) off the coast of Bermuda in a heavy steel bathysphere equipped with thick windows made of quartz. But the record for such dives occurred some time later and is held by the Swiss physicist Jacques Picard, who reached a depth of 35,300 feet (10,863 metres). With this achievement, many of the question-marks surrounding the ocean abysses were cleared up. Contrary to what had once been supposed, the sea-bed showed itself to be inhabited by large numbers of creatures suitably adapted not only to the total lack of sunlight but also to fairly high levels of pressure and to temperatures that are considerably lower than in the other zones. A direct consequence of the absence of sunlight is the lack of plant-life and hence the fact that all the abyssal fauna is carnivorous. The abyssal fishes are never very large (it is not known if this is because of the

none too plentiful food available, or the darkness which has halted growth, or because of a certain imbalance in the ingredients of the water in these regions); the shape of these fishes is never typically hydrodynamic either: they have huge mouths, armed with powerful teeth, which can dilate to hold prey which are larger than the actual predator; in some cases the eyes are so large that they protrude, in others so small that they are almost non-existent. But the features which render these fishes so grotesque are the tactile and olfactory appendages and tentacles which they clearly use to find their way around.

Another strange feature is that in some species (seven to be precise) the male is much smaller than the female and lives attached to his mate's body as a parasite, waiting for the moment to fertilize her eggs. But the most spectacular phenomenon presented by the abyssal regions to their explorers is the so-called phenomenon of *bioluminescence*. William Beebe in fact recounts that the dark depths of the oceans are often illuminated by flashes of light, glowing pin-pricks. The luminous organs responsible for this are very complex structures: in some the light is produced by bacteria living within the body of the creature; in others it is produced by special tissues usually situated on the lower surface of the body. The function of this cold light, the colour of which varies from

The Cetacea are mammals which have become perfectly adapted to marine life, to such an extent that they totally resemble fishes; the fact that they are mammals is evidenced by the fact that they are warm-blooded, suckle their young, and breathe air. Left: a dolphin, an animal of great intelligence. Top right: a killer whale, a fierce hunter. Below: a sperm whale, an enormous animal which, unlike other whales, has strong teeth.

A whale-hunt. These strange creatures, whose size is quite unique, live in schools in cold waters where they feed on plankton which they filter through their whalebone or baleen plates; these are huge horny laminae which hang down from the upper jaw.

species to species, may be twofold: when projected in a forward direction, almost as if to explore the nearby area, it is thought to assist the creature to find food; in other cases it might be used as lure for prey or a means of sexual signal in the mating period.

In the tropical and polar seas the types and numbers of forms of life are extremely diversified. In fact if, by accelerating the processes of growth and development, hot temperatures give rise to a greater numerical abundance of species, the polar waters contain a larger amount of plankton, and thus host a larger number of sea-birds (auks, petrels, albatrosses). Life may be more energetic in the tropics, but in the polar regions the high level of mineral salts contained in the water – due to the mixing process which occurs punctually at every seasonal change-over – gives rise to larger shoals of individuals and this explains why the cold seas have a greater quantity of fish than the warm seas.

Closely associated with the sea, with which they share certain physical characteristics (temperature, salinity, tides, and currents) we find two very specific environments: the barrier reefs and the ocean islands.

Coral reefs occur only where the water temperature reaches a least 70°F. (21°C.), and are hosts to fairly intricate living communities. First and foremost,

it should be said that the coral structures are the work of plants (coral algae) and invertebrate animals (coral polyps), which have in common the fact that they secrete calcareous skeletons; and, furthermore, the fact that they have fairly different shapes, sizes and colouring, on which the type of lodger living in them depends. In addition to sponges, sea-anemones, huge bivalve molluscs, sea-urchins and starfishes boasting dazzling colours, thousands and thousands of fishes with bright colourings and strange shapes in fact live in and around the coral reef. Angel-fish, catfish, butterfly-fish, damsel-fish, parrot-fish, just to mention the most common, move gently among the coral outcrops, and often strike up totally exceptional relationships based on co-operation. On the whole, one of the two partners is an invertebrate (almost always a coelenterate equipped with stinging or even poisonous tentacles) which offers a refuge and protection to the fish in exchange for a little house-cleaning!

But, as everywhere, the bright hues of the coral kingdom conceal threats and dangers for animals and man alike: wounds caused by coral do not heal easily; the prick from certain sea-urchins living here is extremely painful; and ambushes laid by sharks and moray eels for fishermen are anything but rare. Now we come to the ocean islands, that is to say those little fragments of dry

Clumsy and hampered in their movements on dry land, penguins are excellent swimmers: in the sea they can reach speeds of 25 m.p.h. (40 k.p.h.). They are common in the southern hemisphere, above all in the colder regions. Centre and left: the Genteo penguin, living in the Antarctic islands. Right: the king penguin, found along the coast of Antarctica.

The vast hordes of sea-birds use the immense food reserves offered by the sea. Man too is working on the possibility of intensively 'farming' the products of the sea, especially nowadays when it seems that the land can no longer meet his needs.

land which rear their head in mid-ocean and which have never been linked to the continental masses, with the result that their inhabitants had no option but to reach them via the sea.

The Hawaiian Islands, the Galapagos Islands, the Azores, and in the opinion of certain scholars New Zealand too, all have these features. In fact, in all these places the fauna includes no amphibians and no land-mammals (except the bat and the mouse which reached these isolated parts thanks to man).

As well as the aquatic creatures which appear in a reasonable variety of forms, *terra firma* is host to some species of reptiles (tortoises, iguanas) and some birds which are usually characterized by not being able to fly very well (if at all) because there are no predatory mammals to flee from.

Today more than ever before, man has turned a hopeful eye towards the sea, seeing in it not only an alternative source of food (fish, molluscs, crustaceans and, above all, algae), but also a new quarry from which to extract raw materials: it would in fact seem that the bottom of the abysses is covered with a strange sort of paving where the cobbles are made up of precious minerals deposited around bone- and tooth-fragments of fishes, like the mother-of-pearl around the tiny intruding speck in the oyster.

THE FRESH-WATER
VERTEBRATES

Ponds and marshes, lakes and rivers, estuaries and every type of watercourse from the torrent to the stream to the tiniest water-hole with to all appearances no movement whatsoever in it: all these habitats, large and small, are perfectly organized and balanced.

Even if fresh-water life is quantitatively far inferior to ocean- and sea-life, it nevertheless fills the habitats listed above, and plays an important rôle: fresh water is in fact closely allied to the existence of a large number of animal species (and, of course, plant species).

Although basically similar in their own adaptation to an aquatic world (which likewise requires an ability to float, swim, and breathe not atmospheric oxygen but oxygen which has been dissolved in the water), the creatures which live in the inland waters are subject to rules which differ considerably from those which govern life in the sea.

A typical feature of these waters is in fact the inconsistency of their physical factors: quite considerable variations between the maximum and minimum values apply not only to temperatures, pressures, and the absence or presence of plant-life – the primary link in any food-chain – but also to the rate of flow. Not to mention the chemical make-up which varies from place to place and from one season to the next, depending on the type of land through which the water flows; and depending on the existence of tributaries or the total isolation of certain given aquatic environments.

On the whole (and of course there are always exceptions) the absolute saline content of the fresh-water habitats is considerably lower than in salt water; and so because the saline content in the protoplasm of the animal cells is, on the contrary, similar to that of sea-water in its composition, all the creatures living in fresh water must have some mechanism which enables them to tolerate this type of difference in levels by preventing their saline content from being diluted by the water entering them from outside.

The fresh-water habitats are traditionally subdivided into two major categories: *running water* and *stagnant water*. The first category includes all sorts of watercourses of the most varied types: from the most clearly defined rivers, to the estuaries where fresh water mingles with salt water. The second category includes lakes, ponds, swamps and marshes.

In any event, the initial problem to be solved is almost always that of managing to resist the flow of the current by means of structures capable of fighting against the sweeping force of the water: thus we find various devices with which life anchors itself to the bottom or to rocks. The salmon is an expert at making its way upstream against powerful currents; this fish does things the other way round and leaves the sea for fresh water for the sole purpose of laying its eggs in the waters where it was born; it can even leap its way up waterfalls thanks to its muscular tail. Many animals have not managed to find a solution to the problems posed by currents, and have opted out, finding refuge in the calmer waters, although these in turn may be periodically distrubed by torrential watercourses.

One of the most curious inhabitants of the watercourse is the beaver, which actually manages to manipulate the water to its own advantage. Widespread mainly in North America, the beavers are rodent mammals which feed on the bark of trees which they gnaw at incessantly, and gather to build nothing short of dams across watercourses. They do this to prevent possible flooding of the complex network of underground tunnels which they construct in the bank to

Tropical waters contain numerous species of tiny, brightly-coloured fishes which lend themselves well to being bred in aquaria.

act as a refuge, a home for the young, and a winter larder. Capable of swimming at considerable speed with just their hind limbs – their front legs are usually folded back under the thorax – and helped by their curiously shovel-shaped tails, beavers manage to build dykes up to 295 feet (90 metres) in length and 9 feet (3 metres) in height, which obviously greatly alter the direction of the flow of water. The brown bears of Alaska likewise feel very much at home in moving water, and are also capable of making their way up waterfalls in search of salmon – their favourite prey – as are various species of otter equipped with a membrane between the toes of their feet, which they use to improve their swimming performance.

Once a watercourse has travelled through its impassable precipitous stages, which produce such a turbulent flow of water, and settles down to a more sluggish, calmer and flatter course, ready to receive contributions from other tributaries, geographers discard the terms stream and torrent, and call it a river.

In this environment the physical and morphological characteristics undergo considerable changes: the bed of the watercourse becomes wider, the current slows down, and the river bottom becomes sandy or gravelly; in addition, the plant-life along the banks grows more luxuriantly, encouraging animal life.

The principal classes of the vertebrates are all represented here: fishes, amphibians, reptiles and mammals.

Fishes, the first and most ancient fresh-water inhabitants, are still present today, and some of their forms, like the sea-dwelling coelacanth, have extremely primitive features: the ferocious bichir, a polypterid, found in African rivers, for example, which, with its bizarre pectoral fins, has not only gills but a rudimentary lung which only functions when the amount of oxygen in the water becomes inadequate; and the calamoichthys found in the River Niger and the River Congo, which has no fins and a snake-like body.

The cartilaginous fishes are somewhat few and far between (we know of just a few species of rays found in the Amazon and Orinoco rivers): the same does not apply to the bony fishes, however, which are present in such an abundance of forms that they are a good match for marine fishes: from the tiny pandaka, less than 1½ inches (38 millimetres) long, living in certain rivers in the Philippines, to the huge sturgeon found in the River Volga which can reach a length of 12 feet (4 metres); from the peace-loving European perch which is so sought-after by anglers, to the fearsome and bloodthirsty piranha found in South America.

Before discussing the amphibians, which are the typical tenants of and live

An example of Astronotus ocellatus, *a gaily coloured cichlid found in South America. The cichlids are fishes which often keep a very watchful eye over their young, as illustrated by the fact that the eggs are incubated inside the mouth.*

An ancient popular myth has it that the Salamandridae, lizard-like amphibians, are poisonous animals. In fact they are perfectly harmless. Left: the yellow and black salamander. Top right: the spectacled salamander, found in Italy. Below; the newt; the crest on the back is peculiar to the male during the mating season.

almost solely in fresh water, it is worth mentioning a group of animals which are thought of as fishes but whose bodies point to the bodies of the amphibians. These are the so-called dipnoids or dipnoans which have real gills and real lungs, and can thus breathe equally well in water or on land. The best-known forms are the barramunda and the protopterus (or African lung-fish), the former being found in rivers in Australia. Both species are hunted by the indigenous people who are fond of their tasty flesh.

The barramunda has just one lung, which is put to work every half-hour when it goes to the surface to breathe; it hunts by night and when bothered emits a very distinctive rumble.

The protopterus has two lungs and uses its pectoral fins as feet for walking along the river bed. When the river where it lives dries up, it digs a hole and buries itself in it, sometimes living in its dry mud hideout for as long as six months: during its period of dormancy, it breathes air through a small tube also made of mud hidden between its lips, and feeds on the muscles in its tail.

Frogs, toads, newts and salamanders often live in the aquatic habitat provided by a slow-moving river (although they are more often to be found in lakes or marshes), but only at specific times, namely when the eggs are laid and

in the subsequent larval stage. Snakes, tortoises and turtles and crocodiles alternate aquatic life with periods on dry land.

As a general rule birds are not especially attached to rivers: in fact they prefer the calmer waters of lakes or marshes. The same cannot be said of the mammals: although they have never completely adapted themselves to the fresh-water environment (as whales, seals and dolphins have done in the sea in due course), their history has in many respects become linked with the history of a watercourse. Suffice it to mention a species of dolphin found in the Amazon and Orinoco rivers; certain fish-eating bats, using the claws on their hind-feet, can catch small fishes swimming beneath the surface of the water; the whole array of Procyonidae (raccoons) which go down to rivers to clean themselves and wash the fruit on which they feed; bears, tapirs, shrews whose legs sometimes have fringes of rigid hairs which help them to swim; and lastly, the sluggish and colossal hippopotami which can stay totally submerged in water except for their eyes and nostrils which are situated on top of the head.

The undisputed worker/playboy of the river is nevertheless the otter. As agile and quick-witted as a cat, the otter darts through the water like a fish, and plays with anything it comes across: this activity can keep it busy for most of

Covered by warty skin, toads (above left) are among the best known of the anuran amphibians. Below: a toad which lives among reeds. Above right: the green toad. Below: the common toad.

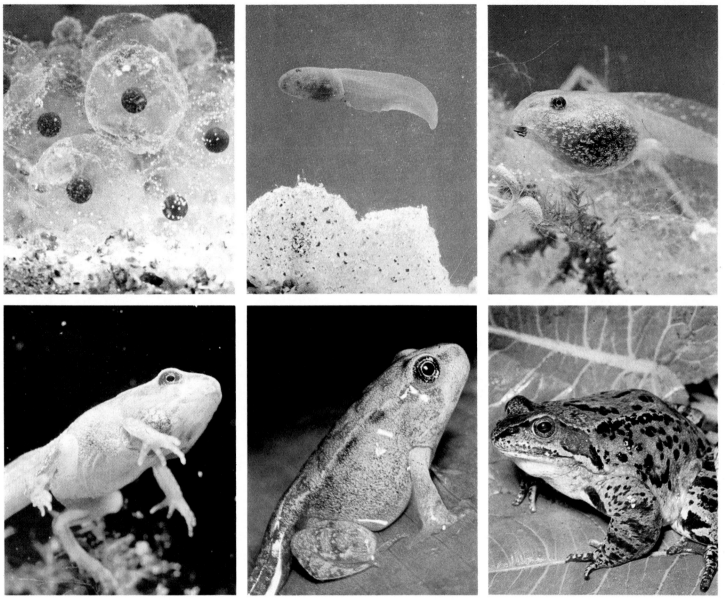

Successive stages in the development of an anuran amphibian. Top, left to right: submerged eggs in a gelatinous substance; a legless, long-tailed tadpole; appearance of the front legs. Below, left to right: appearance of the hind-legs; a successive stage in its development, still with the tail; and lastly the adult form.

the day. And it hunts by night: crayfish, snails, fishes, frogs and water-snakes are its favourite prey; but if these are not forthcoming, it will also stoop to mice or ducks, and even poultry. The otter always eats its meals on dry land, and it is also here that it builds its den, keeping it constantly dry and covered with grass, where it raises its young.

As between land and sea, there are always transitional areas between the river and the land; fresh- or salt-water (or brackish) marshes and swamps, estuaries where the water flows sluggishly in various directions, creating sorts of intermediary physical conditions in which life may flourish or even disappear altogether, giving rise to the so-called *aquatic deserts*. The fertility of these zones – commonly known as *humid zones* – is due to at least three things: 1) the *tides*, which, with the periodic variations caused to the sea-level, promote a rapid exchange of food and rapid removal of waste; 2) the variety and wealth of the *plant-life*, with the consequent exchange of oxygen in the water; 3) relatively constant *light*, *temperature* and *water composition*. Many fishes, like the mullet for example, pass untroubled from salt to fresh water; others, like the salmon already mentioned, and the sturgeon and shad, enter fresh-water environments from the sea only to reproduce. The behaviour of the eels is quite the opposite; at maturity, they swim downstream towards the Atlantic Ocean from inland waters, and head straight for the Sargasso Sea. Here they lay their eggs which hatch into larvae, and these complete their whole cycle of development, retracing, step by step, the route taken by their parents, until they reach the original river.

The pink flamingo is one of the most frequent inhabitants of the low-lying lagoon and banks with little vegetation. This is a large bird with a wingspan of up to $5\frac{1}{4}$ feet (1.7 metres). The beak, flat and curved at the tip, is longer than the actual head and is used to sift through the mud where it finds, and greedily devours, shrimps and molluscs. When the time comes for it to lay its eggs, the flamingo piles mud up into a fairly low mound, like an upturned flower vase with a hole in the middle, where the female lays her eggs which are then covered by both parents, taking turns to do so.

The origins of stagnant aquatic habitats are often to be found in the presence of some obstacle or other in the flow of a watercourse. Caused by natural phenomena (and the geological processes responsible for them are quite plentiful: old ice-floes thawing; movements of the land settling; the filling up of craters belonging to extinct volcanoes) or artificial phenomena (dams built by man, but also those built by beavers!), these obstacles have filled a whole host of aquatic areas with life, such areas being classified as *lakes*, *ponds* and *marshes* depending on their size and depth.

Lakes owe much of the life in them to the temperature: in fact the temperature uses the peculiar property of water of becoming heavier at 39°F. (4°C.) to determine the circulation and hence the exchange of the organic matter. For this reason, temperate lakes are subject to a complete recirculation of their water, first in winter, and then again in summer. With the cold, the heavier surface water sinks; when the warmer weather returns, the whole mass of water is reshuffled, because it tends to reach a balanced temperature, and the whole environment is 'rejuvenated'.

In the extreme climates of the polar and tropical regions such turbulence is only recorded on the surface. In the former, the lake water freezes (which causes even certain animal inhabitants to pass through a period of frozen hibernation, like the Siberian blackfish which can survive a period of real hibernation for as long as several weeks); in the latter, where the temperature never drops as low as those fateful 39°F. (4°C.), there is just a slight surface movement, caused by the motion of the swell and currents, with a subsequent scarcity of oxygen and, hence, fauna in the deeper zones.

As far as the organization of life is concerned, lakes are comparable to the sea. Here, too, we can in fact talk in terms of three zones: the *shore zone*, rich in plants with roots (it is incidentally worth noting that the flora of lakes consists mainly of plants with seeds, in other words the higher plants, which have found their way there from dry land); the *limnetic zone* with free-moving water dominated by suspended plankton; and the *benthonic zone*, where we find animals and plants which are anchored to the bottom.

Lakes are mainly populated by fishes which generally prefer the shallow areas at the edge where the plants which many of them feed on can grow well. There are also numerous species which live attached to the muddy bottom, constantly digging in search of possible prey (worms, molluscs and crustaceans). To do so, these fishes are almost all equipped with suitable devices: the barbels of the sturgeon and carp are like tiny antennae which detect the presence of any source of food; and both these species of fish have the odd ability to evaginate their mouths; the long snout of certain species living in African waters has a special proboscis; and the blind electric eels are equipped with complex electric organs designed solely to find food by avoiding any obstacles in the way.

Almost like a reminder of that far-off day when water began to flow over the earth which was slowly cooling, lakes, ponds and marshes are inhabited by a large number of species of animals which lead a so-called double life: the amphibians. The amphibians have gills during the larval stage (which is usually the aquatic stage) and lungs as adults; and if need be, they can breathe quite

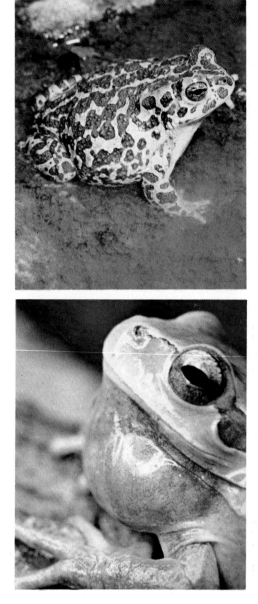

Amphibians which emit sounds. Above: the green toad. Below, the tree-toad showing its sac attached to the throat and digits covered with adhesive pads.

Features peculiar to the beaver are the webbed digits of the hind-feet, the shovel-like tail and the ability to build dams and dykes connected to the nest. Left: a beaver feeding. Top right: dragging a trunk. Below: a dyke formed by tree trunks, made by beavers.

simply through their skins which, being thin and unprotected, can absorb the oxygen required directly. Those members of this class (aquatic frogs and salamanders) which, even when mature, remain allied with the aquatic environment, commit themselves constantly to this type of breathing.

Incessantly preyed upon by snakes and birds, the amphibians all have effective defence mechanisms: glands which secrete irritant or poisonous substances; the ability to assume a mimetic coloration; the capacity to inflict painful bites; great agility when jumping and swimming, thanks to the powerful musculature of their hind-legs; and other special features (interdigital membranes, webbed feet).

If some toads, frogs and salamanders spend much of their life on dry land, usually in the branches of trees or vegetation not too far from water, they are, however, always bound to lay their eggs and spend their initial larval life in water. Of course they are not free from threats here, either: fishes and even certain aquatic insects are extremely fond of such titbits. This is why the parents prevent a total destruction of their young not only by laying very large numbers of eggs, but also by protecting them by constructing intricate and well-concealed nests, or even, in some cases, keeping their young on their backs

321

during the whole incubation period.

The South American toothed bufonid (*Leptodactylus*), for example, forms a mud enclosure near the bank within which it lays its eggs: this enclosure is loosened and broken down by seasonal rains, and the tadpoles, already quite large, and capable of striking out on their own, are released. The Surinam toad carries its eggs on its back; and the male of a frog found in western Africa keeps them in its throat.

The salamanders and newts are on the whole more attached to water than the frogs and toads are: the most remarkable salamander is the giant species found in Japan (3¾ feet [1.25 metres] in length) which never leaves the water, but rises to the surface just to take in air. Stagnant waters are the undisputed realm of the birds: either permanent or more often intermittent (or *passing*) inhabitants depending on their periodic migratory movements, using the water only to search for food or because they have made their nest in this environment in the reeds along the bank, the water-birds (or waterfowl) can be classified in two major categories: the *swimming* birds and the *wading* birds.

With the former, the ability to fly has been more or less completely replaced by the ability to swim; indeed, to be more accurate, the ability to hover in the

The Mississippi alligator (top), a reptile common in American fresh-water habitats, measures no more than 10 inches (25 centimetres) when it leaves the egg (below left); but when fully grown it may reach a length of 14 feet (4·5 metres). It is frequently found in southern parts of the United States (below right: an alligator on the banks of the Mississippi river); in spite of its appearance it is a shy creature and will flee from man; it feeds mainly on fishes.

The anatids are perfectly adapted to a life spent on the water, thanks to their wide, flat beaks which can sift through mud, their webbed feet and waterproof plumage. Top left: a duck taking off over a cane-brake. Below: an African anatid. Right: a flight of swans.

air has been almost totally lost in the tropical species, whereas it has been retained in species living in the cold regions: these latter in fact have to be able to migrate once winter arrives. Although unused their wings remain, and the swimming strokes are made possible with the help of special structures on the feet and toes. The (webbed) feet have membranes between the toes, and the toes are equipped with lobes which considerably reduce friction with the water. The waders, on the other hand, have especially long legs and, consequently, long necks and beaks. Ibises, herons and jacanas can all tread wherever there is shallow water by placing their long, splayed toes on the leaves of floating plants. The supple necks of the waterfowl end in beaks of very varied shapes: the snakebird uses its beak as a spear with which to impale its prey; the spoonbills poke about in the mud with their flat beaks; ducks, on the other hand, filter the water and extract the plankton from it.

Talking of unusual beaks, it is worth mentioning the pelican's beak. Pelicans are large birds which live on the large lakes in the Old World. About 6 feet (2 metres) in length, with a wingspan which may reach $7\frac{1}{2}$ feet (2.5 metres), they are conspicuous for their buccal apparatus. The beak is in fact straight, very strong and almost 19 inches (50 centimetres) long; a sac made of skin

stretches from the tip of the beak to the throat; here the pelican can store up to $4\frac{1}{2}$ pounds (2 kilos) of fish. When it is hungry, it throws all the fish from its sac into the air one after the other, catches them in mid-air and eats them. Excellent swimmer and accomplished in the air, the pelican has an awkward gait on the land, where its very short legs make it slow-moving and unsure. The pelicans spend the hours of darkness on land, perched on the branches of trees or on rocks protruding from the water. Similarly, their nests, made of reeds and grass, are built on dry land: they lay their eggs in May, and these are covered by the female, while the male keeps watch.

For centuries fresh-water ponds, swamps and marshes have fostered health hazards and malaria: nowadays the notoriety of these humid environments has changed to some extent, both because much has been done to combat the diffusion and development of the mosquitoes which carry this disease, and also because they are oases where many species in danger of extinction take refuge and rest. The animal population which inhabits these environments is fairly like that which inhabits lakes, with the addition of certain species of crocodiles and, among the mammals, otters and minks.

There are three environments where water is still a basic element for the life

The sheer elegance of the flamingo (left) contrasts oddly with the shape of its beak which is thick and bent; however, it is perfectly designed for feeding its owner by sucking up mud and water and extracting the plankton. Right: a blue heron; the body structure resembles that of the flamingo, but the shape of the beak indicates that this bird has a different diet; in fact it feeds mainly on frogs, fishes, and other small aquatic animals.

The herons are gregarious birds; they are also migratory and excellent fliers. They have very long legs and splayed digits which stop them sinking into the mud. There are sixty-four species of heron. Top left: the blue heron. Right: the grey heron. The spoonbills (below left) are quite closely related to the herons (both belong to the order Ciconiiformes); spoonbills are named after their distinctively shaped beak.

there: namely, two micro-environments, underground caves and rotting hollows in trees, and artificial lakes created by man.

Caves have a constantly high level of humidity, indicated by the continual dripping of water, with the formation of fairly deep water-holes and small ponds; they offer their denizens conditions of life totally unaffected by seasonal changes. Similarly, the animals living in such habitats have very similar features, often unidentifiable in analogous forms of life in the world outside.

Cave-dwelling creatures are very small: the skin covering, which is thin and colourless, usually helps with respiration; there are hardly ever any eyes in these animals, or if they do have them they work badly; on the other hand their organs of touch and their sense of smell are well developed. With the exception of certain species of bats, there are few cave-dwelling mammals. Among the birds, we should mention the South American guacharo, although this bird can leave its cave at night to look for food; among the reptiles there are some species of snakes; among the amphibians, a form of blind salamander and the olm (*Proteus anguineus*) which, even when mature, breathes by means of gills; among the fishes we find the Kentucky blind fish, just $2\frac{1}{2}$ inches (7 centimetres) long, which is eyeless and has a transparent body.

The rotten hollows of trees, together with the funnel-shaped leaves and flowers of many tropical plants, constitute a host of tiny habitats. It is to some extent the limited size of these habitats which decides what fauna is suited to live in them; this is why, among the invertebrates, the insects are the favourite tenants, along with certain small frogs and humming-birds among the vertebrates.

By building artificial lakes or ponds man has changed the face of the landscape in areas with no natural water reserves. The waters of such lakes and ponds are before very long populated by a micro-fauna which may then become the source of food for fishes and amphibians introduced there by man.

The risk for these environments lies in the possible process of ageing to which they may be subject after a short period of time: this mainly occurs as a result of changes taking place because of the erosion of the basin of the lake which is generally poorly protected. One remedy is periodically to drain the lake or pond and make sure that it is then re-colonized. In Japan, as a complement to agriculture, these water reserves are used for so-called *aquaculture*: algae, molluscs and fishes are bred in large numbers in basins periodically inspected with a view to their physical and chemical characteristics.

The anatids (right: an Australian black swan) give birth to precocious offspring; their young can walk and feed themselves soon after hatching from the egg (left top and below).

THE VERTEBRATES OF THE TROPICAL FORESTS

'... Here we are surrounded by thousands of species of trees which are so tall they seem to touch the sky ... and sheer green, like in Spain in May. And each one of them is at a different stage in its life, depending on the type of tree. ...'
Way back in 1492, Christopher Columbus used these words to describe the equatorial forest of America, but today, five hundred years later, the appearance of the tropical forests (so called because they occur in the tropics) has remained unaltered: a dense and permanently green mantle which seems unaware of the seasons because it is practically always summer in these parts. The leaves fall from the branches and are instantly replaced before they even have time to turn yellow; the buds flower and the flowers turn into fruit without awaiting the right season; and all the while the ground is being covered by a soft carpet which will be the future habitat for a variety of shade-loving species.

Constant heat, abundant light, and a large amount of available water: these three conditions combined make the tropical forests one of the most fertile founts of life on our planet. In fact the tropical forests (which geographers also call rain forests) occur in regions where the annual rainfall exceeds 6 feet (2 metres), with a minimum of 4 inches (10 centimetres) in the driest months. In the rainy season, as a result, the roof of the forest is continually dripping even when the rain has ceased and the consequent humidity (the heat of the sun has a job to penetrate the barrier formed by the foliage) turns the forest into a sort of hothouse where the plant-life grows all the more luxuriantly of its own accord. The tropical forests cover all those areas of the earth's surface which the Equator runs through, except for the eastern seaboard of Africa where the climate is arid and the terrain predominantly desert.

The greatest expanse of tropical forest is in South America where, among other things, the effect of man has been least felt; the tropical forests of Africa are the least extensive, being confined to the basin of the River Congo and its tributaries and stretching north-west along the Gulf of Guinea as far as Liberia; the tropical forests of Indonesia and Malaya are the most fragmented, covering areas of the islands of the East Indies (Sumatra, Borneo, the Philippines) and, on the continent, small areas of Asia as well as the Indo-Malayan peninsula.

We should also include the forests of New Guinea and Queensland in Australia; perhaps because of their isolated nature, however, both these areas have somewhat different fauna and flora – different because these are very specific.

The lush green sea of the forest can be subdivided horizontally into three *strata* in which (as in the sea) the varying intensity of the light also determines a varying distribution of the different living species.

The *roof* or *ceiling* is where we find the process of active photosynthesis: it is in fact here that the rays of the sun touch the forest directly, and of course the entire community relies on the sun for its survival. This uppermost stratum consists principally of trees between 147 feet (45 metres) and 196 feet (60 metres) in height – the greatest height found is 295 feet (90 metres) in a fairly narrow range of species.

The *intermediate zone* may in turn be subdivided into two strata, again on the basis of the height of the vegetation which as a rule grows to one of two well-defined levels: 66 feet (20 metres) and 130 feet (40 metres).

The *ground* or *floor* (known as the *benthos* in the sea) is dimly lit; but the fairly high temperature and level of humidity together with an abundance of decomposing organic matter give rise to a lush plant-life.

As elsewhere, here, too, nature has had to devise various ways and means

Opposite: a typical tropical forest. The incredible tangle of plants provides a habitat for a large number of animal species, whose habits and relations are to date virtually unknown. Above: various species of parrot, typical inhabitants of the forest. Left: the black-throated lovebird. Top right: Australian parrots (Trichoglossidae). Below: parakeets, small parrots bred for decorative purposes.

to make life as uncomplicated as possible: in the oceans we find devices for swimming and floating, and in the green sea of the forests nature has had to solve the problem of flying and to an even greater extent that of moving nimbly through the dense tangle of trees, trunks, branches and leaves. Thus we find prehensile claws and feet, tails which can hang from branches and creepers (lianas), toes equipped with suckers, and long, strong arms.

The prehensile tail, which is found only in the monkeys of South America, is a valuable safeguard in a world where the daily round is spent mainly in trees and well away from potential enemies. This type of tail, with its lower surface devoid of fur, is used like a fifth limb.

Monkeys are the most typical inhabitants of the tropical forests. They are utterly reliant on the hot, constant climate of the tropics: they cannot migrate in the possible event of a cold season; they cannot gather obvious food reserves to be eaten during the winter months; they have a fairly limited diet (they feed mainly on insects, fruit and seeds, all of which are more or less exclusive products of the hot season). As far as classification is concerned, the monkeys are subdivided into four groups: the *prosimians*; the *monkeys of the Old World*; the *monkeys of the New World*; and the large *anthropomorphic monkeys*.

The toucans (left) feed on fruit; but it is not clear what they use their huge curved beaks for. There is no doubt, however, about the function of the strong beaks of the parrots (right), with which they can break open tough seeds, and which they use to steady themselves as they move.

The prosimians include creatures which differ from one another quite considerably, but have the common denominator of leading a predominantly nocturnal life (this does not apply in the case of several species of lemurs living in Madagascar). Tree shrew, lemur, loris and potto are the names of some of the most typical representatives of this group, about whose ways and habits not very much is known. Small and ugly in appearance, with a usually poorly developed brain and huge, glowing eyes designed for life in the dark, the prosimians live in trees, clinging to the branches with their strong feet, the toes of which are equipped with long, curved claws, and feeding on fruit or, even insects.

The real monkeys differ sharply from one another depending on whether they belong to the Old World or the New World. This is possibly explained by the hypothesis that these two animal groups have evolved independently in terms of time and in different ways. In fact, whereas, as already mentioned, the monkeys of South America have prehensile tails, only the monkeys of the Old World have pouches in their cheeks for storing food on a temporary basis: thanks to these pouches the amount of time they spend on the ground – with all its attendant dangers and threats – is greatly reduced. Instead of finding, chewing and swallowing food, these monkeys fill their expandable cheeks with

Top left: an albino example of an entellus (or langur), an Indian monkey which lives by day on the ground and sleeps and rests by night in trees. Below: a young baboon with its mother; these animals live in groups with strict hierarchical structures. Right: a hamadryad (or sacred baboon), a species akin to the ordinary baboon and equipped, like the latter, with powerful teeth.

food and retreat to the safety of some lofty branch to eat their fill, first emptying their pouches and then sorting through the contents which are then consumed in small mouthfuls.

Among the most distinctive features of the monkeys is the brightly coloured buttocks of the African and Indo-Malayan species, designed to attract a mate.

One of the most fearsome natural weapons of the monkeys is the large canine tooth with which the males of many species are equipped. Flattened on the front and sharpened on the rear face like a knife, this tooth reaches a considerable size in some African baboons.

This group also presents another set of habits worthy of mention: the capacity to produce sounds which are to some extent intelligible and classifiable depending on the type of emotion that the monkey wishes to express. Pain, fear, the summons to gather, and the order to be on the alert are communicated to the members of the group in clearly identifiable ways; and this repertoire of sounds is accompanied by a whole series of extremely meaningful facial expressions.

The noisiest species of all is the howler (or howling) monkey, in which the actual structure of the throat enables this animal to emit sounds which are truly

remarkable for their duration and . . . unpleasantness (not least because these shrieks are always issued not by just one monkey, but by a whole community).

All the monkeys of the New World prefer the arboreal life, and live mainly in the so-called roof of the forest.

The anthropomorphic monkeys (the gorilla and chimpanzee in Africa; the Asian gibbons; the orang-utans of South-east Asia) are likewise closely allied to trees, with a particular liking for very tall ones.

As suggested by the very name of this group, these creatures are considered the highest on the zoological ladder, and even akin to man himself. The shape of the skull and the body in fact recalls the features of those distant ancestors and progenitors of our species; the brain is fairly well developed and in addition these are the only mammals to have hands with *opposable thumbs*.

Even when they are not particularly large, gorillas, chimpanzees, orangutans and gibbons are extraordinarily strong in relation to their weight. As typical forest-dwellers, they use their eyes as the primary and fundamental means of identifying the world around them, and are also able to distinguish colours (a feature which otherwise, apart from being a human one, is peculiar to just a few birds and fishes). The anthropoid species do not build refuges or

One of the best-known anthropomorphic monkeys is the orang-utan. It lives alone or in pairs in the mountainous forests of Sumatra and Borneo, where it roams in search of food, consisting of fruit, leaves and small animals. Indolent and gentle by nature, this creature will only do battle if bothered, and then it is a formidable and strong foe.

The gorilla is certainly the largest ape on earth, and possibly the largest that has ever existed. In fact the male can reach a height of 6 feet (2 metres) and weigh between 420 and 660 pounds (200 and 300 kilos). It lives in the forests of central Africa and western Africa in families of five to ten individuals. Despite its appearance and the pose it assumes, beating its chest when annoyed, it is rarely aggressive.

fixed abodes; they do not store food and generally tend to destroy their sources of food.

The gorillas in particular lead a strictly regulated and organized group life under the watchful eye of an old leader. Their day is spent in search of food (which varies in quality and quantity depending on the time of day, in other words depending on the type of meal), and any remaining time is used for moving quickly from one place to another and for building flat pallets hidden among the foliage, for resting.

Like all the monkeys in general, these species also take great care of their young, keeping them close to their breasts during the suckling period and even in the months following.

In addition to the monkeys, the uppermost regions of the forest are host to two other major categories of inhabitant: sloths and birds. To this list we can also add certain porcupines with prehensile tails and long dart-like spines which can envelop the animal in a formidable crown of thorns; squirrels which leap nonchalantly from one branch to the next; certain species of reptile, such as a tiny lizard which resembles a chameleon and can jump upwards, thanks to the small adhesive discs on its feet, and the frightening-looking iguana (which

is in fact an easily tameable creature) which can leap across gaps 75 feet (25 metres) up.

But now we come to the slow-moving sloths, of which two species are known: the three-toed sloth (also known as the aye-aye) and the two-toed sloth (*Choloepus didactylus*). The former has a short tail, and the front legs longer than the hind-legs; the neck is long and supple so that it can be turned towards the ground even when the animal is hanging back downwards. The latter is tail-less. Both species have fairly poorly developed teeth: twenty identical teeth in all, quite well spaced and hollow inside. The sloths live suspended from the branches of trees, to which they cling with their strong, curved claws. They are slow-moving, and chew leaves continually. They are the only mammals with an unusually green-coloured coat: their mantles consist of two types of fur, the one long, coarse and more superficial; the other shorter and finer, and hidden by the former. But the long hairs are often covered with microscopic green algae which help to hide the original colour of the animal and enable it to assume the colour of the surrounding environment.

With their heads permanently hanging downwards, the sloths are only rarely active: if attacked, when their claws turn into dangerous weapons; if they find

The chameleons are saurian reptiles, their heads surmounted by crests, tubercles or small horns, their eyes almost completely covered with skin, and their tails prehensile and capable of being rolled up; but they are best known for their ability to change colour, not so much for mimetic reasons as in relation to the light or as a demonstration of annoyance.

Top left: a slow loris, a South-east Asian prosimian which lives in the forest feeding on small mammals, insects and fruit. Below: a gibbon, a very nimble monkey which moves among the trees using just its arms. Right: a male mandrill with its brightly-coloured face. The females do not have this colouring and are smaller.

one of their kin on the same branch (and when this occurs there is a battle for the territorial rights in question which ends only when the intruder is put to flight); when the young are born, and once a year when they migrate beyond the nearest river in search of a new home.

The other typical inhabitants of the tropical forest are the magnificently coloured birds. Found for the most part in the tropical regions of America, toucans, humming-birds and the vast array of parrots are among the most curious types.

The toucans with their cumbersome yellow beaks live mainly in groups which fly about in the uppermost foliage; their oddest habit consists of building a mud nest in the hollow of a tree in which they imprison the female for the entire incubation period. Only when this period is over is the nest re-opened and the female allowed to come out once again.

The humming-birds have very long, thin beaks which can be introduced into the corolla of flowers to suck the nectar for which they have a weakness.

The parrots with their well-known ability to make imitative intelligible sounds deserve mention because of the varied colours of their plumage which take on the colours of the plant-life in the area and so protect them from

possible attacks. The parrots do not in fact fly well (they move only to look for food), and for this reason they prefer to live perched in trees.

The so-called *intermediate* zone of the forest has a fauna which in many respects resembles that of the uppermost regions: here, too, in fact, animals must be able to move nimbly through the dense tangle of green. As well as certain opossums with long, prehensile tails, this part of the forest houses two very special creatures: the anteater and the pangolin.

Although the anteater at first glance may look clumsy and awkward (it weighs a little less than 110 lb (50 kilos), it can move among the branches of trees with great speed and agility. It has no neck and its snout ends in a sort of thin, toothless proboscis; and as its name suggests, it lives on ants and termites which it flushes from their nests by introducing its extremely long, sticky and protactile tongue. Its body ends in a similarly long tail where the hairs, arranged like a fan, help the animal to keep its grip.

The structure of the pangolin is somewhat different, although it, too, lives off ants. Like a large lizard a metre or even two long, the pangolin's body is covered from head to tail with extremely hard horny scales, arranged like the tiles of a roof and articulated in such a way that, if need be, the animal can roll

In the African forests south of the Sahara, in Madagascar and the Philippines, we find the prosimians which are the least evolved of the primates, the class of mammals to which man also belongs. These animals have distinctive long limbs, slender toes with wide mobile areas of digital pulp, and huge eyes suited to life in the dark (centre below). Left: a mongoose lemur. Centre top: a ringtailed lemur from Madagascar. Right: the lesser galago, found on the African continent.

Vertebrates found in the South American forests. Top left: the iguana, active by day, with sharp vision, which despite its frightening appearance is a vegetarian. Below: the capybara, the largest living rodent which lives on river banks. Centre top; the anteater; its name is somewhat misleading in as much as it feeds on termites rather than ants. Below: the armadillo, an armoured mammal which escapes from predators by burying itself. Top right: the puma, widespread throughout the American continent. Below: the saimiri, a platyrrhine monkey.

itself up in a spiral.

The forest floor, which is covered with a not particularly thick layer of fallen leaves and rotting matter, is the realm of the invertebrates.

Naturally enough we find quite large animals here too, starting with the elephant which, despite its bulk, manages to move quite freely among the creepers and hanging branches, then buffalo, wild boar and leopards in the African forests; and the constantly warring jaguars and long-nosed tapirs of the American forests.

Whether carnivores or herbivores, all these animals are capable of moving with nimble, swift leaps and bounds. In many cases their hides are spotted or striped so as to camouflage them with the light and shadow of the forest around them. Some deer and antelopes have horns on their heads which they use to defend themselves or to win the heart of a female in battle with another male.

The species which are most closely allied to the forest floor have mouths designed to rummage through the ground which they systematically sweep clean with great speed: this applies to the wild boar and the pig-like peccary which both live in herds; and the same goes for certain rabbit-like rodents (for example, the aguti which is found in tropical America) which leave their lairs at

night and go hunting. Among other things, the forests of South America are host to some age-old mammals, the opossums, which are among the few marsupials to have propagated in America.

We also find a curious mammal on the forest floor: the armour-plated armadillo.

There is a limited number of species of armadillo, ranging in size from the giant armadillo (4½ feet [1.5 metres] long, including the tail) to the tiny pichicaigo of Argentina which barely measures 6 inches (15 centimetres). The hide of these animals consists of bony scales which together form a hard shell covered with skin protecting the entire body. The scales are arranged in strips which alternate with soft sections so that the armadillo can roll up on itself: the head and legs are enclosed in the armour and the whole ball is kept hermetically sealed. Despite this armour, the armadillo's life is constantly fear-ridden, the threat coming from jaguars and pumas, life-long enemies, which can prise open the armour with their claws, and wild dogs which can break it with their strong jaws. This is why the armadillos dig quite deep lairs to escape from their enemies, and only rarely venture forth from them.

There are also many birds to be found on the forest floor, as well as those

The sloth (left) is a mammal which lives in the South American forests; it is extremely sluggish, and of limited intelligence. It lives in trees where it hangs by means of its strong claws. Another odd creature found in South America is the cock of the rock (right), with its bright plumage; only the males have the crest covering the beak.

The dangerous and aggressive cobras are widespread in eastern Africa and Asia as far as the Philippines. If bothered, they expand the ribs in their neck and take on a terrifying appearance (top and below left). Below right : the cobra, which here is swallowing a toad, looks like any other snake when its neck is not opened like a fan.

which fill the vault of the forest with their cries: partridges, curassows, trumpeters and tinamous to mention just the best known, as well as the famous scarlet ibis which lives in colonies in marshy areas and ransacks the mud with its curved beak in search of worms, and so on.

But the truly original and also the oldest inhabitants of the forest – where the vertebrates are concerned – are the reptiles and amphibians.

There are two principal types of reptile, the lizards and the snakes (poisonous and otherwise). There is an extremely large variety of lizards: ranging from the iguanas measuring up to 6 feet (2 metres) found in tropical America, to the tree lizards of Malaya which can fly by means of scaly membranes between their front and hind-legs.

A rather special inhabitant of the forest, because of its unique features, is the chameleon. This is a slow-moving reptile which protects itself by being able to change its colour with great speed, and, above all, in a very specific way so as to imitate perfectly the nature of the ground on which every single part of its body rests. The chameleon also has a telescopic tongue which is as long as its whole body with the tip covered with a sticky substance designed to catch its prey. Steadying itself with its prehensile tail, the chameleon moves slowly among

branches, and its large round eyes are constantly on the alert for possible prey to be caught with its extremely swift tongue.

The snakes can be classified on the basis of whether they have poisonous teeth or not. Those that do bite are fully aware of the power of their teeth; those that do not often put on a fierce display by assuming the threatening posture of their more dangerous relatives.

Some of the harmless snakes can move very fast; others, found in the forests of Malaya, are called flying snakes: these in fact live mainly in trees and if attacked by a bird of prey, throw themselves off the tree to the ground where they cunningly hide themselves.

In Africa, where the fearsome and extremely poisonous black and green mambas are found, other arboreal species take on a similar coloration; the same can be said of tropical America, realm of the notorious *fer-de-lance* snake, whose features are imitated by many non-poisonous species.

Among the non-poisonous snakes, mention should be made of the so-called constrictors: up to 32 feet (10 metres) in length and weighing sometimes more than 220 lb (100 kilos), these reptiles are often strong enough to kill fairly large mammals. The aquatic anaconda and the various boas live twined around low

Despite the fact that it weighs 660 pounds (300 kilos), the tiger has a perfectly designed body and is very agile. It lives in the forests of central-eastern Asia (excluding Tibet, Ceylon and Borneo) where it hunts large mammals and often presents a serious threat to man.

The green boa shown here reaches a length of $4\frac{1}{2}$ feet (1·5 metres) and lives in the forests of Guiana and Brazil. It hunts not very large prey, lying in ambush in branches, and has strong but not poisonous teeth.

branches waiting for some victim to pass by; like a streak of lightning they pounce on it and squeeze it to death with their coils, in a deadly embrace.

Among the poisonous species the most widespread in the tropical forests are those belonging to the groups of vipers and cobras. The first group includes the green viper (or adder), found only in the Old World, and the rattlesnake which abounds mainly in the New World. The bodies of all these reptiles are generally short and thick; the head is wedge-shaped, depressed in the rattlesnakes with their so-called *sensory pit* which enables them to detect the presence of warm-blooded bodies in the vicinity. All these serpents kill their prey with the venom which they inject through specially designed *grooved teeth* placed in the foremost part of their upper jaw.

Despite the fierce nature of these snakes, certain small mammals will not hesitate to confront them, being immune to the effects of the poison: this applies to the mongoose and the ichneumon, well-matched opponents of the Indian cobra and the South American *fer-de-lance* respectively.

But the enemies of the poisonous snakes are not limited to just these mammals and certain birds. There are even certain non-poisonous snakes belonging to the same class which attack certain poisonous species, for whose

flesh they have a weakness, and render them powerless. In America, for example, we find the mussoruna, one of the fiercest foes of the New World vipers, which it kills by constriction.

The hot, damp climate of the tropical forests (especially in South America) is undeniably a perfect environment for the amphibians. In fact there are large numbers of frogs and toads with a wide variety of defence mechanisms.

The tropical frogs actually have extremely toxic venom on their skin: as soon as they are brushed against, specially designed glands in the skin emit an irritant substance which does not cause any serious reaction in the case of an absolutely intact skin or hide, but which will show how deadly poisonous it is once it comes into contact with mucus or penetrates the body of the victim through a wound.

Most of the tropical frogs lay quite a small number of eggs (about a dozen) either on curiously funnel-shaped floating leaves or hidden in the hollow of a tree trunk which has been already waterproofed with a waxy substance. The eggs are often large and full of water: in this way the tadpole can develop perfectly within it and only emerges when it is a small frog.

The anaconda is a huge South American snake, up to 27 feet (9 metres) long, which leads a semi-aquatic life. A fine swimmer, it lives on river banks when not in the water, and often climbs up trees. Top left: the head of an anaconda emerging from the water. Below: catching a cayman. Centre and right: an anaconda being caught, these snakes being popular in zoos.

THE VERTEBRATES OF THE SAVANNAH AND THE PRAIRIES

Among the factors which govern and condition the life of an environment, *rain* is certainly of prime importance. If in fact – as we have seen in earlier pages – its occurrence in large enough amounts combined with a hot climate gives rise to the formation of the so-called tropical forests, its scarcity, or a prolonged absence of it (although still combined with a fairly hot climate), may similarly determine the emergence of a new type of world: the *savannah*.

The savannah is in fact a special form of *open country*, dominated by plants belonging to the large group of grasses, undisputed masters of such environments.

Depending on the location in respect of the Equator (and thus in respect of the tropical forests) and the presence or absence of tall grasses, trees and shrubs, low grasses, the open country goes by various names: *prairie* or *grassland*; *savannah*; or *steppe*. We shall mainly consider the savannah, so as to continue on from the remarks we have made about the tropical forests and their inhabitants.

The basic feature of the savannah is thus the presence of *herbaceous plants*, the seeds of which are scattered in abundance and over a wide area by the wind and by animals. These plants have fairly low stems, the leaves are often spiny, the flowers usually fairly dull in colour and the roots are firmly anchored in clods of earth; here and there, usually near watercourses, we find long rows of bushy plants which thrive on moisture. As far as the flora is concerned, this is the aspect of the savannah. And this green world invariably goes hand in glove with a special kind of fauna: like the prairie and steppe, the savannah is dominated by *grazing* animals, which eat the leaves of the plants, and, where the invertebrates are concerned, by the so-called *destructive* insects (locusts, termites and grasshoppers).

In this environment too, the animals have naturally had to face and try to solve certain problems which are basic to their survival: the diet, which is mainly vegetarian; the need to defend themselves in an open area which is thus comparatively short of natural refuges; the necessity of moving on continually in search of new sources of food. Even if the number of vegetarian animals is far greater than that of carnivorous animals (remember that these are the first link in the food chain and thus the first rung of the so-called pyramid of life), not many of them like leaves. The problem of eating leaves is in fact closely allied with that of managing to digest them so as to extract from them the substances necessary for growth, first, and healthy existence, second. Like all the other cells which make up the various vegetable organs, those which make up leaves are also impregnated with a complex substance, *cellulose*, which keeps the walls hard. But those animals whose cells do not have a cellulose structure have great difficulty when it comes to digesting this substance. The devices which they resort to (and in the savannah there is a whole sample selection) are numerous and very varied.

Some herbivores house in their digestive systems micro-organisms (usually bacteria) which can produce a juice capable of dissolving the cellulose; others have teeth specially designed to grind up leaves, so as to obtain a pulp which is more easily digestible; others are *ruminant*, which in a sense is a little like a double digestive process; and still others have the habit of eating their faeces.

A factor common to all the herbivorous forms is that they have a considerably longer intestine than the carnivores, both because of the longer time taken to digest food and because, in terms of quantity, the amount of food

required by a herbivorous animal is far greater than that needed by a carnivorous predator, which can extract from its victim an abundance and variety of all the substances which a living being may need.

If the problem of food is the first concern of the inhabitants of the savannah and, of course, the prairie, an equally serious problem is posed by defence measures against possible hostile predators in an environment which is so open and so exposed to danger.

One of the oddest tactics is that used by the prairie dog, a sort of land-squirrel with a short tail which lives in the plains in the United States. The prairie dogs live in large and well ordered colonies, and have lairs which stretch underground for miles, plunging almost vertically into the ground to a depth of between 9 and 15 feet (3–5 metres) and then opening out parallel with the surface. The network of these tunnels which, like real apartment blocks, have areas earmarked for the storage of food for the winter, areas for the young and nuptial bedrooms lined with grass, is constructed by a pair of animals which work feverishly until the underground residence is finished off with small low walls outside, designed to protect the whole construction from rain.

Given the possible arrival of enemies and the colony's need to gather roots

Once common even in Europe and in Asia, the lion is now confined to the savannah between the Sahara and the Transvaal and to a small area in India. Next to the tiger, it is the largest cat of all; in fact it may reach 440 pounds (200 kilos) in weight. Top left: a lioness hunting. Below: lion cubs. Right: the impressive appearance of a male lion.

Despite its powerful look, the male lion (below left) does not usually go hunting, but leaves the task of finding food to the female (right). After a gestation period of fifteen to sixteen weeks the female gives birth to between two and four cubs (top left), which she looks after with much affection.

and grain, sentries watch over the life of the community day and night. The guard is periodically changed, and no sooner has one sentry gone to eat than his place is taken by a substitute. From time to time the whole area resounds with excited cries: this is the alarm, signalling the appearance of danger. The whole community takes refuge in the holes from which they will only emerge when the sentry has announced that the coast is clear.

Tradition has it that in this environment there is a three-sided friendship between the prairie dog, the rattlesnake and the owl. In reality the two latter species are no more than exploiters who squat in abandoned prairie-dog lairs on the edge of a colony. The owl, in particular, installs itself on the raised areas where the sentries are placed; the rattlesnake, on the other hand, hides in the deepest parts of the lair not only to spend the winter there but also to digest its food, which usually consists of very young prairie dogs.

Another defence method is based on the opposite system to that described above for the prairie dogs. Instead of digging lairs and burrows in which to hide, very many animals (and this applies to large herbivores) gather in very large *herds* made up of adult males, females and young of all ages, down to the newborn. Depending on the region, and hence the situation of the prairie and

the savannah, we thus find herds of elephants and bison, gazelles, antelopes, deer, guanacos, zebras and even kangaroos. The number of animals in the herd is a measure of strength and the basis of the defence potential. Of course, because a group is something more than a mere gathering, it must be well organized; and thus its members must have precise and separate roles. The commonest allocation of tasks is that of males and females for the reproduction and care of the young.

Beyond this level the next step is the establishment of a *society* in which a hierarchy is created: we find one leader (and sometimes several leaders) in command of the group; it is the leader's job to defend the herd's territory; he has first say when it comes to obtaining and sampling food, and has first choice when it comes to a mate. After the leader, the hierarchy descends rung by rung down to the weakest members of the group who often get no more than the leftovers and are the laughing-stock of the group.

The baboons, which are the only monkeys which live in open country, enjoy precisely this kind of perfect social organization. The groups of baboons are usually quite large (up to a maximum of two hundred individuals) and clearly separated from one another in terms of territory. Every large family has

Several species of zebra live in the African savannah, all with the black-and-white striped hide, in herds. Because they are hunted by the large carnivores, especially lions, they have acute senses of smell, hearing and sight; their only defence mechanisms are flight (below) and kicking.

The elegant and often very fast-moving gazelles and antelopes inhabit the savannah and include many species. They usually live in herds (below right: a group of antelopes), often together with other mammals such as zebras and buffalo. Top left: a dorcas gazelle. Below left: Grant's gazelle. Top right: impala gazelles.

various elderly males which are obeyed by all the other members with a natural respect for the rules of life.

The leaders decide when the group should move on or rest; they choose the place best suited for eating and drinking; they settle disputes; and have themselves cleaned of the parasites which infest their coats by females or the weakest young members of the group. Relying on their extremely sharp eyes, baboons will not even retreat when attacked by large cats which lie in wait for them. They leap swiftly into tall trees and shriek and scream until the enemy has moved off. At night they leave the ground and retire into trees.

Sometimes strange alliances are stuck up between animals which are quite different from one another, always for reasons of defence: the baboons, for example, will sometimes join up with groups of impalas, using these animals' extremely sensitive sense of smell where any dangerous scent is concerned in exchange for their own keen vision.

In the same way, antelopes and rhinoceroses live together with oxpeckers, tiny birds which cling to the backs of large herbivores and squawk and shriek to warn them of danger in exchange for being able to feed on the insect larvae clustering in the skin of these mammals.

There is a similar alliance between zebras and ostriches. The African ostriches (like the rhea [or nandu] of South America and the Australian emu) are large birds which, having lost the ability fly, have highly developed legs which can carry them at high speed. In addition they are extremely suspicious by nature and difficult to approach because they have extraordinarily keen vision. But during the day the ostriches are often constantly accompanied by zebras. The zebras are shy and highly sought-after as prey by carnivores, and they entrust themselves totally to the keen awareness of the ostriches which they follow in flight as soon as the large birds quicken their pace when they sense danger. But here, too, there is mutual advantage to be had: the ostriches benefit from the proximity of the zebras because the latter trample the grass with their hooves and flush out hosts of insects, of which the ostriches are fond.

The carnivores living in the prairie and savannah generally belong to two main systematic families: the Canidae and the Felidae (dogs and cats). The former include wolves, jackals and hyaenas; the latter, lions, tigers, leopards, cheetahs and jaguars.

All these creatures, which seem intent on reminding us of those far-off days when our planet was inhabited by fearsome carnivorous monsters, have had

Massive and thickset, with one, two or, in rare cases, three horns on their head, rhinoceroses are usually shy, stand-offish animals, but if disturbed can be quite dangerous. They generally feed on plants, and can eat hard, leathery types of plant, too. Rhinoceros hunting, often carried on to obtain the horn which is supposed to contain an aphrodisiac, has reduced several species to such an extent that they are in danger of extinction.

The ostrich is the largest living bird. Up to 8 feet (2.5 metres) in height and weighing as much as 187 pounds (85 kilos), it has lost the ability to fly, but to make up for this, it is a strong and fleet-footed runner. It lives in quite large groups in dry regions in Africa or Arabia; at one time it was also found in parts of Europe and Asia.

their numbers considerably reduced, but still retain the task, given by nature, of controlling the diffusion of the herbivores, keeping them within limits compatible with the vegetation offered by the open country.

The simple horns of the antelopes and bison, or the branched horns of the pronghorn and deer of the Argentine pampas are adequate defensive weapons against attacks by wolves (whose claws are unable to tear flesh), but are quite ineffectual against the larger carnivores with their well-sharpened claws. For this reason other devices have been elaborated by these animals to escape from predators: an acute sense of smell; speed; and camouflaged colour and markings of the hide. Some antelopes, for example, can run at speeds in excess of 62 m.p.h. (100 k.p.h.); and some types of deer have become so agile and good at jumping that they can take any obstacle in their stride, leaving their enemies hot on the trail but with all four feet firmly on the ground.

Then the hide merges perfectly with the natural colours round about: the various colours and designs of the back go well with a white belly which stands out less than a dark belly would. The striped antelopes, zebras and giraffes likewise seem bent on echoing the alternating light and shadow in their vicinity in their markings.

The prairie environment is furthermore responsible for an interesting biological phenomenon which scholars have termed *convergent evolution*. In effect this complex term describes a very simple fact: more than anywhere else the animals inhabiting the prairies throughout the world have in time acquired similar features although they belong to originally profoundly different groups. On our planet, in fact, there are prairies on every continent which enable a large number of grazing herbivorous mammals to exist. And in every continent the method of locomotion (as we have seen) is basic to survival in the face of possible attacks from carnivorous predators and takes on three different forms: *running*, for pursuit and flight; *jumping* over plants to keep pursuers at a distance; and the *capacity to dig* holes at great speed into which to disappear at the timely moment. The inhabitants of the open country (taking only mammals and birds into account) fall into the following six categories:
(1) *Jumping herbivorous mammals:* the Californian hare (black-tailed jack rabbit); the Asian diplopods (millipedes); the African jack rabbit; and the Australian red kangaroo.
(2) *Burrowing herbivorous mammals:* the prairie dog; the viscacha and cavy (or guinea-pig) of the Argentine pampas; the African ground squirrel; and the

In Greek, hippopotamus literally means 'river-horse'. Although its likeness to a horse is not self-evident, its familiarity with the aquatic environment causes it to spend much of the time in water; it only ventures on dry land to feed (it lives on aquatic plants, roots, shoots and fruit), and spend the night.

The extremely long neck of the giraffe enables it to reach places which are inaccessible to other animals and feed on shoots high above the ground. When the giraffe goes to drink, however (below left), its long neck and front legs pose quite a problem, and render the animal particularly vulnerable to predators.

Australian wombat.

(3) *Mammals which burrow and live in the subsoil:* the North American geomyds (pouched rats); the Argentine tuco-tuco; the Asian mole rat; the African golden mole; and the Australian marsupial mole.

(4) *Running birds:* the Argentine nandu; the African ostrich; and the Australian emu.

(5) *Running herbivorous mammals:* the pronghorn and bison of North America; the guanaco and deer of the Argentine pampas; the Asian saiga; and the African zebra and antelope.

(6) *Running carnivorous mammals:* the North American coyote; the African leopard and lion; the Tasmanian wolf; the Argentine maned wolf; and the Asian manul cat.

In discussing the prairie and savannah we have limited ourselves as far as the vertebrates inhabiting them are concerned to just those animals in the mammal and bird classes whose way of life is of somewhat particular interest.

As far as the amphibians are concerned, it can simply be said that these are spasmodic visitors to the open country, and then only during the rainy season or where a nearby watercourse enables them to lay their eggs and the eggs

351

The two photographs on the left show the difference between the Indian elephant (top) and the African elephant (below). The former has quite small tusks and ears; these are larger on the latter which has no frontal protuberances. Right: African elephants drinking.

subsequently to develop.

As is invariably the case, there are also exceptions to this rule: in fact there is a prairie toad which seems to have completely forgotten what it means to be an amphibian. Indifferent to the presence or absence of water, it spends its whole life underground, laying its large gelatinous eggs in the darkness of the subsoil; these eggs contain within them all the alimentary substances required by the young tadpoles for their development until they undergo their metamorphosis.

Unlike the tropical forests which are rarely visited by man (only the odd native tribe manages to survive on the edge of the forest), the open country environment has always housed a reasonable number of people who, as we have seen in other cases, may have once played the role of simple consumers, and scrupulously observed the rules of nature, but who subsequently, in about 1600, began to move outside the vital chain of which they had to be part, and seriously promoted the destruction of the very environment itself.

The technique of cutting and burning is as old as the hills. Having seen the way in which shoots appeared thicker and faster than ever in the prairies and savannah as a result of fires started by lightning striking the withered brushwood, the indigenous peoples in such areas have learnt how to use

The surly appearance of the hyaenas, and their habit of eating carcasses, make these animals somewhat unprepossessing. In fact their role in the ecosystem of the savannah is valuable in as much as they eat the leftovers of the carnivores. Top: two pictures of the striped hyaena. Below: the spotted hyaena.

controlled fires as a means of maintaining the prairie itself. In this way the grazing herbivores are provided with a fresh supply of food by the new shoots, and the growth from them.

The same cannot be said of the way man has carried on in North America, where at one time there was a thriving community of grazing animals (mainly pronghorns and bison) together with a large number of carnivorous mammals (wolves and coyotes), small burrowing mammals, birds and insects.

Before the white man came to the great plains of North America, nature down the centuries had developed a perfect state of balance. There were herds of bison and pronghorns with abundant supplies of fresh grazing; there were the wolves who not only stopped the herds from becoming too large, but also forced them to keep on the move, thus helping the plant-life which was never completely eaten down to the ground; there were the families of prairie dogs, along with the owls and rattlesnakes.

The first people to arrive were the Indians who began to hunt the bison on foot with what we might call natural weapons: stones, spears and arrows. But the Indians were happy to kill just a few head, and they put every part of the bison to good use. When the Spanish conquistadores arrived in America in the

seventeenth century, they gave horses to the Indians. For the bison this was the beginning of the end. Before long, in fact, Indians and white men unwittingly joined forces in one of the greatest animal slaughters of all time, with the aim of obtaining and then marketing and selling the hides of the bison.

In seventy years, between 1820 and 1890, the bison were systematically hunted, sometimes just for fun. We are told that special trains were arranged for hunters who were allowed to shoot the animals directly from the windows and that the number killed in that period was 60,000!

With the slaughter of the bison came a further harmful development for the prairie: the herds of bison were replaced by herds of cattle and flocks of sheep: the predators and prairie dogs, responsible for the burrows which became traps for the animals introduced by man, were poisoned. And their end also signed the death-warrant for the birds and mammals which ate the carcasses. As a result, all those insects and small rodents, which were constantly preyed upon by the birds, multiplied out of all proportion and caused the full-scale and uncontrolled destruction of the plant-life.

At the present time the situation with the herds of bison is slowly improving thanks to the strict protective measures adopted by the Canadian and

Not as strong as the hyaena and thus exposed to being bullied by it, the vultures spot their food from high up as they wheel in the sky and usually manage to reach it first and eat their fill before other animals arrive. Left: the Egyptian vulture. Top right: a griffon vulture. Below: African vultures.

These pictures show typical inhabitants of the American and European prairies. Top left: the male sage grouse which opens out its tail feathers like a fan. Below: the pronghorn, a swift runner. Centre top: the prairie dog which digs intricate networks of tunnels. Below: a ground squirrel found in the Eurasian prairies. Top right: a herd of American bison. Below: the European bison.

American governments: new livestock is being introduced into the herds; strict checks are periodically made to limit the numbers to a level compatible with the environment; and strict laws allow hunters to kill just one animal per person, and that just once in a lifetime.

If in the case of the massacre of the bison man has tried to repair with his own hands the damage done to nature, who showed herself to be over-generous, in other cases the damage done by mankind has proved to be beyond repair.

This is true with the extinction of the prairie-fowl and the quagga – a South African horse which resembles the zebra. In fact the prairie-fowl disappeared off the face of the earth for good in 1931, finding it impossible to survive in the habitat which man had so radically altered; the quagga, on the other hand, died out rather like the bison, and became extinct for good and all when the Boers slaughtered them indiscriminately for their meat and hides.

Before turning to the inhabitants of the so-called temperate zones, it is worth mentioning an environment which, for reasons of space, we cannot describe in detail, although it covers large areas of the world: the *desert*.

The desert environment is by definition confined to regions where the annual rainfall is less than 10 inches (25 centimetres) and where as a result the

vegetation is very sparse.

In fact there are only four forms of plant-life which have adapted to such hard conditions:

(1) certain *annual plants* which only grow when there is enough humidity in the air, which stops their withering; (2) *desert shrubs and brush*, which, if need be, can shed all their foliage to reduce transpiration to a minimum; (3) the *succulents* which store water in their tissues; (4) that whole host of *microscopic plants* (mosses, lichens, blue-green algae) which explodes into life as soon as the climate cools and there is more moisture in the air.

As far as the fauna is concerned, on the other hand, the number of creatures which have tried to face up to the situation which is dominated by the scarcity of water and the high temperature is quite large.

Some animals, like the reptiles in the case of the vertebrates and the insects in the case of the invertebrates, are in a certain sense pre-adapted to the desert. Their bodies are covered with a thick, weatherproof integument; their need for liquid is extremely slight because, as carnivores, they extract all they need from the blood of their victims; the fact, furthermore, that they are cold-blooded frees them from the need to control the temperature of their own bodies by

Top left: a warthog, a strong wild pig. Below left and centre: two species of monkeys belonging to the family of cercopiths. Top right: a cheetah's head, which is quite small compared to the rest of its body. Below: the ground squirrel, an African rodent.

356

The crocodiles have adapted themselves well to the aquatic life; but they have many features which liken them to the dinosaurs, those fearsome reptiles which inhabited the earth millions of years ago. They hunt at night, catching fish, small mammals, and even young crocodiles.

evaporation; in the event, during the daytime, they move to warmer or cooler places, as need be.

Other animals, on the contrary, and this applies above all to the mammals which are fairly scarce by way of desert forms, have found subsequent solutions to this type of life.

In order to restrict as much as possible the production of sweat essential to the lowering of the body temperature, they live mainly by night, and retire during the hottest hours of the day to cool underground lairs. Some of them (the best-known example and the most closely observed example is that of a species of mouse called a rat kangaroo), manage to manufacture themselves the water which they need to survive, extracting it from the digestion of the dry seeds on which they feed. What is more, they avoid all forms of useless dissipation except for sweating, and eliminate a very concentrated urine.

Other creatures, like the camels, dromedaries and certain donkeys used in the desert certainly need to drink, but are very thrifty in their use of the water which they have stored. During long treks, they use their body liquids and may lose up to 25 per cent of their body weight, which they then restore to its proper level by drinking their fill when they stop.

THE VERTEBRATES OF THE TEMPERATE ZONES

So far we have dealt with those environments which in a certain sense are the furthest removed from our day-to-day lives: vast expanses of water, salty or otherwise, our knowledge of which is often limited to a summer holiday; tropical forests jealously guarding their many secrets; the endless savannah, dominated by its green grasses; and lastly the arid and apparently uninhabited deserts. But our planet still offers a fairly large number of environments which, because of their geographical location, can be grouped under the heading *temperate zones*.

Flat or mountainous, exposed to the beneficial effects of the ocean and sea currents, or quite independent of them, the temperate zones have in common a fairly harsh climate during the winter months and a fairly warm one during the summer season. The climatic conditions of the intermediate seasons vary considerably from region to region, passing through a whole range of subtle gradations: from the so-called Mediterranean regions, where there is a mild spring-like climate for almost all the year, to the cold northern regions where the harsh Atlantic current affects both spring and summer.

There are two temperate zones on earth: the *northern temperate zone* in one hemisphere, and the *southern temperate zone* in the other. Differing in area (the former measures 27 million square miles [70 million square kilometres], the latter only 10 million square miles [26 million square kilometres], and in the way the land is distributed (the northern zone is a continuous whole except for two small gaps, the Bering Straits and the Atlantic Ocean, whereas the southern zone is split between three continents: Africa, America and Australia), these regions also have quite different types of vegetation.

In fact, if we consider the predominant flora (exceptions occur in those areas which, because of their altitude or latitude, have a more uniform and generally colder climate, in other words, the mountainous regions and those which border directly on the Poles), we can say that the temperate zone is dominated by a *forest-like* vegetation.

Of course, the most northerly forests, i.e. those which form a belt immediately to the south of the tundra – which borders directly on the polar regions – consist almost exclusively of evergreen conifers (silver fir and Norway spruce); the forests lying more to the south of these, with a humid-temperate climate, have a larger number of species, ranging from the evergreen conifers (here we find numerous varieties of pines) to deciduous broadleaf trees like oak, beech, etc. This is the case with the northern hemisphere.

In the southern hemisphere, on the other hand, we do not find deciduous forests like the above. In fact here the high degree of humidity caused by abundant rain during the summer, together with a warmer climate, have given rise to a lush type of vegetation, more akin to that of the fully fledged tropical forest, and this type of flora is called the *temperate rain forest*.

From the point of view of fauna, given the great variety of environments in these temperate zones, it will be as well to proceed zone by zone, as follows: first, the inhabitants of the northern evergreen forests; then of the deciduous forests; and lastly of the temperate rain forests. Having so done, we shall deal with those species which, unbothered by the cold, forge their way across the tundra close to the polar ice-floes, and those species which have adapted themselves to life at high altitudes in the various mountain ranges.

The northern forests stretch across vast areas of the Asian continent where they form the huge forests of Siberia, and to a lesser extent across the American

The long ramified horns or antlers give the deer an elegant and regal appearance. Only the males have them and each year, in February, they drop off and sprout again immediately, adding another point. The illustrations here show three different species: the red deer, a species peculiar to Europe (left), the wapiti of North America (top right), and the spotted deer of India, Sri Lanka (Ceylon) and Nepal (below).

continent, where they form the Canadian forest. As we said earlier, the predominant trees are the silver fir and the Norway spruce with, in more southern parts, various species of pine trees and willows and poplars near watercourses and lakes.

Contrary to how it may appear, the northern forest houses a fairly large number of animal species. Generally belonging to the mammals and birds (although there are various reptiles and amphibians among the vertebrates, and numerous insects where the invertebrates are concerned), these different creatures have found ways and means of surviving even where the winters are long, because even here there is a reasonable amount of food – the seeds of the evergreen trees, for example – and safe refuge.

The presence of a white mantle of snow for many months obviously creates difficulties. First and foremost that of movement. This may not be much of a problem for the birds, many of which are excellent fliers, but the mammals which move across the white and often soft covering must avoid sinking in: the American elk, or moose for example, has long, stilt-like legs; and the polar hare has longer feet, to give it more support than the corresponding species which live where there is no snow on the ground. The second problem is that of

359

surviving periods of extreme cold; here, too, we find various types of adaptation.

Some large mammals, like the musk-ox, find it enough to thicken their coats; others, on the other hand, prefer to retire to lairs burrowed beneath the insulating layer of snow where they have stored up enough food to live on (this habit is usually peculiar to small rodents and insectivores which have feet and snouts designed for digging); others leave the bitterly cold regions and head southwards (like some members of the large family of Cervidae, for example). The third and final problem posed by this white world is that of keeping out of sight of enemies. Here again it is the small mammals which are most affected by this problem, in particular certain rodents and one family of carnivores, the weasel family (or Mustelidae), which includes a large number of creatures which are often unfortunately known only for the value of their fur: ermine, sable, marten, etc. Nowadays found mainly in the remote forests of Scandinavia, Russia and the Rocky Mountains in America (where they are protected by very strict laws), the Mustelidae are typically nocturnal animals. Fierce hunters and just as fierce when defending themselves, they eat mice and other small rodents, as well as birds and eggs. They usually lead a solitary life, except for the brief reproductive period. Many members of this family can

Mustelidae, usually quite small mammals, are carnivorous predators. The existence of many such species is threatened by the fact that their fur is much sought-after, and that they are consequently much hunted. Top left: the wolverine. Below: the badger. Centre: the marten. Top right: the ermine. Below: the weasel.

Popular tradition has it that the wolf is fierce and cruel, but it is unfair to attribute human labels to the wolf as to any other predator. In effect, strength and cunning are simply the means with which natural selection has equipped this creature to enable it to survive. Top left: a common wolf. Below: a wolf's meal. Right: its typical grey coat.

change the colour of their fur with the seasons. The weasel (*Mustela nivalis*), for example, periodically changes the colour of the fur on its back; at the beginning of spring dark fur begins to appear on its back (the belly is always white, however) and by the height of summer covers its whole body. In about October the reverse process commences: patches of white fur appear first on its flanks and then on its back, and these give its coat a typical spotted appearance. At the end of November it is completely white and blends perfectly with the surrounding environment.

The polecats are particularly well known for their defence mechanism. In fact they are not bothered by their bright coloration and keep predators at a safe distance by emitting from anal glands an evil-smelling substance which can be squirted as far as 12 feet (4 metres).

Typical mammal inhabitants of the *taiga* (the name given by scholars to the northern forests), among the herbivores, are deer, and, among the carnivores, wolves. Depending on the region, we find many species of deer. The males live in groups under the command of an old individual; during the reproductive period, when it is time to have a family, they try to mate with as many females as possible in order to be sure of as many offspring as possible. A typical feature

of these animals is the fact that all the males of the genus *Cervus* (and both sexes where the reindeer and caribou are concerned) have horns which are deciduous. These horns (or antlers) are bony growths which emerge from the skull and are not covered by skin (unlike those of sheep and cattle) and drop off during the winter, growing again in the following spring, and increasing annually in size and in the number of points. The points are used for defence and for sexual prowess, females being won after fierce battles to the accompaniment of the clash of antlers. In some cases the antlers grow so large that they actually complicate the animal's life: this happened in the case of the Irish deer, whose extinction was possibly even caused by this feature, the horns growing to a width of some 11 feet (3.5 metres).

If the deer family defends itself with its horns and by galloping, the wolves entrust their defence and their ability to hunt to their teeth. Their favourite prey are in fact the caribou, the mountain sheep and the American elk, which they attack by a strategy of encirclement. The wolves live in groups – called *packs* – made up of about ten individuals: two or three males with their respective families. Usually only the adult members of the pack will hunt; the younger members simply observe the various tactics used. Having reached

Few other animals in the temperate forest can match the squirrel's ability to move through the trees. This small rodent lives essentially by day and feeds on plants, and nuts in particular. The photographs show three different species: the common squirrel (left), found in Europe and Asia; the red squirrel (top right) and the grey squirrel (below right), both of which are American.

The problem of survival is particularly acute for a timid and defenceless creature like the hare (below left: the common hare), a favourite prey of carnivores. The only ways it can escape from its predators lie in the speed with which it can run and jump (top left), reaching as much as 44 to 50 m.p.h. (70 to 80 k.p.h.), and camouflage (right: the Arctic hare with its white winter coat).

maturity, they will then be able to help attack prey, but they will not actually touch it until the most experienced wolf has killed it.

As far as the birds are concerned, the taiga is undoubtedly a favourable environment for them during the summer, when insects abound. In the colder winter months it is inhabited mainly by vegetation species which find adequate amounts of seeds on the evergreen conifers. This is why, in winter, the northern forests are abandoned by several of their habitual winged lodgers, and left with just the crossbills, rock ptarmigans and certain varieties of hawks and owls.

The crossbills live in flocks of about a hundred individuals: their oddly cross-shaped beak is the key to their survival. Feeding on pine-seeds, thanks to the shape of their bills they are able to break open the ligneous scales and extract the seeds. The rock ptarmigans, unlike the other birds, also live on the ground; to enable them to do so they have long legs and wide feet with the dual function of stilts-cum-snowshoes for moving across land which is covered by snow for much of the year. Another specific adaptation to this world is that achieved by the owls, typical nocturnal creatures which also hunt frequently by day: in fact the short summer nights do not enable them to gather enough food.

If the environment described above has the cold as its major obstacle to the

A series of typical inhabitants of the temperate forests. Top left: the lynx; below: a Canadian chipmunk; centre: the dormouse; top right: the capercaillies; below: a young goldfinch.

survival and propagation of the various species living in it, life in the warmer deciduous forests is no simpler. In fact if on the one hand the greater variety of plant-life enables a larger number of animals to live here, on the other hand the dormant period which most of the plants undergo when winter arrives (with the consequent loss of their leaves) makes the search for food a tricky business.

But as always there are many ways out of this kind of situation: the cold-blooded creatures, that is the reptiles and amphibians among the vertebrates, who cannot remain active when the temperature drops below zero, spend the winter months in a state of lethargy buried in a safe refuge protected from the world outside; as far as the warm-blooded animals are concerned, some simply carry on their daily activities; others (bears, skunks and opossums) fall into a sort of prolonged sleep from which they awake to go for a short stroll or have a quick snack; other types of animal, on the other hand, undergo a kind of torpor (called *hibernation*) during which the body temperature and the breathing rhythm and heartbeat all drop considerably. This applies to hamsters and marmots, dormice and squirrels. During hibernation all these animals consume the fatty reserves in their bodies in order to survive.

Curled up in a tight ball, they seem quite lifeless: in the case of a squirrel, for

The Procyonidae are mammals which in appearance resemble small bears or dogs, and they are closely akin to these latter. They are found mainly on the American continent. Top left: the red coati which uses its tail when moving in trees. Below: a raccoon, a member of the genus Procyon (right).

example, it has been recorded that the body temperature is just a few degrees above zero; that the heartbeat, which is between two hundred and four hundred a minute in normal conditions, drops to a minimum of five; and that the breathing rhythm drops from two hundred a minute to four or less. This period is usually spent in specially prepared dens; the badger's set is a model of cleanliness and perfection. Situated at the end of long tunnels are actual rooms with smooth walls and floors lined with dried grass to make them soft. Appropriate apertures ensure a constant flow of air. The dormouse, on the other hand, builds its straw bed, lined and upholstered with moss, in the hollow of a tree. It retires here at the end of autumn, well fed and fat so as to cope better with its period of lethargy, and remains hidden in its nest until the fine weather returns. The marmots, lastly, prefer to spend the winter months in company. Having prepared a deep lair which they cover with hay and having eaten and drunk their fill beforehand, ten or fifteen of them gather in their hideout from which they will not emerge for six months.

The most decisive alternative to the seasonal problem is that chosen by those who decide to leave: some animals, birds in particular, seem to reckon that there are no on-the-spot solutions and prefer to leave the forest, which has

become so inhospitable, and short of food, on a temporary basis, and head southwards where survival is easier. In quite large flocks, wild geese and cuckoos, tits and swallows, as well as an infinite number of tiny passerines leave the forest as winter approaches and head for warmer climates. Depending on the species, the birds travel either in orderly formations or with little organization; they make short and extremely long flights, flying day and night or sometimes just at night, covering great distances without stopping, or stopping here and there to rest and refresh themselves. One of the favourite destinations is Malaya where, from October to March, animals from all over northern Asia are taken in and put up for the duration (cuckoos, plovers, etc.); and between April and September the same process applies to species fleeing from the winter in the southern hemisphere.

In general the animals living in the deciduous forest live a less arboreal life than those living in the tropical forests, and this is easily explained if one realizes that, basically, this environment only offers a refuge for a limited part of the year. The squirrels are the most common and competent tree-climbers; they manage to shin up tree trunks clinging on first with their front feet then with their hind-feet. When they come down, unlike most similar creatures, the

Popularly notorious as a particularly wily creature, the fox is common in Europe, Asia and North America. Even if its favourite food consists of small mammals and birds, it will, if need be, content itself with insects, carcasses and even plants.

All the animals depicted here are large herbivores which live in different environments. The caribou (top left) lives in the Arctic tundra; the roe deer (below) in the forests of Europe, the ibex (top centre) in the Alps; the elk (below) in coniferous forests; and the mouflon (right) in the Mediterranean forests.

squirrel sprints head-first using its tail as a parachute. Moving swiftly from one branch to another, they store up shoots, nuts, acorns and so on, and are not averse, in some instances, to taking certain birds' eggs.

The so-called flying squirrels are 'cousins' of the true squirrels; these are tiny animals which have wide folds of skin between their front and hind-legs and hurl themselves into space, gliding to a preselected landing-place.

The bear's way of scaling a tree trunk is as clumsy and awkward as the squirrel's is nimble and graceful. Clinging to the bark with the sharp curved claws on all four feet, it lets itself slither down, or, if being pursued, clambers cumbersomely upwards.

When moving in trees the slow-moving Canadian porcupines use not only their well-developed claws, but also the rough pads on the soles of their feet, as well as the hard, modified spines at the base of the tail.

Some forest-dwelling mammals are of interest because of their way of dealing with their young. Let us take the opossum as an example; this is one of the few marsupials living in America. After a gestation period of less than a month, the baby opossums are born. Blind and defenceless, weighing just two grammes, they nevertheless have well-developed small front legs with small

claws. Using these, they clamber up the mother's abdomen, following a path marked out by her with saliva, and finally reach the incubator pouch of the marsupium.

Within the marsupium there are thirteen nipples, no more and no less: each baby opossum has its own private teat, to which it attaches itself and does not let go for two whole months. If there are more than thirteen young, the surplus have no option but to die of hunger: this is nature's spontaneous way of controlling her populations. At the end of the suckling period, with the coat now complete and the sense organs all working, they are too large to remain in the mother's pouch; but at the same time they are still too inexperienced about the ways of life outside and thus prefer to stay as close as possible to their mother who carries them everywhere with her on her back for about forty days.

Like all the monkeys, so the macaques inhabiting the mountainous forests of North Africa and Gibraltar are very concerned about the family unit and caring for their young which they keep constantly attached to their bellies when they travel long distances with the arrival of the cold time of year.

In the raccoon family it is the mother who for a whole year assumes the task of looking after the young, not only those to which she has given birth, but also

The merciless laws of nature force certain animals to be cruel predators and others to play the extremely unfortunate role of possible prey. The first group includes the owls (top right); the second the small passeriformes (left: the wren) and rodents (below right).

Among the passeriformes some species are mainly granivorous, and others feed mainly on insects. The first group includes the hawfinch (top left), the second the blue-tit (below) and the blackbird (right).

any orphans in the area. In fact after mating the males' only task is to hunt, and they leave all matters concerning the growth of their offspring and the defence of them from predators by day and birds of prey by night to the females.

As we have already seen with the tropical forests, here, too, there are large numbers of birds. First and foremost we find the woodpeckers, whose feet are so formed as to enable them to climb nimbly up tree trunks, and whose beaks, which are pointed and very hard, enable them to penetrate the wood and flush out insects and larvae hidden in it. It is quite common to see cuckoos, truly parasitic birds which do not make their own nests, do not hatch their own eggs, and do not rear their own young. The cuckoo lays its eggs in another bird's nest, and lets the occupant hatch, feed and rear the young until they are completely developed. The female usually lays her egg with the species which has reared her: if she was born in a thrush's nest, for example, she will lay an egg which has the typical colour of a thrush's egg, in a thrush's nest, while the owner is not at home. The second egg will be laid in another nest, and so on and so forth. Sometimes the female cuckoo will take one of the rightful eggs from a nest and throw it away so that when the mother returns she will find the same number of eggs as when she left. Ornithologists have tried to explain why the

cuckoo behaves in this way by the hypothesis that because she lays her eggs with long intervals between each one, she would find herself having simultaneously to hatch one egg while rearing an already hatched young cuckoo, the two tasks being incompatible with one another.

There are various sorts of gallinaceans living in the forests: wild turkeys, capercaillies, quails and pheasants in great numbers live on the ground where they make their nests, and only perch in trees to rest and sleep. They leave the wood in search of food from cultivated areas. The males of these species have brightly coloured plumage with metallic highlights; the females are more dully coloured.

If the life of the forest by day is at its most lively, with the constant to-ing and fro-ing of large numbers of inhabitants, it is at night, in the silence of darkness, that mammals, birds, reptiles and amphibians put their sense of smell and hearing to the test, and tread their way stealthily through the trees to catch their unsuspecting prey. As well as being the only mammal that can fly, the bat is one of the most frequent nocturnal figures. The subtle perfection of the structure of its body deserves mention. The toes of its front legs which are long and thin are joined together by delicate membranes which run along the

The Australian continent has a very specific fauna, with many forms which are now extinct in other parts of the world. The marsupials are the best represented mammals: here we see the koala (left top and below), the wombat (top centre) and the dasyure (below). Other archaic forms of life include the duck-billed platypus (top right) and the tuatara (below), a reptile which is considered to be a true living fossil.

The best-known Australian animals are certainly the kangaroos, marsupial mammals – forty-five species all similar in appearance but ranging in size from 12 inches (30 centimetres) to about 6 feet (2 metres) in height. Here we see a female with young (left) and one of the small species known as the wallaby (top right). Below: two young kangaroos, not long out of the pouch.

sides of its body and are connected to its hind-legs. Here the toes are free and clawed and enable the animal to hang from all kinds of supports to rest (which it does head downwards!). Between its hind-legs there is another strip of skin which acts as a net in which food is gathered.

The bat hunts by night, guided by a complex and still somewhat mysterious radar system. In fact while in flight the bat emits very high-frequency squeaks which bounce off objects and keep it constantly informed about everything in its vicinity. In this way the bat can fly in safety without having to rely on its eyesight and sense of smell, which are totally inadequate, to catch nocturnal insects on which it lives.

Similarly the owls cannot rely too heavily on their eyesight when darkness comes. Having let out their raucous shriek, almost as if to warn other animals of their presence, and having set their highly sensitive sense of hearing in motion, they take off. The soft, silken nature of their feathers enables them to fly quite silently; the short, hooked beak can with a single blow shatter the head of the victim which is then swallowed whole by the greatly expanded jaws. When the hunt is over, it is up to the stomach to select the digestible substances and reject the others: bones and fur are rejected in the form of small pellets.

So far we have had a glimpse of certain aspects of forest life, but we have not gone into too much detail, for reasons of space, but also because the generic heading of temperate forest covers an enormous number of more localized but equally distinctive habitats in terms of both flora and fauna.

The temperate climate, which is conveniently altered by rain, as we said earlier, is also dominant in those countries (Australia and New Zealand) which, possibly, contain the strangest creatures in the world.

Among these is the duck-billed platypus, an odd character which seems to have been born in error, combining as it does the features of various different species: it has a temperature which sometimes varies with the outside temperature, like reptiles'; its bill is like a duck's, and it lays eggs like a bird; it is covered with hairs and suckles its young like a mammal (it belongs to this class); it can swim, thanks to its webbed feet and the flat tail which acts as a rudder, but it can also walk on land somewhat clumsily and dig long underground tunnels with its claws to reach its lair where each year the female lays two eggs. In addition it is one of the few mammals to have poisonous glands which, in the male, are connected with the movable hollow spur placed at the end of its front legs. As an adult the duck-billed platypus is completely

The duck-billed platypus (top left and below) is an Australian animal which presents nothing short of a zoological paradox. Although a true mammal (it suckles its young), it lays eggs and has a bill and webbed feet which resembles a duck's; a close relative is the echidna (top centre), an animal covered with spines which is shown here with a platypus (top right).

The activity of the lizards, cold-blooded animals, is closely related to the conditions of the ambient temperature; in fact in the morning, to get moving they must reach a body temperature of between 86–98°F. (30–37°C.). A typical feature is their ability to regenerate the tail should it be broken off: but it often grows double or even triple (below left). Top left: a lizard close by its eggs. Top right: the wall-lizard. Below right: the green lizard.

toothless, although almost as soon as it emerges from its egg it is equipped with a set of very simple teeth.

As well as the duck-billed platypus Australia is also host to the so-called spiny anteater, i.e. the echidna. This strange creature has spines covering its back which it can curl up into a ball. It spends its days burrowing in the ground in search of ants which constitute its diet. It is toothless and has a long, sticky, telescopic tongue; like the platypus it lays eggs which, however, it does not sit on and hatch, but arranges in a sort of marsupium where the milk glands emerge.

All the marsupials belong to the group of less-evolved mammals, that is the kangaroos, the koalas which eat the leaves of certain special varieties of eucalyptus; and the various marsupial rats, mice and wolves. These animals differ sharply from one another, but are all characterized by the fact that they are born in such an incomplete state that they have to remain for quite some time in their mother's ventral pouch, called a marsupium (whence the name marsupials) where, as well as being suckled, their development is completed.

The kangaroos have a somewhat poorly developed brain, short front legs, mainly used for tearing up plants and carrying them to the mouth, and very

strong hind-legs designed to bear their 198 lb (90 kilos) bodies when jumping. They also use their long, strong tails when jumping, and can reach speeds of 28 m.p.h. (45 k.p.h.).

We are intentionally leaving until last the description of the mountain environments and their inhabitants, because in a certain sense this can serve as a kind of summary of all we have discussed so far.

Latitude and altitude are in fact comparable because a move northwards is rather like climbing to the top of a mountain. Here, too, we find a *stratification* of the vegetation which, starting from the tropical forests (or the prairies or temperate plains) rises up to the snow-level and the eternal ice of the mountain peaks through a strip of deciduous forest, followed by a taiga-like belt, and then by a flora which starts off by being low and herbaceous and then gives way to forms of mosses and lichen similar to those found in the tundra.

This is why, if we are to give a full picture of the earth's environments, it seems vital to mention the world of the *tundra*, which is a sort of Arctic prairie where the subsoil is permanently frozen, which towards the north gives way to the frozen polar regions.

In the tundra, as everywhere else, life is conditioned by a physical factor: in

Of varying colour, as shown by the photographs (left and top right), the common viper (or adder) is characterized by the presence of poisonous grooved teeth (top centre), an elliptical pupil and a raised snout (below centre). This species feeds on small animals, and mice in particular (below right).

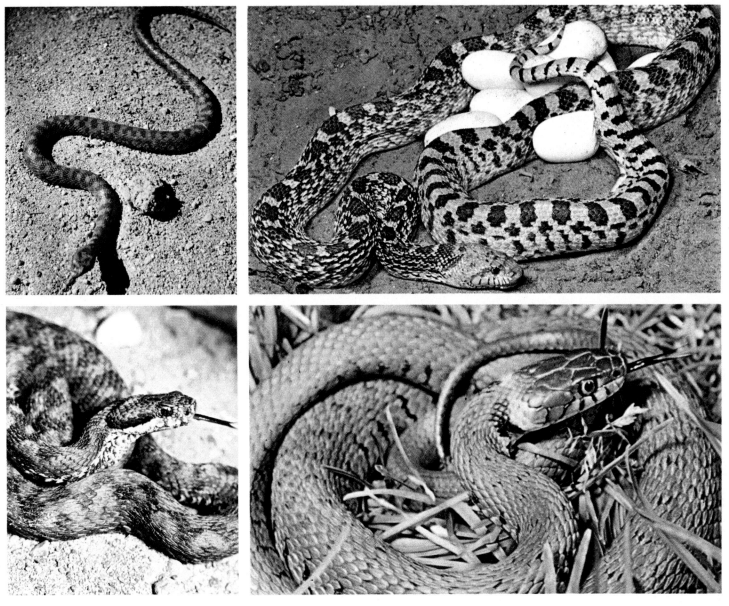

These photographs show how you need a trained eye to differentiate at once between a harmless snake and the dangerous viper, not least because these animals come in a variety of colours. Top left: the checkered water snake. Below: the viper. Top right: a colubrid with its eggs. Below: a ringed-snake. All are harmless.

this particular case it is heat and not water as in almost all the other land environments which we have dealt with so far. Because heat is scarce, precipitation has little effect because evaporation takes place very slowly and thus the existing water reserves are usually adequate.

The tundra is commonly known as *barren land*: in reality the number of species which have managed to adapt to the cold is quite large both in the case of plant-life, which occurs in the form of tough herbaceous species which can make the most of the brief summer warmth to produce lush growth (these include sedge, mosses and lichens) and where animals are concerned, too (mammals, birds and, during the summer, insects as well), which manage to extract the maximum possible nourishment from whatever plants are available. The best-known inhabitants of the tundra are the caribou (also called American reindeer), the true reindeer and the musk-ox. These are all large herbivores which, during the long winter, often head into the northern forests below the tundra to stock up.

The musk-ox lives in herds of twenty to thirty head in Greenland and in the northernmost regions of the American continent. They usually remain in the tundra throughout the year, thanks to the way they have adapted themselves:

their large bodies are covered with a heavy, hirsute layer and their thickset legs end in wide hooves which give them good support when moving over snow or ice. They have a characteristic defence mechanism which they adopt when faced by wolves, their constant foe: as soon as wolves are sighted, the adult oxen form a tight circle with their heads facing outwards, as if to display – for the purposes of intimidation – their strong curved horns. The calves and newborn animals who have not yet learnt how to defend themselves and are the object of the wolves' attack are kept in the centre of the circle.

The caribou never remain in one place. Day and night they roam in large groups along the tracks which link summer and winter pastures in search of food.

Among the smaller creatures, which are the main source of food for the predators living in the tundra, the lemmings are certainly the most curious because of their habits. Because of a distinctive cyclical arrangement, which is renewed every three to four years, the lemming population undergoes a more or less total change. Having increased in number in an amazing way, helped by an extraordinary abundance of food, thousands and even millions of lemmings abandon their habitual territory and head off in all directions in search of

The insectivores are considered by zoologists to be the most primitive mammals, having a placenta. Their name derives from the fact that they feed mainly on invertebrates and more often than not on insects. Left: a shrew, one of the smallest mammals. Top right: the hedgehog with its erectile spines. Below: the mole, a glutton for earthworms.

The Talpidae are insectivorous mammals which have adapted to a specific way of life. The moles depicted in the two photographs on the left and below right have a physical structure which makes them quite at home underground, and is most evident in the form of their front legs and feet; the desman or muskrat (top right), with its toes joined by a membrane is, on the contrary, designed for swimming.

better conditions. First of all their journey is undertaken in small orderly groups but as soon as they encounter the first natural obstacle in their path (a watercourse for example) one can witness the chaotic crush which brings death to much of the population. The reduction in the numbers of the lemmings causes a reduction in the numbers of the carnivorous species which live on them, namely the Arctic foxes and owls.

Talking of Arctic foxes, this is the right moment to mention another interesting and typical phenomenon of the species which live in cold climates. In order to avoid any useless dissipation of heat, all the Arctic mammals are smaller than their counterparts living in warmer parts. Furthermore they have smaller feet, tails and ears which are all completely covered with fur.

The birds which live in the tundra are mainly aquatic: ducks, plovers and skuas only periodically visit this environment, and they do so in summer when the food is more varied and plentiful. The rock ptarmigan, whose plumage changes colour three times during the year (white in winter; grey in autumn and brown in summer) lives here all the year round, however.

At this juncture it only remains for us to deal with the last habitat: life in the mountainous regions. If the problems of high altitudes may affect almost all

Closely related to the domestic pig, the common wild boar, depicted here, which inhabits large areas of Europe, Asia and Africa, is a strong creature with keen senses, and feeds on tubers, roots, acorns, chestnuts and small animals. Note (left) the speckled coat of the young which acts as camouflage in the light and shadow of the undergrowth.

the vegetable and animal species which inhabit the most remote and disparate mountains of the world, the life in the regions at the foot of these mountains and in the areas which might be called half-way up is on the contrary extremely varied. In fact the morphological, physical and climatic nature of these areas can be utterly different, not only from one mountain range to the next, but even on the two opposite sides of one and the same chain of mountains, which may be arid and bleak on one side and lush and green on the other.

By way of example we might mention the rather special case of a mountain chain which rises up from a flat expanse through the series of most typical vegetable stratifications (*deciduous forest, evergreen forest, herbaceous plants, mosses and lichens, and permanent ice* with no flora) to the highest peaks.

The animals which are at home in the flat open country with its temperate climate are above all creatures which live on the ground and in the subsoil. The mammals are represented mainly by the insectivores, rodents and lagomorphs, that is, hares and rabbits. These are small animals which live in burrows and have the curious habit of accumulating provisions for the harsh months.

Moles, hedgehogs and shrews among the insectivores dig burrows in the ground using their front limbs which are appropriately designed for the

Three species of birds which are often considered 'unpleasant'. Top left: a (grey) crow. Below: a rook, a typical bird of ill omen. Right: the cuckoo, a bird which has the nasty habit of laying its eggs in other birds' nests.

purpose; while squirrels, beavers, and whistling hares which belong to the two second groups, gather, respectively, nuts, wood and hay and hide them under the cover of a plant or in the hollow of a tree trunk.

Other typical ground-based inhabitants include certain reptiles, namely the quick-moving lizards and the slower-moving snakes, slower to all appearances at any rate, but still handy in combat. A distinctive feature of the lizard is the way its tail breaks off in the face of an aggressor: this is not a bloody business, but occurs at a point predetermined by a split in a vertebra and also involves the musculature in that area which is designed so as to break off cleanly as well. In Europe at least the most common snakes are the harmless grass snakes, which live mainly in damp habitats, and the considerably more dangerous vipers which defend themselves and obtain their daily bread by means of their powerful poison.

The plains gradually give way to the lower slopes of the range, and the greater diffusion of trees, as compared with the low-lying and bushy plants lower down, combines with a different distribution of animal species. Bears and foxes, wolves and wild boar, as well as badgers, dormice and porcupines, to mention the most common, inhabit these regions, which are generally wooded.

379

Things do not begin to change until the 6,500-foot (2,000-metre) mark: the lower atmospheric pressure with its consequently lower amount of available oxygen, together with a lower temperature, limits the diffusion of plant- and animal-life. Among the vertebrates only the mammals and the birds have managed to adapt themselves to and survive in these conditions.

Ever on the alert for fierce birds of prey, capable of moving agilely over rocky and often precipitous terrain, and not too fussy or demanding about their diet, goats and sheep, ibex and chamois all once inhabited the upper regions of the Alps in large numbers. Today they are reduced to a few strictly protected flocks.

Akin to the ibex and chamois of the European mountains for their agility, caution and resistance to cold and hunger (in summer a chamois will content itself with munching a shrub and a few shoots and buds, and in winter just a little moss and lichen) there are four species of Camelidae which inhabit the upper reaches of the Andes in South America (where there are no species of goats). The llama, alpaca, guanaco and vicuna are all animals hailing from these parts; nowadays they are generally raised for their meat and wool. Their hooves have sharpened edges, are split into two toes and have a thick sole,

Awkward on the ground (below left) but graceful and majestic in flight (top left), the hawks play an important rôle in the various ecosystems on the land, in as much as they control the numbers of many species of animals on which they prey, such as rodents (centre) and snakes, which would otherwise multiply indiscriminately. Top right: a kestrel. Below: a bateleur eagle, found in Asia.

The ibex lives in places inaccessible to other large land animals; it inhabits the mountainous and rocky parts of Europe. It has strong gnarled horns (right) which it uses mainly in battle against other males during the mating period (below left). Top left: a group of Alpine ibex.

slightly concave in the lower section which functions to some extent like a sucker.

The vicuna never leads an isolated life: usually they form two types of group, the males who are free and independent until the reproductive period, and the females and young, with just one elderly male to lead them. If bothered, the vicuna has the odd habit of spitting foul-smelling saliva, chewed vegetation and small leftovers of food over its shoulder at the aggressor.

The constant enemies of the mountain mammals are the large birds of prey, for whom the thin air of high altitudes poses no problem as far as flying is concerned.

Eagles, condors and vultures are completely unbothered by the absence of trees on which to perch or in which to make their nests. Barren crags and cliffs which jut into the void are in fact the best observation posts for these birds, and also offer them inaccessible places where they can live undisturbed. Relying on their speed, and on the strength of their talons and strong curved beaks, these large birds of prey move without warning from their hideouts to pounce in one fell swoop on the prey they have sighted and carry it back to their rudimentary nests where they can tuck into it in peace and quiet.

So far then, we have offered an overview of life as it is distributed across the earth: it is of necessity incomplete, but we have tried to show how the biosphere – that thin layer of life which covers our planet like a halo just 7½ miles (12 kilometres) thick – is not a compartmentalized arrangement.

Even if, to all appearances and above all for our own convenience, we have singled out a whole host of different and separate habitats (deserts and forests, mountains and rivers, open sea and fresh-water marshes) the biosphere is in fact one large network of animals, plants and microscopic organisms which depend on each other in the closest possible way.

The only landscape which we have so far not dealt with is that created, or, perhaps better stated, altered by that other animal: man.

And because man's interference with nature is becoming so far-reaching – because of man's requirements but also because of his rashness – that there is a risk that he will totally dominate our planet, it seems proper that we should also direct our attention to this aspect of the problem to see how animal life may or may not find a place in this new, artificial environment.

From time immemorial the eagle has been surrounded by myths and this bird which lives a solitary life in the most inaccessible of places justly deserves this reputation. Right (top and below): the golden eagle, found across the globe except for Australia and New Zealand. Left: traditional hunt in Sin-Kiang with the eagle.

THE VERTEBRATES AND MAN

Man is an animal with a spinal column and an internal skeleton (or endoskeleton); and like fishes, amphibians, reptiles and birds he must also be considerered a vertebrate. His anatomical structures have further induced zoologists concerned with classification to place him in the class of the mammals, and, more precisely, in the group of the primates: man, like the gorilla, the orang-utan and the other monkeys is thus probably descended from a single original ancestor. Despite this, there are certain features which undeniably make man so different from the other animals that he finds it hard to consider himself an animal and conceives of himself as a higher being, and the highest form of life on our planet.

These differences are usually summarized by the term *intelligence*: it is commonly said that man is intelligent and animals are not. Intelligence is a combination of factors, namely: the capacity to *memorize*, that is, to retain in the mind information supplied by the senses; and then to *associate* this information and thus to *abstract* it. These capacities do not pertain exclusively to man, but are present in a much more limited form in the other animals. By what means has man managed to attain this level?

First and foremost is his brain, the largest and most highly developed of all animal brains. This enormous warehouse of usable information is further assisted by man's physical structure, by the structure of the hands, for example, which can touch, grip, handle and move objects which supply via the senses of touch, sight and hearing a vast array of data to be conveyed to the memory.

Other advantages enjoyed by man as an intelligent being are the facts that he does not become independent of his parents for a very long time and that he lives in a society: it is not hard to see how both these facts hugely increase the quantity of data supplied to his brain. This is mainly because man is equipped with another very remarkable privilege, a language made up of *words*, again assisted by the particular structure of the organs which emit sounds, which offer him an infinite variety of tones and expressions.

If man did not have these features, his physical structure would not basically be very appropriate for survival in the fearful struggle for existence: unable to run fast because of the very shape of his legs, his senses not exceptionally keen, he has managed to make up for these shortcomings by the infinite ways he has found of taking from his own environment all the energy which he himself does not possess. All this has come about in a series of gradual transitions, from fire to the atom.

The relationship between this being, who is not so much superior as different, and the other animals, has always been varied and complex. First for the affection they promote in us, we should recall those vertebrates which, without striking up any direct relationship with man, have kept him company by living in his actual home: the swallows, discreet creatures quite happy to build their nests in the shelter of eaves. The seasonal coming and going of these birds accompanies the biological rhythms of nature and reminds man of these rhythms from which he stubbornly tries to detach himself. Then there are the swifts, distinguishable at a glance by their black breasts and larger size: in the air they are just as unpredictable and tireless as the swallows. Have you ever seen a swift still? Its legs are so short that zoologists have named it *Apus*, which means footless.

Man's relationships with the other vertebrates have nevertheless almost never been limited to simple cohabitation or coexistence. After a period in

which men and animals found themselves out of common necessity sharing the same caves and the same prey (the bones discovered in the caves dating back to those days tell of the first battles fought between the two parties, often ending with the death of both combatants from the wounds mutually inflicted), man managed to assert himself over his fellows, not by force, but because of his intelligence and creativity which enabled him to invent tools.

In fact this phase in his history includes the first rudimentary sticks used to dig the soil and flush out small wild animals; the first sharpened stones which were flung at any particularly large and savage prey; the first cutting tools to carve up the victim and make it easier to carry back to the cave and distribute to the various members of the group. Gradually, and precisely because of increasingly refined tools and techniques, the number of animal species of interest to man also increased considerably. Not only did our forbears hunt mammoths, horses and reindeer, they also hunted birds, reptiles and even seals and whales.

Swords, spikes, hooks, harpoons, bone and stone needles, bows and arrows and hides were nonetheless simply mechanical improvements of the most varied techniques whereby man, by using to the full the strength of his own

Man's dwellings house numerous species of animals which are often considered harmful or at least bothersome. This is not the case with the swallow (top left, centre and below), which has always been appreciated for its carefree good humour and as a harbinger of summer. City skies, on the other hand, are dominated by the swift (top right), a tireless flier with completely black plumage.

Probably derived from fertile hybrids of various forms of Canidae, the domestic dog is one of the animals which has undergone the most thorough selective process by man, to such an extent that nowadays there are more than two hundred races. It is worth noting that individuals belonging to different races show different instinctive attitudes, as when defending themselves for example (below left) and hunting (right).

muscles, managed to obtain food and clothe his body. In other words, all these inventions were aimed at allowing man to save his own energy. The next step, in addition to fire, was the discovery of alternative sources of energy. This came about, for example, when man domesticated his first animal: the dog.

Mankind's first contact with dogs was very probably associated with the rubbish produced by his own home. Man could have chased the dogs away but before long he became aware of the dog's ability to run, follow a scent and perceive sounds and noises beyond the range of man's own ear. This is why an alliance sprang up between the two species, based on hunting, which was advantageous to both parties. With the help of an appropriately equipped man, the dog was more certain of being able to catch game than on its own, and in addition it would receive a good ration of bones and innards. For its own part the dog could contribute to this happy arrangement by tracking down small prey, pursuing fast-moving herbivores, and pursuing and lying in wait for large animals until the master's arrival. At night it could keep its master warm with its thick coat, warning him of any imminent danger; in some cases it could even help men to drag heavy loads. And even today, in the less-developed societies, the relationship between man and dog differs according to the task

entrusted to the latter. In fact, although in agricultural societies the dog is more or less ignored, and allowed to live only on the very edge of the human community, sheep dogs are well treated everywhere because not only do they protect the flock from attacks by wolves, but they also keep the flock together and thus save the shepherd from using a considerable amount of energy.

If in the course of history the welcome extended to the dog has been slight compared to the treatment of dogs nowadays (it was only in the late Middle Ages that dogs were introduced to court alongside nobles and princes; whereas the fashion for dogs as companions, living with their master on the basis of affection and nothing else, was introduced during the Italian Renaissance), the cat was certainly not received any more warmly.

Apart from a brief spell of glory enjoyed by the cat in ancient Egypt, it was never greatly favoured in Europe. Brought there by the Arabs during their invasion, it was used during the Crusades purely as a servant to keep down the vast numbers of mice. Interest was only really shown in the wild cat which supplied hides to be tanned. The common cat was relegated to the kitchen and the inn because it was considered too asocial; it could be neither tamed nor even trained for contests or games. The vogue for cats was born in Paris at a

The fact that the dog has followed man into almost every region inhabited by him owes much to its pronounced ability to adapt to very different environments, more or less in the same way as man does. Thus while the races found in the temperate zones are used mainly for hunting (top) and defence, there are dogs which also offer invaluable help to man by pulling his sleds in the Arctic (below).

When man was still ignorant of the laws of genetics, he managed to apply them unwittingly and created canine races with surprising physical structures. This is the case with the extremely elegant dogs depicted in the photographs here.

much later date.

In the nineteenth century in fact, as a result of the then recent discoveries in the field of hygiene, there was widespread terror about any and every possible source of contagion. Where animals were concerned, dogs and horses represented possible sources of disease and as such they were kept somewhat apart. The same did not apply to the cat which, to all appearances at any rate, came to be considered an extremely clean creature which respected the most modern hygiene standards because of its habit of continually licking its fur to remove from it all trace of dust and dirt.

Cats were thus allowed into the home, but unlike dogs did not offer companionship and demand in exchange affection and caresses: even within the four walls of the home cats usually retained their independence and their basically solitary nature. To such an extent is this true that scholars consider that where cats are concerned one should talk in terms of self-domestication in the sense that these animals have not been domesticated by man but have domesticated themselves by entering human homes and adapting to the life there.

The third and last animal with which man has established a more friendly

and intimate relationship is the horse. But in this case the basis is more one of collaboration than of a simple communication of affection: horse and rider must in fact form a harmoniously functioning unit together.

To grasp the importance of these creatures, it is enough to consider that in earliest times the horse was used in armies and that cavalry appeared for the first time in Greece in the seventeenth century BC. Then in ancient Rome, with Caesar, the cavalry was used for what in modern times was to become one of its main tasks: reconnaissance.

The Arabs, who had always carefully raised and selected tall, agile horses which were high spirited and extremely swift in the field – and are still highly prized today – won many major military victories with their cavalry.

At an even later date the mounted warriors of Ghenghis Khan conquered Europe. In his writings Marco Polo relates in fact that the Mongol leader had organized a very efficient mail service to keep tabs on all the territory he had conquered.

In the West the horse became the basis of political power: the whole feudal system in Europe in the Middle Ages was in fact based on the efficiency of and respect for an armed horseman: land and wealth were always considered to

Many races of dogs once used for hunting are nowadays appreciated mainly as companion animals. This is the case with the terrier, a dog which is English in origin. Top left: the fox-terrier. Below left: the hunting terrier. Centre: an Airedale terrier. Top right: the Boston terrier. Below right: the Bedlington terrier.

If man has managed to mould the dog's instinct by exploiting its advantageous qualities, the same success certainly does not apply in the case of the cat, an animal with an extremely independent nature which is hard to train and condition. The photographs show three typical pictures of the domestic cat.

take second place. It was the existence of large, heavy suits of armour which made it necessary to select stronger and tougher varieties than the swift and nimble Eastern steeds. Of course the development of these features sacrificed the quality of speed and the horses' resistance to climates other than Western; and in fact the European horses were unable to tolerate the hot climate of Palestine where during the Crusades they died in great numbers.

In times of war, victory or defeat depended largely on the strength and persistence of the cavalry; in peace time horses represented the most common and surest means of transport, as well as the basis of jousting tournaments and contests which were then very fashionable.

The beginning of the end of the horse as an animal for use in war and as a means of transport was the invention by James Watt of the steam engine. But it should not be thought that man's machines instantly and straightaway replaced the work that horses had carried out for him for centuries (and the work carried out by donkeys, camels and oxen). The way for the mechanization of land vehicles was opened in 1830 when the first steam railway was opened in England between Liverpool and Manchester to transport people at the giddy speed of 15 m.p.h. (24 k.p.h.), as opposed to the 9 m.p.h. (14 k.p.h.)

Left: the African tawny cat, a wild form which is considered to be the ancestor of the domestic races. Right: two of the most highly prized domestic races, the Persian cat (top) with its long fur, and the Siamese cat (below) with its typical two-tone coat.

of the French stagecoaches which were considered to be among the fastest.

If the horse, from the viewpoint of its practical use, suffered considerably from the rapid advances made by man where vehicles and machines in general were concerned, it simultaneously acquired, or rather promoted, a rôle which it had already been playing for ages: namely, that of a race-horse. Carefully selected and bred and trained, race-horses started to attract large crowds of both sexes and gave rise to a new and thriving industry: betting.

Today, quite rightly, horses can be considered as 'jobless'. They are no longer used as means of transport (modern asphalt roads are not even suitable for their hooves); even as a source of entertainment they have lost their edge (now that football has monopolized the attention of millions of fans all over the world); and they have also lost their original importance in the military field (in the last great war they were used almost exclusively in the rear as auxiliary draught animals).

Fate has not treated the horse's cousins – donkeys, mules and hinnies – quite so drastically. Less delicate and demanding, and used only for transporting goods on poor and tricky roads by societies too poor to have mechanized transport, they have managed to keep their rôle as man's helpers almost intact.

Originating from wild species probably from China and Siberia, the present-day race of horses are divided into the dolichomorphic (long and slender), the mesomorphic (stronger, but still swift) and the brachyform (very strong indeed and large). Top left: Icelandic ponies. Below: horses from the Fijian Islands. Right: a pair of horses with a foal.

Before concluding this glimpse at the animals which, like the dog, the cat and the horse, share part of their existence with man's, we should briefly mention the carrier pigeon.

More economically viable than trains and vehicles, carrier pigeons can travel as far as 625 miles (1,000 kilometres) in an uninterrupted flight of fifteen to twenty hours at an average speed of 55 m.p.h. (90 k.p.h.). As a result their use as an 'air-mail' service was not limited just to war time, but saw peace-time use as well. Tradition has it that at the beginning of the nineteenth century certain influential businessmen used these carriers to send valuable and secret economic bulletins from one continent to another.

But here, too, machines put an end to a career which had known moments of true glory (carrier pigeons which were responsible for commendable missions were regularly awarded medals for gallantry).

As we have already mentioned on several occasions in this discussion, man is still strictly reliant for many of his requirements on the natural world in which, in this respect at least, he plays the rôle of consumer. Mankind still turns to the animal kingdom (and the world of plants) for many of the ingredients of his diet.

The daily requirement of protein, vital first for his growth and then for the maintenance of his body, has always been supplied to him by the flesh of many sorts of vertebrates: we could in fact say that species of all the classes of the vertebrates are used in one way or another, depending on the tastes, customs and menu of people in the various parts of the world. Of course at the dawn of their history men did not have many demands: but whatever came within range of their arrows or spears was always and certainly welcome. Then their demands gradually became more specific and refined: they went in search of animals whose flesh was tastier and tenderer, eating only the choicest parts of them, and leaving the rest for their dogs.

Archaeologists (who are concerned with man's earliest history) maintain that there are basically two premises for the domestication of animals with the purpose of using them at a later stage: the spread of farming practices which had turned the nomadic peoples into sedentary and attentive observers of the phenomena and laws of nature; and the invention of the axe made of smooth, dressed stone which was capable of working wood and thus enabled man to build enclosures to stop his animals from going astray and to protect them from attacks by carnivorous wild animals, which roamed the earth in large

Few animals have been as important to the development of man's civilization as the horse: for many peoples in fact it has been the only form of beast of burden and war-machine (left: Moroccan warriors). Nowadays, with the advent of mechanization, the horse is used in Western societies almost exclusively for racing and other spectacles (right).

In regions where a great deal of hunting is carried on, the impoverishment of the fauna obliges man to raise animals which are not easily domesticated so as to repopulate the denuded areas. This applies to pheasants (left and top right: the common pheasant) which are raised in special pheasantries (below right).

numbers in those far-off days.

The first animals to come under man's sway were essentially four in number, all belonging to the so-called palaearctic fauna living in the northern temperate zone of the Old World from England to China.

Scholars disagree somewhat about the origins of these four species; most of them do agree, however, that the forbear of the present-day goats was the wild goat of Turkestan and Afghanistan; that the ancestor of the sheep was the mountain argali of northern Iran; that the parent of the cow was the great long-horned *Bos primigenius* found in the southern plains of Russia together with possibly another smaller species; and that the pig is derived from two species, the wild boar, which is still found in Europe, Asia and Africa, and a wild pig found in South-east Asia.

It is hard to say how the first herdsmen managed to capture their first head of livestock: perhaps one of the ways was to raise and feed young specimens caught during hunting forays until they were mature adults.

Whatever the case, and independently of the methods used to capture them, goats, sheep, cattle and pigs formed a perfectly balanced quartet.

The goat is a tough animal which can graze on any kind of inaccessible and

hard terrain. It supplies meat, milk and a skin which is particularly suited to making receptacles; its hide, which is smooth and waterproof because of the fatty substance which makes it soft, is excellent for making garments and tents.

The sheep supplies milk and, above all, wool; it is not a fussy eater, its favourite food being green or dried grass.

The pig provides a good supply of meat; it grows rapidly and produces large numbers of young at a time. It, too, is happy with the simple things of life, feeding on acorns and tubers. Its teeth make fine knives and necklaces.

Unlike the three animals above, cattle are of real use to man as a means of carrying heavy loads. Even today, along the Persian shores of the Caspian Sea, in the Caucasus and in the Basque provinces, oxen are frequently harnessed to ploughs, harrows and carts. As well as supplying meat, the cow also supplies milk, an extremely important food because it contains all the essential ingredients of man's food and is the raw material for the manufacture of other products (butter and cheese) which can be preserved for longer periods. The hide of cows is used for making straps and belts, and in particular, footwear; the horns are used to make receptacles and household objects.

Man's first violations against nature occurred as early as the distant

Goats are often kept in poor regions, precisely because these animals manage to live in such barren places (top left) to which other grazing animals, such as cattle, cannot adapt. Below: goats in a Peruvian village. Top centre: goats in a Swiss setting. Below: herds of goats in Georgia (USSR). Top right: a nanny-goat with her evident fetlock and wattle. Below: a European goat.

Found in the wild state in almost every corner of the world (left top and below), cattle are among the most sought-after livestock because of the quality and abundance of their meat and milk. Centre: the guar, an Asian cow which can be domesticated. Top right: cattle of the Maremma in Italy. Below: cattle in the Oise region of France.

Neolithic era. Selecting those plants and animals which were of most use to him, he limited and finally destroyed some of these species and obviously seriously damaged nature's stock. Changes were noted in the size and shape of the body of the animals which had become domesticated: the horns and teeth, no longer used as defence measures against enemies, became much smaller; the legs, now no longer used for running from predators, grew shorter and squatter; the bodies became more thickset and fatter, offering man more meat and other products.

Among modern peoples it was the Portuguese who introduced animal colonizaton into newly conquered lands. The initial discovery of the New World was immediately followed by surprise at finding this world uninhabited by those animals which were by then part and parcel of mankind's daily life. The natives of South America had never heard of goats or sheep or cattle or pigs or horses which were all introduced first by the Portuguese and then by the Spanish so as to ensure their own survival in these new lands where they had come to settle.

In discussing domestication and animal colonization, we should not forget two mistakes made by man in his attempt to get round the rules of nature: the

first is called 'goat' and the second 'rabbit'.

Totally heedless of the laws of nature, man had in fact introduced these two animals into the islands of Saint Helena and Australia respectively, secretly hoping to reap great rewards from raising them. But in both cases the result was simply the destruction of the host-territory by the two species in question which simply upset the balance in both places. In fact the goats destroyed the mantle of vegetation which was not designed to stand up to their hard, sharp hooves, and thus destroyed the lives of many tiny native creatures; the rabbits multiplied so copiously, in the absence of their natural carnivorous enemies, that they too wrought havoc all across Australia.

Just as the South American countries were rich in exotic fauna, which the natives treated in much the same way as the primitive Neolithic people treated their fauna (that is, they simply hunted the animals according to their requirements without really trying to breed and raise them), so the cold regions of North America presented a fauna comparable to that of Europe with an additional abundance of animals with fur coats. (In South America hides and wool were supplied in plentiful amounts by llamas, alpacas and pumas.)

The cold regions of Canada – like the frozen wastes of Siberia – contained a

Where the environmental conditions are too hard for the horse, its work is done by Camelidae, extremely strong animals; in the cold deserts of Asia the camel is mainly used (top left: a picture taken in the Gobi desert), while the dromedary is used in the hot deserts (below left and right) and the llama in the Andes (top right). Opposite: a herd of a type of alpaca, a species raised mainly for very fine hide.

large number of foxes, beavers, elks, martens and deer whose hides and coats were of no particular value in commercial terms for the people who inhabited those places in those days. It was the English first of all, followed by the French (matched in Russia by the Cossacks) who launched one of the markets – for furs and hides – which meant that many animal species were subsequently exterminated.

Today, with thousands upon thousands of animals slaughtered and hundreds of species extinct for ever, fairly strict measures have been taken to safeguard the lives of those animals which have managed to escape man.

In other cases (the most widespread today is that of the mink), attempts have been made at controlled and almost totally mechanized breeding. As happens for different reasons with battery chickens (which are virtually 'artificial' creatures because their life and food are so totally ruled by precise schedules and clearly defined synthetically prepared feed), so the mink too is reared 'at home' to obtain large amounts of fur with the qualities specifically required by the market.

The vastness of the sea has always intrigued man, as have its moods which can switch from being utterly calm to inexplicably tempestuous, and its

The raising of farmyard animals has always been a useful addition to agriculture in rural areas: but recently it has assumed a truly industrial aspect and is thus of considerable economic importance.

Salmon-fishing is based on the exploitation of one of the most remarkable forms of reproductive instinct known to us. This fish, born in mountain rivers and then descending to the sea (centre), returns when sexually mature to the rivers of its youth to lay its eggs (below left). Accustomed to overcoming all sorts of obstacles (top left), it nevertheless fails to avoid the fisherman's net. Right (top and below): two stages of salmon-fishing.

inhabitants which live in huge numbers in the uppermost waters and in the deepest abysses.

And man has always turned to the sea (as he has to other forms of aquatic environment) as if it were an inexhaustible and prodigious larder.

This is why some of man's first inventions were tools which were vital for catching fish: hooks, nets and harpoons appeared at the same time as the first rudimentary boats. And man tried to put every part of the catch to use: the flesh, which was nutritious and tasty, was eaten; the skin was dried and used to make tools; the bones and other remains which were casually thrown away on the ground proved to be excellent fertilizer for plants which grew all the more luxuriantly. And still today this is what man expects his catch to provide, and still today man expects his catch to be as large as it once was, be it in salt water or fresh water.

In reality, here, too, the theme of man's blame comes to the fore. Either directly or indirectly polluted, the various waters of the world are for the most part dead. Much fish has vanished; some bob to the surface, poisoned in their own environment; others, which have managed to survive, accumulate in their bodies such high amounts of toxic substances that they are themselves lethal if

eaten. As in the case of the vertebrates, man has also tried to find ways of breeding fish artificially. Fish-farming usually applies to just certain species (trout, eels, salmon and sturgeon) which live completely, or at least partly, in fresh water, where the physical and chemical factors are easier to reproduce and keep constant than the same factors pertinent to salt water.

Apart from the food and clothing which man obtains from animals, he also owes many of his scientific and technical advances down the centuries to animals. Anatomy, physiology, physics, chemistry and pharmacology, for example, are all disciplines which have employed animals for their most decisive experiments, and usually those animals which could offer man neither food nor clothing. These animals (and they include in particular frogs, tortoises, mice, guinea-pigs and rabbits among the vertebrates, as well as a truly vast array of insects, crustaceans and molluscs among the invertebrates) had, and still have, to put up with everything that man has invented but is unable to try out on himself or his fellow-men.

By way of example, let us mention Galvani's frog used to demonstrate the existence of animal electricity; the ram on which Guillotin tried out his fatal guillotine; the trio consisting of a sheep, a cockerel and a duck which were the

Cod-fishing is very common in northern Europe, and north-west Canada and America. It is carried on mainly in winter when the shoals of cod gather to mate and reproduce in the vicinity of the Lofoten islands, off the coast of Newfoundland and south of Iceland. Below left and right: cod hanging up to dry. Top right: stages of cod-fishing.

The massacre is the last stage of tunny-fishing as still practised in Sicily and Sardinia. Driven by a complex system of nets, the tunny-fish are forced to enter an area known as the 'death-chamber' (below left). Here they are beaten with hooked poles and once worn out hoisted into the boats. Top left and right: two stages of the tunny massacre.

first to try out the thrill of flying in the paper fire-balloon launched at Versailles in 1783; and the numerous dogs and monkeys which, in modern times, have been shut up in space capsules for flights into space both real and simulated.

To animals we also owe the first discoveries and applications of vaccines and various types of serum to combat fatal diseases. Closely associated with the advances achieved by the new mechanical means of transport, the nineteenth century was in fact a period in which disease was very widespread, and because of the new transport systems many diseases managed to become international. If cattle, sheep, dogs, horses and camels were decimated by anthrax, many human beings fell victim to rabies, smallpox and diphtheria. The technique of preventive inoculation (e.g. vaccination) adopted and experimented with by Jenner on cows was perfected by Pasteur to deal with anthrax.

At this point the way was wide open: once the agent of the disease was discovered, a small quantity was inoculated into a healthy animal which might be a horse, a cow, a monkey or a pig. Thus infected, the organism released into its blood the antidotes capable of combating the disease. When appropriately recovered and purified, these substances so produced represented for man and other animals the means of preventing or combating an infection of that order.

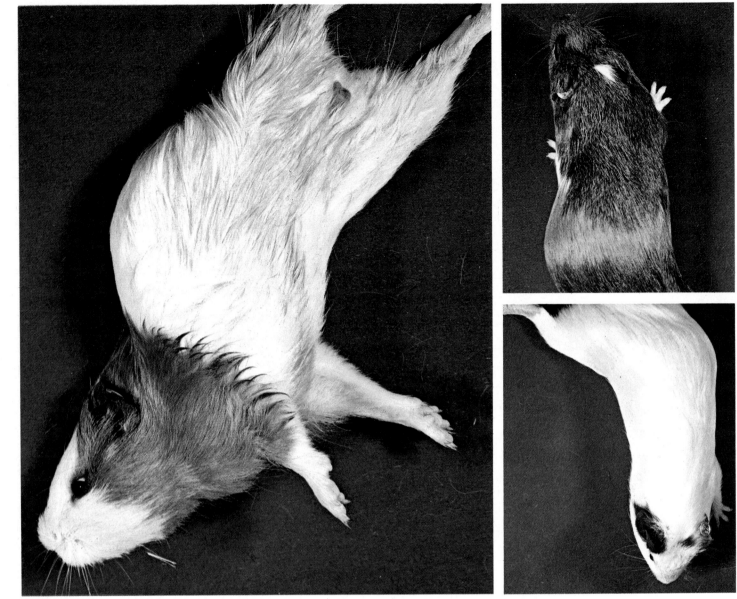

The history of human civilization is full of such examples. Exploited and revered, destroyed and domesticated, feared and imitated, animals have known a history that is no less fascinating than man's. But this has not meant that, apart from what man's vital needs have been and will continue to be, mankind has treated the animal world in any way other than as a selfish master. And this is largely due to the continuous and dizzy increase of the population, which is numerically the largest of all large living beings. This is why, in a certain sense, it is up to man to decide how many animals he needs for food, how many from which to obtain the material for the manufacture of his clothing, and how many may be used to transport his goods and chattels and his own person, and how many may be used as 'guinea-pigs' in his experiments. And there will never be a surplus of animals in the world: why, for example, if wool can be obtained by biochemical means in a laboratory, should man continue to raise large flocks of sheep? If one day a synthetic replacement for meat is discovered (and in certain plants, such as the soybean, quantities of protein have been discovered which do not fall far short of the amounts contained in meat), why should we continue to breed livestock with all the economic and organizational problems that this entails in a world of 'machines'?

The cavy or guinea-pig is a rodent originating from Peru and can be considered the typical laboratory animal. Subjected to often cruel experiments (the photos show guinea-pigs under the influence of tetanic toxins), they have supplied man with extremely important physiological and pathological data.

Man's habit of clothing himself with the hides of slain animals dates way back to prehistoric times: nowadays many kinds of fur can be bred, as is the case with the chinchilla (right), thus avoiding the danger of extinction. It is inconceivable, on the other hand, that the existence of wild species such as the ermine (left) should be threatened for reasons which are basically purely ornamental.

And so if man has the power to decide whether those animals which have been of use to him in the past should live or die, what principles should he adopt in the face of that whole multitude of species which, to all appearances at least, do not benefit human society, and even 'rob' it of precious space and in some cases represent an irksome danger?

The immense concrete jungles of man's cities have grown up and out to the detriment of those green expanses of plants and trees inhabited by animals which, when not killed or when they do not die naturally because they have been unable to adapt to a world which is not theirs, have been pushed further and further towards the edges, leaving man with just an ever-growing swarm of disease-bearing mice and rats as inseparable companions.

The forests of the Amazon and India, the far-flung deserts of Africa and Asia, hitherto considered uninhabitable by man, and, naturally, the deepest abysses in the sea and the unscaleable mountain peaks still represent oases where life has remained uncontaminated by man's interference.

Must we therefore prepare ourselves to bid farewell to the animals, and also resign ourselves to losing those which we consider part of our lives (not forgetting that life is like a net which involves every single sort of creature in its mesh)?

Perhaps, now, we can entertain a ray of hope for greater human prudence. Following the example set by England, where animals enjoy more respect and consideration than anywhere else in the world, National Parks and Reserves have been established in almost every other country where the animals can find both the environment most suited to their nature and, if need be, adequate concern for their survival, in such a way that species which still survive today need not follow in the footsteps of many of their less fortunate fellows at least in the immediate future. Two facts give an immediate idea of the magnitude of the problem: in the last two thousand years about two hundred species of vertebrates have vanished from the face of the earth, but at least a third of these have done so in the past fifty years. If one considers that there are currently at least three hundred and fifty species in danger of extinction, it is not hard to see that, without adequate preventive measures, their fate is sealed.

This would undoubtedly be a grave loss: every animal that becomes extinct takes away with it a small part of the history of life on earth.

It is well known that man's activities are endangering the existence of an ever larger number of animal species. Given this state of affairs, the role of the national park becomes increasingly important; the Yellowstone National Park in the United States is a typical example. Here in fact together with incomparably beautiful natural phenomena, numerous animal species have found a safe place in which to live (below left).

sylvatica) 97, 126
Forsythia 124
Fossils *25–31,* 25–29, 50, 84, 100, 101, 137, 139, *140,* 143, 145–6, *146–7, 154,* 176, 242–3, *295–6,* 299, 305, 370
Four o'Clocks (*Mirabilis*) 18
Fox *366,* 379, 298; Arctic 377
Foxglove, purple (*Digitalis purpurea*) 97, 127
Fraxinus – see Ash
Frigate-bird *65, 284*
Frog 60, *64, 155, 208,* 231, 289, 298, 317–18, 319, 320–2, 326; tropical 342
Fumarioideae 117
Fungus – *see also* Mushroom; toadstool 16, *35,* 37, *42,* 43, *56,* 68, 78, *79,* 80, 82, 232, 236, 245–6, 248, *249;* Dutch elm 111–12; gill 81; imperfect 81–2; malefic pore (*Boletus satanus*) *80;* pore (*Boletus edulis*) *80;* ringworm 82; sac 80; rust 81; smut 81; slime 79; true 80

Gabon 28
Galago *336*
Gametophytes *83,* 83, 85, 94, 95
Gardenia 128
Garlic 132, *133*
Gastropacha quercifolia 233
Gastropods 28, 184–5, 197, 201, *208, 212,* 216, *256,* 260
Gavial 296
Gazelle 46, 346, *347;* dorcas *347;* Grant's *347*
Gecko 288
Gentiana (gentian) 96, 125; *G. aculis* (Alpine) 125; *G. andrewsii* (bottle) 125; *G. asclepiadea* 125; *G. crinita* (fringed) 125; *G. lutea* (yellow-flowered) 125
Gentianaceae 125
Gentianales 125
Genista tinctoria (Dyer's greenweed) 119
Geometridae 225, *233*
Geomyd 351
Geraniales 114
Geraniaceae 114
Geranium 114
Gerridae (*Halobates*) 176
Gibbon 332, *335*
Ginger, wild (*Asarum*) 123
Ginkgo biloba (maidenhair tree) 101
Ginseng (*Panax quinquefolium*) 96, 122
Giraffe 24, 293, 296, 349–50, *351*
Gladiolus 132
Globigerinae 180
Glycyrrhizia lepidota (liquorice plant) 119
Gnat, water 176
Gnetales 101
Goat 155, 380, 393–4, *394,* 396; wild 393
Golden club (*Lysimachia vulgaris*) *123*
Golden rod (*Solidago*) 130–1
Goldfinch 364
Goose, wild 366
Gorilla *23,* 296, 332–3, *333*
Gossypium herbaceum (cotton plant) *120,* 120
Graminales 132
Gramineae 132–3
Grape (*Vitis*) 89, 269

Grapefruit 115
Grass 132, 133, *134;* marsh 154
Grasshopper 159, 169, 171, *174, 226,* 227–8, *230, 233,* 233, 343
Grebe 288, *294*
Ground-cherry *126*
Grubs (pupae) 169, 218, *224,* 224–5, *232,* 265
Guacharo, South American 325
Guanaco 346, 351
Guar *395*
Guillemot *293*
Guinea-fowl 296; -pig 351, *402*
Gurnard 306–7
Gymnocladus dioica (Kentucky coffee tree) 119
Gymnosperms 71, 86, 88, 94, 100–5, *101,* 106, 107
Gymnothorax mordax 302

Hackberry 112
Halobates (Gerridae) 176
Hamadryad *331*
Hamamelidaceae 118
Hamamelis mollis (witch-hazel) 118
Hamster 364
Hare 378; Californian 350, common *363;* polar 359, *363;* whistling 379
Harvestman *252*
Hawfinch *369*
Hawk 363, *380*
Hawkmoth bumble-bee *37*
Hazel (*Corylus*) *108,* 110
Hazelnut (filbert) *99*
Heather (*Calluna vulgaris*) 124
Heaths (*Erica*) 124
Hedera helix (English ivy) 122
Hedgehog *376,* 378–9
Helianthus annus (common sunflower) 130
Heliotrope (*Valeriana officinalis*) 128
Helix polatia 171
Helobiae 136
Hemipter, cryptocerate *215, 232*
Hemlock (*Tsuga*) 102, *122;* poison (*Conium maculatum*) 123
Hemp (*Cannabis*) 112
Henbane, black *126*
Hen 34, 319
Hepatica 115
Hepaticae (liverwort) *56, 83,* 83
Heron 323, *324–5*
Herring 301
Herring gull 292
Hevea brasiliensis (Brazilian rubber tree) 115
Hibiscus (rose mallow) *120,* 120–1; *H. esculentus* (okra) 121
Hickory (*Carya*) 111
Hinny 390
Hippocastanaceae *121,* 121
Hippopotamus *282,* 296, 318, *350*
Histiophoridae 306
Holly (*Ilex*) 121
Hollyhock (*Althaea rosea*) 120
Holothurian (sea-cucumber) 160, *185,* 187–8, 193
Homarus americanus 191
Homopter *271*

Mussoruna 341–2
Mustard *99*, 117; greens 117
Mustela nivalis 360, 360–1
Mustelidae *360*, 360–1
Mustelus canis 302
Mycena 35
Myosotis scorpioides (forget-me-not) 126;
  M. sylvatica (garden forget-me-not) 126
Myriapods 171
Myricales 124
Myrtaceae 120
Myrtales 120
Myrtle, common (*Vinca minor*) 125–6
Myxomycophyta (slime fungus) 79

Nandu 351
Narcissus 90, 132; *N. jonquilla* 132; *N.
  poeticus* 132; *N. pseudo-narcissus* 132
Nasturtium officinalis (watercress) 117
Nautilus *28, 177, 183*; paper *183*, 184; pearly
  183, 183–4
Nematodes 158, 217, 229
Nemertines 214
Nepeta cataria (catnip) 127
Nettle, stinging (*Urtica urens*) 112
Newt 62, 298, *316*, 316–17, 322
Nicotiana tobacum (tobacco) 96, 127
Nightingale 292
Nightshade *126–7*
Nitrobacteria 236
Notonectid (back-swimmer) 209
Nummulite *146*
Nymph 213
Nymphaeaceae (water-lilies) *44, 47*, 115

Oak – *See Quercus*
Oak, poison (*Rhus toxicodendron*) 121
Oats (*Avena sativa*) 96, 133
Ocelot *279*
Octopus 172, 181–3, *183*, 272
Odontoglossum grande (orchid) *135*
Oenothera (yellow evening primrose) 120
Oil palm 135
Okra (*Hibiscus esculentus*) 121
Olea europa 102, *124*, 124, *125*; *sativa* 124
Oleaceae 124, *125*
Oleales 124
Oligochaetes 168
Olive tree (*Olea europa*) *102, 124*, 124, *124,
  125*; wild (*Osmanthus americanus*) 124
Olm 325
Onagraceae 120
Onion 54, 90, *91*, 132, *133*
Ophioderma (Ophiuroidea) *192*
Opossum 296, 338, 364, 367–8
Opuntia (prickly pear cactus) *111, 113*
Opuntiales 113
Orange 115, 230, *271*
Orang-utan 296, *332*
Orchid (*Cypripedium*) 95, 106, 134–5, *135*;
  green adder's mouth (*Microstylis unifolia*)
  135; lady's slipper, pink (*Cypripedium
  acaule*) 135; lady's slipper, white (*C.
  candidum*) candidum) 135; lady's slipper,
  yellow (*C. parviflorum*) 135; yellow snake
  mouth (*Pogonia ophioglossoides*) 135

Orchiaceae 134
Orchidales 134
Orchesella 239
Oregano (*Origanum vulare*) 127, *128*
Oribatidae 236
Oriole 293
Orosphaera serpentina 181
Orpine (Crassulaceae) 117
Orthopter (*Tettigonia viridissima*) *169*, 227
Oryza sativa (rice) 132, 133
Oscarella celata 186
Osmanthus americanus (wild olive) 124
Osmylidae 233
Ostracoderm 274, 280
Ostrich 296, 348, *349*, 351
Ostrya (hop-hornbeam) 110
Otaria 309
Otter 289, 315, 318–19
Owl 113, *290–1, 293*, 345, 353, 363, *368*, 371
Oxpecker 347
Oyster 24, 158, 159, 164, 169, 176, 180–1,
  185, 197, 198, 199, 204, 264–5, *268*

Padina pavonia (brown algae) *77*, 77
Paederus baudii (rove beetle) *244*, 251
Palmaceae 135
Panax quinquefolium (ginseng) 96, 122
Pandaka 316
Pandanales 136
Pangolin 336
Pansy, garden (*Viola tricolor*) 114; wild 114
Papaveraceae 116
Papaver radicum 116; *P. somniferum*
  (common poppy) 116
Papilio trolius 234
Papilionoideae 119
Paradise, MacGregor's bird of *293*
Parage aegeria 223
Parakeet 329
Paramecium 155, 164, 204, *208, 210*
Parietales 114
Parmelia coperata 82
Parrot 296, *329*, 335–6
Parsley (*Petroselinum crispum*) *99*, 122, 123;
  hemlock- (*Conioselinum*) 123
Parsnip (*Pastinaca sativa*) 123
Partridge 338
Pastinaca sativa (parsnip) 123
Patella vulgata 171
Pawpaw 116
Peach *117*, 118
Peacock *23*, 296
Peanut (*Arachis hypogaea*) 119
Pear tree 220
Pea, garden (*Pisum sativum*) *18*, 18, 20, 21,
  99, 119, 119; sweet 95
Pecan (*Carya*) 111
Peccary 337
Pecten *173*, 173, 199
Pelican 323–4
Penguin 31, *38, 158, 293*, 312
Penicillium notatum 81, 81; *P. roqueforti 81*
Pepper, black (*Piper nigrum*) 112
Peppermint (*Mentha piperita*) 127
Perch *289*, 316
Periwinkle (*Vinca minor*) 125–6

413

PICTURE SOURCES

Museums, Libraries, Research Institutes and Learned Societies

British Museum (Natural History), London; Chaumenton Collection; Musée d'Histoire Naturelle, Paris; Museo Civico di Storia Naturale, Milan; National Museum of Natural History, Leyden; National Museum of Natural History (a branch of the Smithsonian Institution), Washington, D.C.; University of Mainz; U.S.I.S., Rome

Commercial, Industrial and Other Sources

Canadian Aereoservice Ltd.; Coboi, Trieste; Embassy of Japan, Rome; Guide Agricole Philips; Metropolitan Water Board, London; Praga Rapid; Shell; Syndication International

Artists

Giorgio Arvati; G. B. Bertelli; Luciano Corbella; Piero Cozzaglio; Raffaele Curiel; Giuliano Minelli; Sergio Rizzato; Remo Squillantini

Photographers, Photographic Agencies, Publishers, Literary Agents, Photographic Archives, Magazines

A.F.A. Agency; Afsen; F. Albergoni; Aldus Books; Alpha Photo, New York; Annan Photo Features; Antonini; Archivi Mondadori; Atesa, Geneva; R. Atkinson; Ayeiff; Russel Barnett; Bertelli; Bevilacqua; Bonatti; Bonzi; Stanley & Kay Breeden; Fred Bruemmer; Bucciarelli; Burton; Buzzini; J. L. Calderon; Camera Press; Carrara Pantano; Carrese; Casale; J. Allan Cash; Central Press; Chaumenton Collection; Christiansen; Church; Ciapanna; Cirani; Bruce Coleman Ltd.; Collins Publishers; Gene Cox; Cozzi; Craighead; Crespi; Dal Gal; Dalton; Walter Dawn; De Biasi; Sergio del Grande; De Wit; Dei; Devez; © Walt Disney Productions; Dossenbach; Drury; Dulevant; George Evans; M. Fantin; Fantinelli; A. Fatras; Douglas Faulkner; Ferri; Erwin Fieger; Werner Forman Archives; E. Foss; Foto Pictor, Milan; Foto Platt; Fototeca Est; Freelance Photographic Guild, New York; Freson; Frezzato; Fumagalli; Gasberg; Giacomelli; Giankas; Giussani; W. Green; Michel Guy; Editions Hachette; Hammacher; Hansel; Harper; Hermes; G. Holton; Iwasaki; Jasinsky; B. B. Jones; Kerneris; Russ Kinne; Edizioni Kister, Geneva; Klotz; Albert Krafczyk; Yves Lanceau; Frank W. Lane Agency; Mark J. Lann; Lanza; Larson; Latouche; George Laycoevi; Nina Lean; L.H.N.; Lod; Lombardi; S. Lorrain; Lotti; Lucien Louvgenies; G. Lunardi; A. Macchiavello; Magnum Photos; Maltini; Margiocco; Mariani; J. Markham; Martinerie; Mati; Mazza; Arborio Mella; Jane Miller; Miller Services; Milotte; Robert Mitchell; Edizioni Scolastiche Mondadori; Monkmeyer; Kelly Monterspaugh; Walter Mori; Motto; F. Moulen; David Muench; Josef Muench; Nardini; Newmann; Nilsson; Noailles; Novosti Press; Okapia Tierbilder; Orion Press; P2 Agency, Milan; Panicucci; Pasotti; Paterniani; Paul Popper, London; Pedone; L. Perkins; Philips & Thomas; Photocil; Photo Researchers Agency, New York; Pinna; Poroniey; Prato; Pucciarelli; Quilici; Agence Rapho (photographer Guillumette); Dean Ray; G. L. Relini; Ricciarini; Ridge Press; Rivolier; Enrico Robba; Rocco Longo; Fulvio Roiter; Roman; Ross; Emilio Rossi; Rutherford Platt; H. Saint Girons, Paris; Agence Satour, Paris; Scala Agency, Florence; Scarnati; Schulthness; Schwabe; S.E.F.; Sella; Kurt Severin; George Silk; Simion; Jan Six; Smythe; Solaini; James Sopper; Stephens Publishing House, Cambridge; P. Summ; Daniel Thase; Time Life Inc.; G. Tomisch; M. Tongiorgi; Maspons Ubiña; Verin; Viola; A. Visage; Ina Wadson; Walcott; Warner; Williamson; D. P. Wilson; S. C. Wilson; Nicholas Wright; Peter Wyss; Y.L.L.A.; Z.E.F.A. Agency, Dusseldorf; Z.F.A.

A special thanks to Walt Disney Productions, Burbank (California), who have kindly allowed us to use photographic material published in their series

We apologize if we have omitted or incorrectly credited any of the above bodies